NANOMATERIALS

Synthesis, Characterization, and Applications

About the Editors

A. K. Haghi, PhD

Dr. Haghi holds a BSc in urban and environmental engineering from the University of North Carolina (USA), a MSc in mechanical engineering from North Carolina A&T State University (USA), a DEA in applied mechanics, acoustics, and materials from the Université de Technologie de Compiègne (France), and a PhD in engineering sciences from the Université de Franche-Comté (France). He is the author and editor of 45 books as well as 650 papers in various journals and conference proceedings. Dr. Haghi has received several grants, consulted for a number of major corporations, and is a frequent speaker to national and international audiences. Since 1983, he served as professor in several universities. He is currently editor-in-chief of the *International Journal of Chemoinformatics and Chemical Engineering* and is on the editorial boards of many international journals.

Ajesh K. Zachariah

Ajesh K. Zachariah has obtained his masters degree in chemistry from Kerala University with a specialization in organic chemistry. He had done his research internship at NIIST, Trivandrum, Kerala, India. Currently he is working as Assistant Professor at Mar Thoma College, Tiruvalla, Kerala, India. Also he is doing his doctoral research in polymer chemistry and nanomaterials at the School of Chemical Sciences, Mahatma Gandhi University, Kottayam, Kerala, India. He has several internationally cited articles and one patent. His research areas are synthesis and characterization of nanomaterials such as photocatalysts and nanoporous materials and polymer nanocomposites.

Nandakumar Kalarikkal, PhD

Dr. Nandakumar Kalarikkal has obtained his masters degree in physics with a specialization in industrial physics and PhD in semiconductor physics from Cochin University of Science and Technology, Kerala, India. He was a postdoctoral fellow at NIIST, Trivandrum and later joined with Mahatma Gandhi University in 1994. Currently he is the Joint Director of the Centre for Nanoscience and Nanotechnology and Assistant Professor in the School of Pure and Applied Physics of Mahatma Gandhi University. His current areas of research interests include synthesis, characterization and applications of nanophosphors, nanoferrites, nanoferroelctrics, nanomultiferroics, nanocomposites, and phase transitions.

Advances in Nanoscience and Nanotechnology

Series Editors-in-Chief

Sabu Thomas, PhD

Dr. Sabu Thomas is the Director of the School of Chemical Sciences, Mahatma Gandhi University, Kottayam, India. He is also a full professor of polymer science and engineering and Director of the Centre for nanoscience and nanotechnology of the same university. He is a fellow of many professional bodies. Professor Thomas has authored or co-authored many papers in international peer-reviewed journals in the area of polymer processing. He has organized several international conferences and has more than 420 publications, 11 books, and two patents to his credit. He has been involved in a number of books both as author and editor. He is a reviewer to many international journals and has received many awards for his excellent work in polymer processing. His h Index is 42. Professor Thomas is listed as the 5th position in the list of Most Productive Researchers in India, in 2008.

Mathew Sebastian, MD

Dr. Mathew Sebastian has a degree in surgery (1976) with specialization in Ayurveda. He holds several diplomas in acupuncture, neural therapy (pain therapy), manual therapy, and vascular diseases. He was a missionary doctor in Mugana Hospital, Bukoba in Tanzania, Africa (1976-1978) and underwent surgical training in different hospitals in Austria, Germany, and India for more than 10 years. Since 2000 he is the doctor in charge of the Ayurveda and Vein Clinic in Klagenfurt, Austria. At present he is a consultant surgeon at Privatclinic Maria Hilf, Klagenfurt. He is a member of the scientific advisory committee of the European Academy for Ayurveda, Birstein, Germany, and the TAM advisory committee (Traditional Asian Medicine, Sector Ayurveda) of the Austrian Ministry for Health, Vienna. He conducted an International Ayurveda Congress in Klagenfurt, Austria, in 2010. He has several publications to his name.

Anne George, MD

Anne George, MD, is the Director of the Institute for Holistic Medical Sciences, Kottayam, Kerala, India. She did her MBBS (Bachelor of Medicine, Bachelor of Surgery) at Trivandrum Medical College, University of Kerala, India. She acquired a DGO (Diploma in Obstetrics and Gynecology) from the University of Vienna, Austria; Diploma Acupuncture from the University of Vienna; and an MD from Kottayam Medical College, Mahatma Gandhi University, Kerala, India. She has organized several international conferences, is a fellow of the American Medical Society, and is a member of many international organizations. She has five publications to her name and has presented 25 papers.

Yang Weimin, PhD

Dr. Yang Weimin is the Taishan Scholar Professor of Quingdao University of Science and Technology in China. He is a full professor at the Beijing University of Chemical Technology and a fellow of many professional organizations. Professor Weimin has authored many papers in international peer-reviewed journals in the area of polymer processing. He has been contributed to a number of books as author and editor and acts as a reviewer to many international journals. In addition, he is a consultant to many polymer equipment manufacturers. He has also received numerous award for his work in polymer processing.

Contents

List of Contributors xi
List of Abbreviations xv
Preface xvii

1. **Synthesis of Oxide Nanomaterials** 1
   Ibetombi Soibam, Sumitra Phanjoubam, and Chandra Prakash
2. **Shape Controlled Synthesis, Characterization, and Optical Properties of Silver Nanostructures** 19
   P. Kumbhakar, A. Maity, R. Sarkar, and A. K. Mitra
3. **Nanocomposites of Poly(Methyl Methacrylate) and Brookite Titania Nanorods** 41
   Niranjan Patra, Alberto Barone, Marco Salerno, Gianvito Caputo, and Athanassia Athanassiou
4. **Biomorphic Scaffolds for the Regeneration of Load-bearing Bones** 51
   Simone Sprio, Andrea Ruffini, and Anna Tampieri
5. **Porous Nanostructure Ti Alloys for Hard Tissue Implant Applications** 79
   J. Jakubowicz and G. Adameks
6. **Synthesis of Water-dispersible Carbon Nanotube-fullerodendron Hybrids** 119
   Toyoko Imae, Kumi Hamada, Yu Morimoto, and Yutaka Takaguchi
7. **A Facile Sol-gel Method for the Preparation of N-doped TiO_2 Photocatalyst for Pollutant Degradation** 133
   P. P. Silija, Z. Yaakob, N. N. Binitha, P. V. Suraja, S. Sugunan, and M. R. Resmi
8. **TiO_2 Nanoparticle Thin Films Cast by Electrophoretic Deposition** 139
   Sameer V. Mahajan, Maria T. Colomer, Mario Borlaf, Rodrigo Moreno, and James H. Dickerson
9. **Transforming Magnetic Photocatalyst to Magnetic Dye-adsorbent Catalyst** 147
   Satyajit Shukla
10. **Synthesis of Metal Nanoclusters in Dielectric Matrices by Ion Implantation: Plasmonic Aspects** 167
    P. Mazzoldi, G. Mattei, G. Pellegrini, and V. Bello
11. **From Single Atom to Nanocomposites** 197
    Beata Kalska-Szostko

12. Electrically Conductive Nanocomposites for Structural Applications..... 239

S. Mall

13. Electrical Properties in CrO_2 Modified C-ZrO_2 of Ferromagnetic Nanocomposite........ 255

A. Sengupta, A. K. Thakur, S. Ram, and R. K. Kotnala

14. Processable Aqueous Dispersion of Fullerene C_{60}: A Nanofluid................ 261

M. Behera, S. Ram, and H. J. Fecht

Index 269

List of Contributors

G. Adamek
Poznan University of Technology, Institute of Materials Science and Engineering, M. Sklodowska-Curie 5 Sq., 60-965 Poznan, Poland

Athanassia Athanassiou
Dr, National Nanotechnology Laboratory and Center for Biomolecular Nanotechnologies of IIT - Lecce - Italy, via Barsanti 1 - 73010 Arnesano - Lecce - Italy, athanassia.athanassiou@iit.it, +39-(0)832-29.57, +39-(0)832-29.82.38,

Alberto Barone
Dr, Italian Institute of Technology (IIT) - Genova - Italy, via Morego 30 - 16163 Genova - Italy, alberto.barone@iit.it, +39-(0)10-71.781.756, +39-(0)10-72.03.21

M. Behera
Materials Science Centre, Indian Institute of Technology, Kharagpur-721 302, India

V. Bello
Physics Department, University of Padova, via Marzolo 8, I-35131 Padova (Italy)
*paolo.mazzoldi@unipd.it

N. N. Binitha
Department of Chemical and Process Engineering, Faculty of Engineering and Built Environment, National University of Malaysia, 43600 UKM Bangi, Selangor, Malaysia,
Department of Chemistry, Sree Neelakanta Government Sanskrit College, Pattambi,
Palakkad-679306, Kerala, India, Ph: +91 466-2212223. Fax: +91 466-2212223, Email:binithann@yahoo.co.in

Mario Borlaf
CSIC, Instituto de Ceramica y Vidrio, E-28049 Madrid, Spain

Gianvito Caputo
Dr, National Nanotechnology Laboratory - Lecce - Italy, via per Arnesano 16 (km 5) - 73100 Lecce - Italy, gianvito.caputo@unile.it, +39-(0)832-29.82.70, +39-(0)832-29.82.38,

Maria T. Colomer
CSIC, Instituto de Ceramica y Vidrio, E-28049 Madrid, Spain

James H. Dickerson
Department of Physics and Astronomy, Vanderbilt University, Nashville, TN, USA
Vanderbilt Institute of Nanoscale Science and Engineering, Vanderbilt University, Nashville, TN, USA

H. J. Fecht
Institut für Mikro- und Nanomaterialien, Universität Ulm, Albert-Einstein Allee-47 D-89081, Ulm, and Forschungszentrum Karlsruhe, Institute of Nanotechnology Karlsruhe, D-76021, Germany

Kumi Hamada
Graduate School of Science, Nagoya University, Chikusa, Nagoya 464-8602, Japan

Toyoko Imae
Graduate Institute of Engineering and Department of Chemical Engineering, National Taiwan University of Science and Technology, Taipei, 10607, Taiwan, ROC
Graduate School of Science, Nagoya University, Chikusa, Nagoya 464-8602, Japan

J. Jakubowicz
Poznan University of Technology, Institute of Materials Science and Engineering,
M. Sklodowska-Curie 5 Sq., 60-965 Poznan, Poland

Beata Kalska-Szostko
University of Białystok, Institute of Chemistry, Hurtowa 1, 15-399 Białystok, Poland

R. K. Kotnala
Magnetic Materials Division, National Physical Laboratory, New Delhi 110012, India

P. Kumbhakar
Nanoscience Laboratory, Department of Physics,
National institute of Technology, Durgapur, 713209, West Bengal, India
*Corresponding author, E-mail:nitdgpkumbhakar@yahoo.com & pathik.kumbhakar@phy.nitdgp.ac.in

Sameer V. Mahajan
Interdisciplinary Graduate Program in Materials Science, Vanderbilt University, Nashville, TN, USA
Vanderbilt Institute of Nanoscale Science and Engineering, Vanderbilt University, Nashville, TN, USA

A. Maity
Nanoscience Laboratory, Department of Physics,
National institute of Technology, Durgapur, 713209, West Bengal, India
*Corresponding author, E-mail:nitdgpkumbhakar@yahoo.com & pathik.kumbhakar@phy.nitdgp.ac.in

S. Mall
Department of Aeronautics and Astronautics
Air Force Institute of Technology
AFIT/ENY, Bldg. 640, 2950 Hobson Way,
Wright-Patterson AFB, OH, 45433-7765, USA
E-mail:Shankar.Mall@afit.edu

G. Mattei
Physics Department, University of Padova, via Marzolo 8, I-35131 Padova (Italy)
*paolo.mazzoldi@unipd.it

P. Mazzoldi
Physics Department, University of Padova, via Marzolo 8, I-35131 Padova (Italy)
*paolo.mazzoldi@unipd.it

A. K. Mitra
Nanoscience Laboratory, Department of Physics,
National institute of Technology, Durgapur, 713209, West Bengal, India
*Corresponding author, E-mail:nitdgpkumbhakar@yahoo.com & pathik.kumbhakar@phy.nitdgp.ac.in

Rodrigo Moreno
CSIC, Instituto de Ceramica y Vidrio, E-28049 Madrid, Spain

Yu Morimoto
Graduate School of Natural Science and Technology, Okayama University, Tsushima, Okayama 700-8539, Japan

Niranjan Patra
Dr, University of Genova and Italian Institute of Technology (IIT) - Genova - Italy, via Morego 30 - 16163 Genova - Italy, niranjan.patra@iit.it, +39-(0)10-71.781.756, +39-(0)10-72.03.21,

G. Pellegrini
Physics Department, University of Padova, via Marzolo 8, I-35131 Padova (Italy)
*paolo.mazzoldi@unipd.it

Sumitra Phanjoubam
Department of Physics, Manipur University, Canchipur, Imphal-795 003, India

Chandra Prakash
Department of ER&IPR, DRDO Bhawan, Delhi-110 011, India
Email: ibetombi_phys@rediffmail.com

S. Ram
Materials Science Centre, 2 Department of Physics and Meteorology, Indian Institute of Technology, Kharagpur 721 302, India

M. R. Resmi
Department of Chemistry, Sree Neelakanta Government Sanskrit College, Pattambi,
Palakkad-679306, Kerala, India, Ph: +91 466-2212223. Fax: +91 466-2212223, Email:binithann@yahoo.co.in

Andrea Ruffini
Ph.D.
Institute of Science and Technology for Ceramics, ISTEC-CNR, Via Granarolo 64 48018 Faenza (ITALY)
tel: +39 0546 699756; e-mail: andrea.ruffini@istec.cnr.it

Marco Salerno
Dr, Italian Institute of Technology (IIT) - Genova - Italy, via Morego 30 - 16163 Genova - Italy, marco.salerno@iit.it, +39-(0)10-71.781.444, +39-(0)10-72.03.21

R. Sarkar
Nanoscience Laboratory, Department of Physics,
National institute of Technology, Durgapur, 713209, West Bengal, India
*Corresponding author, E-mail:nitdgpkumbhakar@yahoo.com & pathik.kumbhakar@phy.nitdgp.ac.in

Daivd Schmool
IN-IFIMUP and Departamento de Física e Astronomia, Universidade do Porto, Rua Campo Alegre 687, 4169 007 Porto, Portugal

A. Sengupta
Materials Science Centre

Satyajit Shukla
Ph.D.
Nano-Ceramics Section
Materials and Minerals Division (MMD)
National Institute for Interdisciplinary Science and Technology (NIIST)
Council of Scientific and Industrial Research (CSIR)
Industrial Estate P.O., Pappanamcode
Thiruvananthapuram 695019, Kerala, India
Phone: +91-471-2535529; Fax: +91-471-2491712; E-Mail: satyajit_shukla@niist.res.in

P. P. Silija
Department of Chemical and Process Engineering, Faculty of Engineering and Built Environment, National University of Malaysia, 43600 UKM Bangi, Selangor, Malaysia

Ibetombi Soibam
Department of Physics, National Institute of Technology, Manipur, Takyel, Imphal-795 001, India

Simone Sprio
Ph.D.
Institute of Science and Technology for Ceramics, ISTEC-CNR, Via Granarolo 64 48018 Faenza (ITALY)
tel: +39 0546 699759;
e-mail: simone.sprio@istec.cnr.it

S. Sugunan
Department of Applied Chemistry, CUSAT, Kochi-22, India.

P. V. Suraja
Department of Chemical and Process Engineering, Faculty of Engineering and Built Environment, National University of Malaysia, 43600 UKM Bangi, Selangor, Malaysia

Yutaka Takaguchi
Graduate School of Natural Science and Technology, Okayama University, Tsushima, Okayama 700-8539, Japan

Anna Tampieri
Ph.D
Institute of Science and Technology for Ceramics, ISTEC-CNR, Via Granarolo 64 48018 Faenza (ITALY)
tel: +39 0546 699753; e-mail: anna.tampieri@istec.cnr.it
fax: +39 0546 46381

A. K. Thakur
Department of Physics and Meteorology, Indian Institute of Technology, Kharagpur 721 302, India

Z. Yaakob
Department of Chemical and Process Engineering, Faculty of Engineering and Built Environment, National University of Malaysia, 43600 UKM Bangi, Selangor, Malaysia

List of Abbreviations

ADC	Azodicarbonamide
$AgNO_3$	Silver nitrate
CdS	Cadmium sulfide
CH_3OH	Methyl alcohol
$CoFe_2O_4$	Cobalt ferrite
CT	Charge transfer
DDI	Dipole-Dipole interactions
DLS	Dynamic light scattering
DMIM	Discontinuous metal-insulator magnetic multilayers
Ea	Activation energy
EDS	Energy dispersive X-ray spectroscopy
EMI	Electromagnetic interference
EPD	Electrophoretic deposition
ESD	Electrostatic discharge
Fe_2O_3	Hematite
FMR	Ferromagnetic resonance
FT	Fourier transform
GDVDP	Glow discharge vapor deposition polymerization
GMM	Mie-Maxwell-Garnett
GMR	Giant magnetoresistance
HA	Hydroxyapatite
Hc	Magnetic coercivity
HRTEM	High-resolution TEM
MA	Mechanical alloying
MB	Methylene blue
MMG	Mie-Maxwell-Garnett
MNCF	Metal nanocluster composite film
MR	Magnetoresistive
Ms	Magnetization
NCs	Nanoclusters
NiCCF	Nickel-coated carbon woven fabric
NiNS	Nickel nanostrands
NPs	Nanoparticles
PMDA-ODA	Dianhydride-4,4 oxydianiline
PPMS	Property measurement system
PVA	Poly(vinyl alcohol)
ROS	Reactive oxygen species
RTM	Resin transfer mold
SAED	Selected area electron diffraction
SEF	Surface enhanced fluorescence
SEM	Scanning electron microscopy

SERS	Surface enhanced raman spectroscopy
SiO_2	Silica
SnO_2	Tin oxide
SPM	Superparamagnetic
SPR	Surface plasmon resonance
Tanδ	Power loss
TEM	Transmission electron microscope
TiO_2	Titania
TOF	Time-of-flight
UTS	Ultimate tensile strength
UV	Ultraviolet
VSM	Vibrating sample magnetometer
XRD	X-ray diffraction
ZnO	Zinc oxide
ZnS	Zinc sulphide
Er	Dielectric constant
Σdcb	Conductivity

Preface

The history of the nanomaterial begins from old days of alchemists. The science of "nano" started in 1959, after Richard Feynman's famous talk at the annual meeting of the American Physical Society describing the molecular machines building with atomic precision titled as "There's Plenty of Room at the Bottom" and he predicted that a revolution in the field of science and technology was awaiting. In 1974, Norio Taniguchi, a Japanese scientist, coined the term "nanotechnology" in a paper on ion-sputtering machining. A technical paper discussing the new technology was presented after seven years from Taniguchi's introduction of new term. It was Eric Drexler who presented the first technical paper on molecular engineering to build with atomic precision. Since then many studies were done in the field of nanotechnology.

The present book is mainly centered on the production of different types of nanomaterials and their applications. The book can be grouped as:

- synthesis of different types of nanomaterials
- characterization of different types of nanomaterials
- applications of different types of nanomaterials including the nanocomposites

The book is intended to be a reference for basic and practical knowledge about the synthesis, characterization, and applications of nanotechnology for students, engineers, and researchers. Chapter 1 deals with the synthesis of oxide nanomaterials focusing on lithium ferrites. They form a very important group of materials. They are magnetic material having varied technological applications that can be fabricated by various synthesis methods such as conventional and nonconventional methods. Ferrites were generally produced by a ceramic process involving high temperature solid-state reactions between the constituent oxides/carbonates. Shape-controlled synthesis, characterization, and optical applications of silver nanostructures are described in Chapter 2. It discusses novel methods for chemical synthesis of silver nanoparticles of different size and shapes, such as nanorod, nanospindle, nanocube, nanosphere, nanoellipsoid, and some plate-like structures.

Chapter 3 discusses the growth and characterization of metal nano-sized branched structures on insulator substrates by electron-beam-induced deposition. Using an electron-beam-induced deposition (EBID) process in a transmission electron microscope, the authors fabricated a self-standing metal nano-sized branched structures including nanowire arrays, nanodendrites, and nanofractal-like trees, as well as their composite nanostructures with controlled size and position on insulator (SiO_2, Al_2O_3) substrates. The techniques such as high-resolution transmission electron microscopy and X-ray energy dispersive spectroscopy and the growth mechanism are discussed.

Bone tissue diseases are among the most disabling pathologies and affect an increasing number of people worldwide. The diseases that most seriously impact people's lives and activities are those involving long bone portions subjected to mechanical load. The impact of such problems is particularly relevant among aged people (i.e. due to osteoporosis), but in the last decade the number of relatively young patients is continuously increasing due to modern lifestyles that include intense sports activity, tendency to obesity, etc. Chapter 4 describes the smart biomaterials obtained by biomorphic transformation. Chapter 5 discusses the porous nanostructure Ti-alloys for hard tissue implant applications. Formation of Ti-6Al-4V, Ti-15Zr-4Nb and Ti-6Zr-4Nb porous nanocrystalline bioalloys is described in Chapter 5.

Chapter 6 discusses synthesis of water-dispersible carbon nanotube–fullerodendron hybrids. Chapter 7 discusses nitrogen doping on TiO_2 via the sol gel method and improvement of properties for N-doped systems when compared to the post preparation N doping processes. Chapter 8 discusses the production of densely packed films of titanium dioxide (TiO_2) nanoparticles via electrophoretic deposition (EPD). Chapter 9 discusses the magnetic dye-adsorbent catalyst synthesized via hydrothermal processing. The magnetic dye-adsorbent catalyst consists of a core-shell nanocomposite with the core of a magnetic ceramic particle (such as mixed cobalt ferrite and hematite, and pure cobalt ferrite) and the shell of nanotubes of dye-adsorbing material (such as hydrogen titanate).

Composite materials formed by metal nanoclusters embedded in dielectric films (Metal Nanocluster Composite Film, MNCF) are the object of several studies owing to their peculiar properties suitable for application in several fields, such as nonlinear optics, photoluminescence, catalysis, or magnetism. Chapter 10 describes the plasmonic aspects of synthesis of metal nanoclusters in dielectric matrices by ion implantation. Chapter 11 discusses different approaches for the development of nanocomposites. Chapter 12 describes electrically conductive nanocomposites for structural applications. Electrical properties of CrO_2 modified ZrO_2 of a cubic (c) crystal structure are discussed in Chapter 13. Chapter 14 describes poly (methyl methacrylate) nanocomposites filled with brookite nanocrystals.

— A. K. Haghi, PhD, Ajesh K. Zachariah, and Nandakumar Kalarikkal, PhD

1 Synthesis of Oxide Nanomaterials

Ibetombi Soibam, Sumitra Phanjoubam, and Chandra Prakash

CONTENTS

1.1 Introduction 1
1.2 Spinel Structure Ferrites 3
1.3 Preparation of Ferrites 5
1.3.1 Conventional Ceramic Method 5
1.3.2 Non-conventional Method 6
1.4 Magnetic Studies 8
1.4.1 Superexchange Interaction in Ferromagnetic Spinel 8
1.4.2 Saturation Magnetization in Ferrites 9
1.4.3 Curie Temperature (T_C) 9
1.5 Mössbauer Studies 10
1.6 Lithium Zinc Nickel Ferrites 12
1.7 Conclusion 17
Keywords 17
References 17

1.1 INTRODUCTION

The nanomaterials have unique properties for some exceptional as well as ordinary applications. In the past few years due to their scientific importance fundamentally and technologically nanoparticle systems have received much attention, In this thrust, area of research lithium ferrites form a very important group of materials. Lithium ferrites which are magnetic material having varied technological applications can be fabricated by various synthesis method such as conventional and non-conventional method. Ferrites were generally produced by ceramic process involving high temperature solid state reactions between the constituent oxides/carbonates. The properties of ceramics like morphology, particle size distributions, magnetic properties, and so on are strongly dependent on the features of the starting powder. This is in particular true for lithium based ferrites. One of the effective means to control the powder characteristics is to modify the powder synthesis process itself. The different methods have been proposed such as sol-gel, auto-combustion, coprecipitation, sol-gel auto combustion, Pechini method, and so on. Among the synthesis methods, citrate precursor method is one which is gaining importance for preparing fine grained ferrite powders for suitable purposes. The method appears to be simple and convenient as it neither requires expensive chemicals such as alkoxides which are used in sol-gel processing nor equipment like autoclaves for combustion process. On the other hand, the final product obtained is in the nanometer range, exhibit high purity, chemical homogeneity, and compositional control. The magnetic properties are immensely enhanced. The preparation of conventional ceramic method as well as the non-conventional method like chemical coprecipitation and citrate precursor is discussed, where emphasis is given to citrate precursor synthesis method. A particular series of samples of Li-Zn-Ni ferrite is presented along with their magnetic and Mössbauer characteristics.

The importance of fabrication and the understanding of the magnetic properties of oxides of metals attracted scientist and engineers from various fields. The lithium ferrite is one of the most used versatile magnetic materials. They are important components for microwave device and memory core due to their high values of Curie temperature, saturation magnetization, and so on [1, 2]. A number of lithium ferrite compositions comprising of the oxide of lithium and iron, and if desired the oxides of other metals such as Mn, Ni, and Zn are known to be useful in applications requiring soft ferrimagnetic properties. The lithium ferrites are usually synthesized and have been prepared by mixing and milling together powder mixture of lithium carbonate, iron oxide, and oxides of other suitable metals. After prolonged milling the powder mixture is calcined at an elevated temperature of about 1,000°C to co-react the individual oxides into the spinel lithium ferrites which is then processed into the desired components [3, 4]. The solid state reaction has some inherent disadvantages like chemical inhomogeneity, introduction of impurities during ball milling, and so on. The high temperature induced lower magnetization due to formation of Fe_3O_4 and α-Fe_2O_3. However, most of the factors degrading the properties are overcome in chemical synthesis method. The art of preparing lithium ferrite by conventional method is not suitable for preparing chemically uniform lithium ferrites of extremely small grain size. The quest for synthesis of small grain size demands the synthesis of ultrafine ferrite materials [5]. Consequently, it led to the discovery of various synthesis methods. Several patents disclose various wet chemical techniques namely sol-gel, auto combustion, coprecipitation, Pechini method, and so on [6-9]. In the chemical coprecipitation process an aqueous solution of suitable salts of iron, lithium, and other desired, suitable materials is mixed with a precipitating agent that will cause the precipitation of a fatty acid salt of lithium with the hydroxides of the other metals present in the solution. The precipitate represents a substantially uniform mixture of precursor compounds of the ferrite metals. The precipitate is filtered from the mother liquor and dried. The dry precipitate is heated in air at an elevated temperature, preferably about 200°C to dehydrate the precipitate and to burn out carbonaceous matter leaving a residue of the oxides of the respective metals. The particles are then heated at a temperature of 400–500°C to co-react the lithium oxide, ferric oxide, and other metal oxides present to form a spinel lithium ferrite. At this stage, the ferrite particles which are extremely small in size about 100 Angstrom are formed. It is then sintered at about 1,100°C. The sintering operation results in some grain growth, although such growth can be advantageously inhibited by the presence of a small amount of bismuth. It is to be noted that the grain size is significantly smaller than the grain size obtained in conventionally prepared lithium ferrites. However, the coprecipitation process takes a very long duration to obtain the required materials. This is to overcome in the sol-gel auto combustion process also known as the citrate precursor method. In this method, the respective nitrates and citric acid are heated making a spontaneous combustion process to occur giving the required ferrite sample in just a few second. A brief discussion of what is spinel structure ferrite, different preparation methods, basic magnetic, and Mössbauer background and some characteristic properties of Li-Zn-Ni ferrites prepared by citrate precursor method is given.

1.2 SPINEL STRUCTURE FERRITES

The spinel structure ferrites have the general formula MFe_2O_4, where M represents a divalent metal ion. The divalent metal ion can be one of the divalent ions of the transition elements like Mn^{2+}, Co^{2+}, Ni^{2+}, Cu^{2+}, Zn^{2+}, Cd^{2+}, and Mg^{2+}, or a combination of these ions, or a combination of ions having average valency of two, like $0.5Li^{1+}$ + $0.5Fe^{3+}$ as in $Li_{0.5}Fe_{2.5}O_4$. The trivalent ion such as Al^{3+} or Cr^{3+} ion and tetravalent ion such as Ti^{4+} or Ge^{4+} ion can completely or partly replace the Fe^{3+} ions in MFe_2O_4.

The crystal chemistry of compounds crystallizing in spinel structure has been extensively dealt with by Gorter [10], Hafner [11], and Blasse [12] although it was first determined by Bragg [13] and Nishikawa [14]. The unit cell having spinel structure is shown in Figure 1.

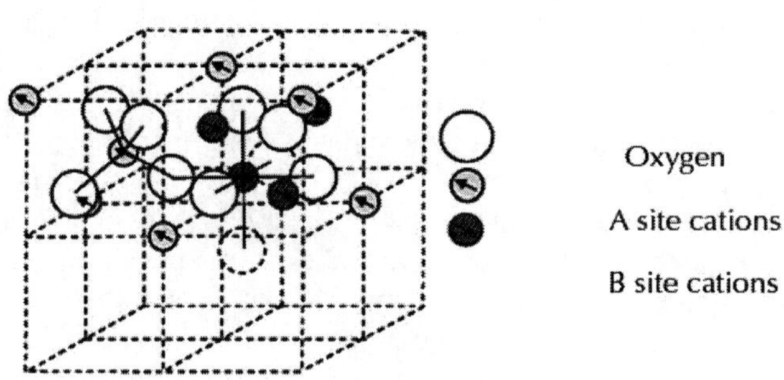

FIGURE 1 Structure of spinel ferrite.

The spinel crystal structure primarily depends on the oxygen ion lattice [10-13], the radii of the oxygen ions (~1.32A°) being several times larger than the radii of the metallic ions (0.6A° and 1.0A°). The crystal structure can consequently be thought of as being made up of the closest possible packing of oxygen ions, that is it forms an fcc lattice leaving two kinds of interstitial sites, the tetrahedral or the A site and the octahedral or the B site. A metallic ion located at A site has four nearest oxygen ion neighbors forming a regular tetrahedron and is said to be in a site of tetrahedral coordination, whereas, a metallic ion in octahedral site has six nearest oxygen ion neighbors forming a regular octahedron and the metallic ion is in a site of octahedral coordination. The tetrahedral site has smaller volume compared to octahedral site and both types of sites are smaller in size compared to oxygen ion diameter.

In the unit cell of a spinel lattice there are eight molecules of MFe_2O_4 consisting of 32 oxygen ions forming a cubic close packed structure, 8 divalent ions and 16 trivalent ions. The 32 oxygen ions leave 64 tetrahedral interstices (A) site and 32 octahedral interstices (B) site. Of these only 8 A sites and 16 B sites are occupied by the metal ions [14, 15].

To describe the structure, the unit cell with cell edge 'a' can be subdivided into eight octants with edge $\frac{1}{2}a$. Two adjacent octants are shown in Figure 2.

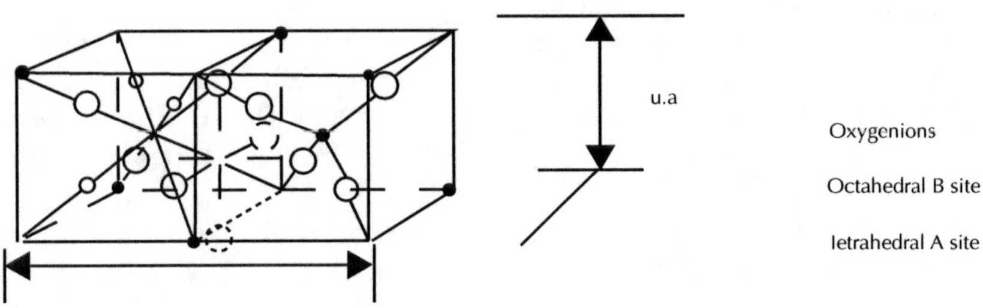

FIGURE 2 Two adjacent cubes showing two octants of the spinel structure.

The oxygen ions have identical positions in all the octants. Each octant contains four anions, which form the corners of a tetrahedron. The interstitial positions occupied in any two adjacent octants are different but are the same in two octants sharing an edge. If the ionic arrangement in any two adjacent octants are given, the unit cell can be constructed by translation of the two octants along the three (110) axes. The surrounding of a tetrahedral ion by the other ions has strictly cubic symmetry but is not the case for an individual octahedral ion. The octahedral ions, of course are cubically surrounded by the oxygen ions but not by the neighboring metal ions.

It has been mentioned earlier that oxygen ions form an fcc structure. But in the real case there is existence of slight deviations due to deformation caused by the metal ions. The tetrahedral sites are too small to accommodate a metal ion, so that when a metal cation fills a tetrahedral A site. It displaces equally the four oxygen ions outwards along the body diagonals of the cube. These four oxygen ions still occupy the corners of an enlarged tetrahedron and the symmetry around each A ion retains cubic symmetry. But the four oxygen surrounding the B site ion is shifted, in such a way, that this oxygen tetrahedron shrinks by the same amount as the first expands. Therefore, the symmetry is no longer cubic but is a trigonal symmetry such that a trigonal field appears on the B site. This field is a function of the oxygen parameter, u, which is a qualitative measure of the displacement. It is given by the distance between an oxygen ion and a face of the cube, which is put equal to au, where 'a' is the cell edge. The ideal fcc structure has oxygen parameter $u_{id} = \frac{3}{8}$.

Spinel ferrites can be classified into three categories, *viz.* normal, inverse, and mixed ferrites on the basis of site occupancy by divalent and trivalent metal ions.

Normal Ferrites

The kind of ferrites in which all the tetrahedral sites are occupied by the eight divalent cations and the octahedral sites by the 16 trivalent metal ions, are termed "normal" ferrites. Examples are zinc ferrite, $Zn^{2+}[Fe^{3+}]O_4$, cadmium ferrite $Cd^{2+}[Fe_2^{3+}]O_4$, and so on, where the cat ions outside the square brackets occupy A site and those within the square bracket occupy the B site.

Inverse Ferrite

In this ferrite eight out of the 16 trivalent metal ions occupy tetrahedral sites, and the octahedral sites are occupied by eight divalent metal ions and the remaining eight trivalent cations. Examples are magnetite, nickel ferrite with the respective compositional formula as $Fe^{3+}[Fe^{2+}Fe^{3+}]O_4^{2-}$ and $Fe^{3+}[Ni^{2+}Fe^{3+}]O_4^{2-}$.

Mixed Ferrites

This kind of ferrites is neither completely normal nor completely inverse since the A and B sites are randomly occupied by both divalent and trivalent metal ions. An example of this category of ferrites is magnesium ferrite, $MgFe_2O_4$ with the general cation distribution represented as :

$$Mg_x^{2+}Fe_{1-x}^{3+}\left[M_{1-x}^{2+}Fe_{1+x}^{3+}\right]O_4^{2-}$$

1.3 PREPARATION OF FERRITES

1.3.1 Conventional Ceramic Method

Ferrites are usually synthesized and have been prepared by mixing and milling together the powder mixture of lithium carbonate, iron oxide, and oxides of other suitable metals [16]. It is shown in Figure 3. The reactants have been weighed out in the required amount and then mixed. For small quantity of less than 20 g manual mixing may be done using agate mortar and pestle. Mostly, agate made of porcelain is used as it is hard and unlikely to contaminate the mixture. Homogenization is made by adding sufficient amount of a volatile organic liquid like acetone or alcohol which gradually volatilizes after 10–15 min of grinding and mixing. If the total quantity is much larger than 20 g manual mixing is not advisable and in such cases mechanical mixing is better using a ball mill but may take several hours. After prolonged milling the powder mixture is calcined at an elevated temperature like 1,000°C to co-react the individual oxides.

Solids do not usually react together at room temperature over normal timescales, so, it is necessary to heat them to much higher temperature often 1,000–1,500°C, in order for reaction to occur at an appreciable rate. The heating schedule to be used depends on the form and reactivity of the reactants. If one or more of the reactants is an oxysalt, the first stage of reaction must be decomposition of the oxysalt and mixture should be heated first at an appropriate temperature for few hours such that decomposition occurs in controlled manner. If the stage is not done decomposition will occur vigorously and may cause sample to spit out of container. The reaction to give final product usually requires hours or even days depending on reaction temperature. However, this art of preparing lithium ferrite are not suitable for preparing chemically uniform lithium ferrites of extremely small grain size.

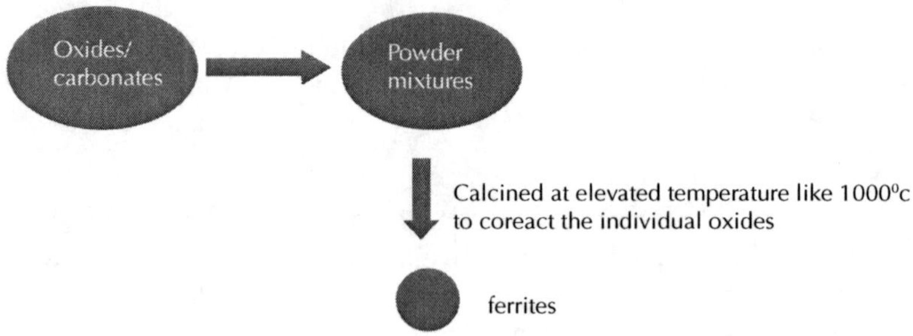

Mixing and Milling

FIGURE 3 Ceramic method of preparing ferrites.

1.3.2 Non-Conventional Method

There are several non-conventional methods but here we give a basic outline of the chemical coprecipitation method and the citrate precursor method.

Chemical Coprecipitation Method

In the chemical coprecipitation (Figure 4) an aqueous solution of suitable salts of iron, lithium, manganese, and other desired, suitable materials is mixed with a precipitating agent that causes the precipitation of a fatty acid salt of lithium with the hydroxides of the other metals present in the solution. The precipitate represents a substantially uniform mixture of precursor compounds of the ferrite metals on an atomic scale. The precipitate is filtered from the mother liquor and dried.

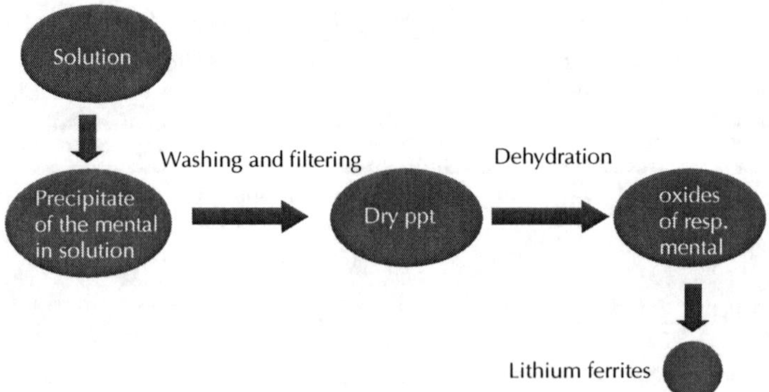

Preciptating agents

FIGURE 4 Chemical coprecipitation method of preparing ferrites.

The dry precipitate is heated in air at an elevated temperature, preferably about 200°C to dehydrate the precipitate and to burn out carbonaceous matter leaving a residue of the oxides of the respective metals. The particles are then heated at a temperature of 400–500°C to co-react the lithium oxide, ferric oxide, and other metal oxides present to form a spinel lithium ferrite. Because of high degree of homogenization, much lower reaction temperature is sufficient for reaction to occur. The overall reaction for formation of lithium ferrite may be written as:

$$Fe_2[(COO_2)]_3 + Li(COO)_2 \rightarrow \text{Lithium ferrite} + \text{carbon monoxide} + \text{carbon dioxide}$$

It is then sintered at about 1,100°C. The sintering operation results in some grain growth, although such growth can be advantageously inhibited by the presence of a small amount of calcium, so that an ultimate smaller grain size is obtained. It is to be noted that the grain size is significantly smaller than that obtained in conventionally prepared lithium ferrites [17]. The method does not work well in cases where two reactants have very different solubilities in water and the reactants do not precipitate at the same rate. However, the coprecipitation process takes a very long duration to obtain the required materials and is often not suitable for preparation of high purity, accurately stoichiometric phases. Therefore other methods such as citrate precursors are used.

Citrate Precursor Method

In this method (Figure 5), the metal ions from the starting materials such as nitrates are complexed in an aqueous solution with $\alpha-$ carboxylic acids such as citric acid.

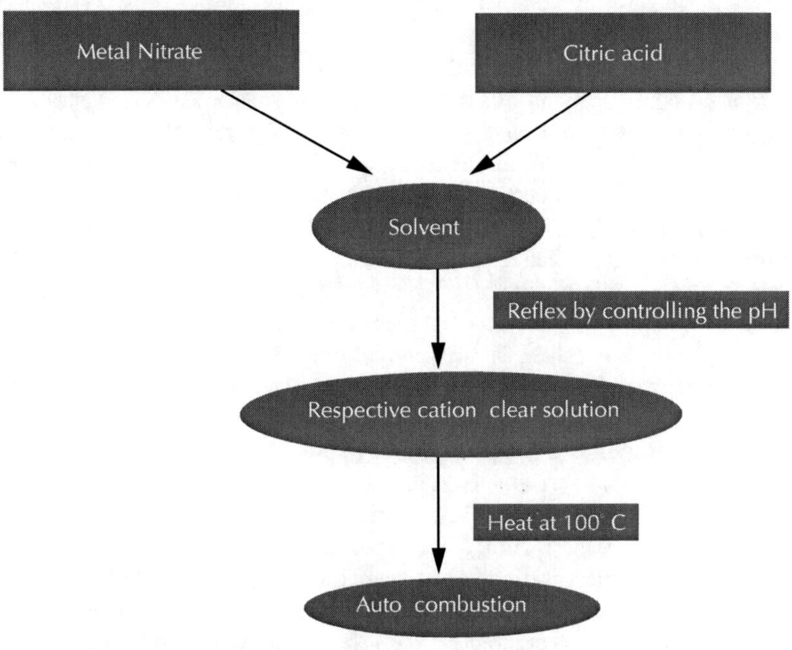

FIGURE 5 Citrate precursor method of preparing ferrites.

The ratio of metal nitrates to citric acid is taken in the ratio of 1:1 in order to make uniformity in the chelation of the metals. The pH of the solution is controlled at 7 by adding ammonium hydroxide. The pH controlled solution is refluxed at 40°C with continuous stirring using magnetic bar agitator and dried in an oven at 100°C. A highly viscous mass is then formed by the evaporation of the solution resulting from metal nitrates and citric acid. Now the citric acid which also act as a fuel provider for redox reaction provides fuel for the ignition step of the process as it is oxidized by the nitrate ions. Finally, auto-combustion or the ignition process takes place just like the eruption of a volcano giving rise to a dark brown voluminous product which is the ash-synthesized ferrite powder. This has been reported [18, 19].

The underlying idea for the reaction is that the metal nitrates react with water producing metal hydroxide and gases like nitrogen dioxide. Then the carboxylic acid functionalities from the citric acid chelant are deprotonated with ammonium hydroxide and the metal hydroxide by removing protons. This results in production of metal ions having a positive valency and carboxylic ions with negative valency together with water molecules. Finally, coordinate bonds are formed between the metal ions and carboxylate ions of the chelant to produce the metal and acid complex which is the ash-synthesized powder obtained. It is then given final sintering. Several characteristic properties of substituted lithium ferrite were prepared by this method and investigated.

1.4 MAGNETIC STUDIES

1.4.1 Superexchange Interaction in Ferromagnetic Spinel

The superexchange interaction of magnetic ion to understand the magnetic properties of oxides of transition metal was first recognized by Kramer [20] and demonstrated later by Neel [21] but the most advanced state was brought by the work of Anderson [22]. Superexchange interaction is the interaction that takes place through a passive intermediary. The mechanism of superexchange interaction can be illustrated by considering two metal ions and an oxygen ion having a purely electrovalent bond, represented by M-O-M where M are the two metal (Fe^{3+}) ions separated by an O^{2-} ion.

The metal–metal distances are sufficiently large and the orbital wave functions of the 3d electrons of Fe^{3+} ions do not overlap the next metal ion but it overlaps the neighboring oxygen ion. Therefore, the direct exchange interaction between the two Fe^{3+} ions is not possible but involves the participation of the intervening O^{2-} anion. The outer shell of the O^{2-} anion has an electronic configuration s^2p^6. Since, the outer shell is completely filled there is no spin coupling. However, in an excited state one of the electrons from 2p shell move out bonding itself to the 3d shell of one of the neighboring metal (Fe^{3+}) cation. As a result, the oxygen ions becomes O^- with an outer electronic configuration of s^2p^5 and a resulting spin moment of one Bohr magneton μ_B parallel to the resultant spin of the metal ion involved in the interaction. These net spin of the O^- ion interact directly with neighboring magnetic ion.

The superexchange interaction has a dependence on the bond separation as well as the bond angle. With increase in the bond separation there is a decrease in the interaction. These interactions are found to be strongest when the bond angle between the two metals and oxygen ion represented as M-O-M is 180° With further decrease in bond angle the interaction decreases and reaches the minimum for an angle of 90°.

According to Neel, there are three types of superexchange interactions present in spinel ferrites—the interaction between neighboring magnetic ions located at A sites (AA interaction), the interaction between various ions in sublattices A and B (AB interaction) and the interaction between neighboring magnetic ions located at B sites (BB interaction) [20, 21]. The AA interaction is very weak because the metal oxygen distance is very large about 3.5 A° and the angle of A-O-A is about 80°. The distance for the BB exchange interaction is reasonable, being about 2.1 A°, but the angle of B-O-B is 90° and hence the magnitude of interaction is minimum. The AB interaction, for which the distance is short ~1.9 A° and the angle A-O-B is favorable, being about 125°, is the strongest of the three. The BB interaction is not, however, completely negligible and plays an important role in certain mixed ferrites like the mixed zinc ferrites. When the AA or BB interaction becomes comparable in magnitude to the AB interaction, the magnetic moments of the sublattice are arranged in a triangular form and are said to be canted. These canted magnetic moments occur because of the magnetic dilution of one of the sublattice by substituting non magnetic ions in the ferrites. Besides, non magnetic substitution, change in temperature can also lead to this competitive situation [23, 24].

1.4.2 Saturation Magnetization in Ferrites

In ferrites, the magnetization is due to the antiparallel uncompensated electron spins of the two A and B sublattices. Hence, the net magnetization is given by the difference between the magnetization of the two sublattices. The saturation magnetization occurs when all its individual magnetic domains are aligned in the direction of the applied magnetic field such that with further increase of the external field, the intrinsic magnetization within the domain increases. This saturation magnetization value is observed to change with increase in temperature. The increase in temperature destroys the magnetic ordering thereby decreasing the magnetization and leading finally to the zero magnetization which gives the Curie temperature.

Ferrite exhibiting ferrimagnetisms is governed by the superexchange interaction. According to Neel, the AB exchange interaction is stronger than the AA and BB interaction. This is because for AB interaction the cation and anion distances are small and the angle A-O-B is about 125° whereas for AA and BB interaction the angles are about 80° and 90° respectively. The saturation magnetization which is the vector sum of the two sublattices can now be written as:

$$\overrightarrow{M_S} = \overrightarrow{M_B} - \overrightarrow{M_A}$$

where M_B and M_A are the magnetizations of the B and A sublattice respectively.

The effect of magnetic interaction on the magnetic moments can be studied considering the different kind of spinel ferrite such as normal, inverse, and mixed ferrites.

1.4.3 Curie Temperature (T_C)

The T_C is the critical temperature at which the aligning effect of exchange interactions on the spins of magnetic ion is cancelsed by the disordering effect of random thermal motion and the spontaneous magnetization of the ferro or ferrimagnetic material vanishes by changing to the paramagnetic state. It is possible to attain perfect alignment of the magnetic spins of all the molecules only at 0 K.

In the case of ferromagnetic materials, the exchange interaction of neighboring electrons produces a strong effective field called the molecular field [25]. This field is responsible for the alignment of the magnetic dipoles parallel to each other. As the temperature is increased the spontaneous magnetization decreases from M_o, its value at t = 0 till the magnetization disappears at the point called the Curie temperature. This disappearance of magnetization at T_C is due to the fact that there is sufficient thermal energy of the lattice to overcome the interaction energy and hence the magnetic ordering is destroyed making the materials paramagnetic. In ferrimagnetic materials, the variation of magnetization with temperature may be more complicated. As in the case of ferromagnetic materials, here also the spontaneous magnetization disappears at the Curie temperature. The magnetic characteristics of a ferrimagnetic material is mostly controlled by superexchange interactions between the Fe^{3+} ions in A and B sites mediated by the oxygen anions. The strongest superexchange interaction is given by the antiparallel alignment of spins between the A and B sites. Hence, the overall strength is determined by the AB interaction which has a direct relation with the T_C [26]. This strength of the AB interaction is determined by the number of $Fe_A^{3+} - O^{2-} - Fe_B^{3+}$ linkages per magnetic ion per formula unit. Hence, the T_C depends on the number of $Fe_A^{3+} - O^{2-} - Fe_B^{3+}$ linkages per Fe^{3+} ion per formula unit.

According to the relationship given by Gilleo, the interaction energy assigned to each linkage may be written as $\frac{kT_C}{n}$, where k is the Boltzmann constant and n is the number of interactions. Therefore, if E_A and E_B are the respective energies needed to invert the spins of ions at A and B sites for a ferrite, the average thermal energy required to create absolutely non-interacting ions is given by [2]:

$$kT_C \simeq \frac{n_A}{n_A + n_B} E_A + \frac{n_B}{n_A + n_B} E_B$$

where n_A and n_B are the number of magnetic ions at A and B sites, respectively.

In a compound of ideal crystal, the chains of such linkages are infinitely long. However, the nonmagnetic ion substitution decreases the number of active linkages weakening the various exchange interactions. Thus, there is a reduction of thermal energy required to offset any spin alignment with the result that Curie temperature is reduced. The study of T_C gives an estimate of the operating temperature limit for the magnetic materials.

1.5 MÖSSBAUER STUDIES

Mössbauer spectroscopy is concerned with transition that takes place inside atomic nuclei. The incident radiation that is used is a highly monochromatic beam of gamma rays whose energy may be varied by making use of Doppler effect. The gamma rays that are used in Mössbauer spectroscopy are produced by decay of radioactive elements such as $^{57}Fe_{26}$ or $^{119}Sn_{50.}$ Under certain condition of recoilless emission all of the energy change in the nuclei is transmitted to the emitted gamma rays and this give rise to a highly monochromatic beam of radiation. This radiation is than absorbed by a sample that contains similar atoms to those responsible for the emission. In practice the energy of the gamma rays is modified by making used of the Doppler effect. The energy of the gamma ray can be increased or decreased by giving positive or negative velocity. If the source is moved with different velocities towards the absorber, behind which the detector is placed, then whenever the effective values of the gamma ray energies are matched at a certain Doppler velocity, resonance will be at a maximum and the count rate in the detector, a minimum. At any higher or lower velocity, resonance will be less the count rate will increase, and at velocity far away from that defining maximum resonance, the resonance will get destroyed and the count-rate a maximum. A plot of relative transmission *versus* series of Doppler velocities between source and absorber is the basic form of Mössbauer spectrum. The absorption line is Lorentzian in shape with width at half height corresponding to 2⁻. The schematic set up used in the Mössbauer experiment in transmission geometry, and the typical Mössbauer spectrum is shown in Figure 6.

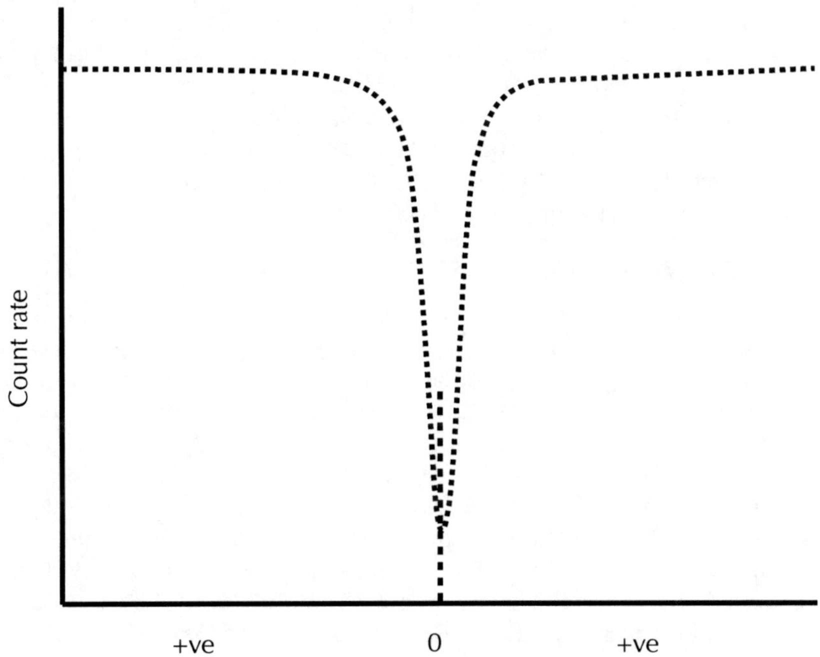

FIGURE 6 Mössbauer spectra.

The Mössbauer effect has been detected in a number of transitions of many isotopes in different elements. The most commonly used Mössbauer isotope is ^{57}Fe obtained from a ^{57}Co source. Suitable lifetime and energy in its first excited state combines to make it widely used for most of the works in Mössbauer experiment.

Considering the example of Mössbauer isotope Fe^{57}, the 14.4 keV gamma ray emitted as a result of transition from the 3/2 state to 1/2 state has a line width of 4.19 $\times 10^{-9}$ eV determine entirely by the lifetime of the excited nuclear state which is $\tau = \frac{1}{\gamma}$ $= 0.98 \times 10^{-7}$ s. This width permits an intrinsic resolving power given by:

$$\frac{\Gamma}{E_o} \cong \frac{4.19 \times 10^{-9}}{14.4 \times 10^{3}} \cong 10^{-13} \text{ eV}$$

It is particularly useful in studying a variety of hyperfine interaction between the positively charged nucleus and the surrounding electron cloud such as (1) isomer shift or the electrostatic monopole interaction between the nucleus and the s-electron density at the nucleus, (2) electric quadrupole interaction between quadrupole moment of the nucleus and the electric field gradient at the nucleus and (3) magnetic dipole interaction between the magnetic dipole moment of the nucleus and the magnetic field at the nucleus. The isomer shift is the coulombic or electrostatic interaction between the positively charged nucleus and the electrons. The s-electron wave functions have non zero electron charge density within the nuclear volume and change in this s-electron density may arise from change in the valiancy. This results in an altered coulombic interaction manifesting itself as a shift of the nuclear levels. This shift also arises due to the fact that the effective charge radius of nucleus is different in the ground and isomeric excited state. Thus, the isomer shift depends on the chemical environment of the nucleus and gives useful information on the interaction of nuclear charge and electron cloud. Mössbauer spectroscopy helps in observing directly the Zeeman splitting of nuclear levels and in determining the hyperfine field.

1.6 LITHIUM ZINC NICKEL FERRITES

Lithium zinc nickel ferrites having the general formula $Li_{0.4-0.5x}Zn_{0.2}Ni_xFe_{2.4-0.5x}O_4$ where $0.02 \leq x \leq 0.1$ in steps of 0.02 were prepared using the citrate precursor method in the following way [27]. The starting chemicals used are lithium nitrates [$LiNO_3$ {Merck Germany}], zinc nitrate [$Zn(NO_3).6H_2O$ {Sigma Aldrich}], iron nitrate [$Fe(NO_3).9H_2O$ {Merck Germany}], nickel nitrate [$Co(NO_3).6H_2O${Merck Mumbai}] and citric acid [$C_6H_8O_7$ {Merck Mumbai}]. A homogenous solution was made such that the ratio of the metal nitrates to citric acid is 1:1. The pH of the solution was then controlled at 7. After controlling pH at 7 it was refluxed at 40°C with continuous stirring for about half an hour. The solution was then put in an oven at 100° where auto combustion process takes place with evolution of large amount of gases. The ash-synthesized product (Figure 7) so obtained is the typical spinel structured lithium ferrite powder with nanocrystallite size.

The powder is mixed with polyvinyl alcohol and pressed into pellets and torroids by applying 50 kN of pressure. The samples are then densified by giving final sintering at a temperature 1,080°C for 5 hr heating rate being controlled at 5°C/min. The samples are then furnace cooled to room temperature.

FIGURE 7 Ash-synthesized powder.

The phase analysis was confirmed from the simple XRD technique. A typical XRD pattern for $Li_{0.38}Zn_{0.2}Ni_{0.04}Fe_{2.38}O_4$ is shown in Figure 8.

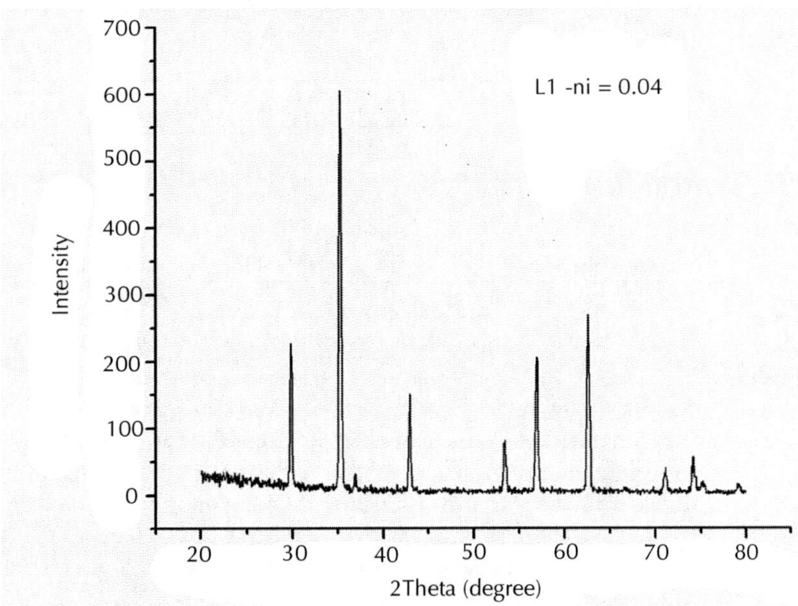

FIGURE 8 Typical XRD pattern.

From the XRD data the lattice parameter ‘a’ was calculated and a decrease with increasing Ni^{2+} ion concentration is observed (Table 1). It had been learnt that lattice expansion takes place if the doping ion has larger radii than the displaced ion. In the present series of ferrite samples Ni^{2+} ion with ionic radius 0.078 nm substitute Fe^{3+} and Li^{1+} ions with radii 0.067 nm and 0.078 nm respectively, and the lattice parameter is expected to increase. However, the observed decrease, contrary to expectation may be attributed to the fact that the contribution to lattice parameter in certain cases of substitution does not solely depend on ionic radii but also on the force of interaction which results from the equilibrium between long range attractive coulomb forces and short range repulsive interaction between the ions [27-29]. This requires precise knowledge of shape and surface structure of the particles as well as existence of any surface charge; moreover the properties are complicated for small size material. The work has been reported [27].

The crystallite size was estimated from XRD analysis using the Scherrer formula which is given as:

$$D_{hkl} = \frac{0.89\lambda}{\beta\cos\theta},$$

where λ is the incident wavelength of CuK_α radiation of XRD β is the full–width at half–maximum and θ is the diffraction angle.

An increase in the crystallite size with increasing Ni^{2+} ion concentration is noticed. The values are shown in Table 1.

TABLE 1 Lattice parameter, Crystallite size, Saturation magnetization, and T_C for $Li_{0.4-0.5x}Zn_{0.2}Ni_xFe_{2.4-0.5x}O_4$.

Concentration (x)	0.02	0.04	0.06	0.08	0.1
Lattice parameter (A°)	8.437	8.494	8.457	8.414	8.408
Crystallite Size (nm)	26	37	40	49	54
Saturation magnetization (emu/gm)	94	80	102	78	79
Tc (°C)	584	573	571	566	563

The compositional variation of room temperature saturation magnetization is given in Table 1. It depicts a decrease with the progressive substitution of Ni^{2+} ions, although an anomalously high value is obtain for Ni = 0.06 concentration. The variation in the saturation magnetization can be explained by considering the Neel’s molecular field model and the cation distribution. According to Neel’s model AB superexchange interaction having the most favorable A-O-B angle of about 125°C dominates over intrasublattice AA and BB interaction. Therefore, the net magnetization is given by the vector sum of the magnetization of the two sublattices M_A and M_B. Again the cationic distribution has been assumed to be [27]:

$$(Zn_{0.2}Fe_{0.8})_A[Li_{0.4-0.5x}Fe_{16-0.5x}Ni_x]_BO_4$$

It is observed that Li^{1+} ion and Zn^{2+} ion being non magnetic they do not make any contribution to the magnetization, and the contribution comes only from Ni^{2+} ion with magnetic moment $2\mu_B$ and Fe^{3+} ion with magnetic moment $5\mu_B$. Hence, the net magnetization will be given by:

$$M_S = M_B - M_A = [(2x + 8.0 - 2.5x) - 4.0] = 4 - 0.5x$$

The saturation magnetization is thus expected to decrease, and has been observed experimentally.

The variation of the T_C with Ni concentration, determined using the Soohoo's method [30] and a fall in its value is observed. This can be understood in terms of the magnetic superexchange interaction proposed by Neel, which has a direct relation with the T_C [26, 31, 32]. According to this model, the inter sublattice AB superexchange interaction dominates over the intra sublattice AA and BB superexchange interactions. The strength of this A-B interaction is determined by the number of $Fe_A^{3+} - O^{2-} - Fe_B^{3+}$ linkages [26] per formula unit and the overall strength of this linkage determines the T_C of the ferrite samples.

From the cationic distribution of the present series of ferrites expressed as $Li_{0.4-0.5x}Zn_{0.2}Ni_x Fe_{2.4-0.5x}O_4$ the magnetic Fe^{3+} ions are being replaced by less magnetic Ni^{2+} ions. Thus, with progressive substitution of Ni^{2+} ion the number of active $Fe_A^{3+} - O^{2-} - Fe_B^{3+}$ linkage decreases, thereby weakening the AB superexchange interaction. Hence, the thermal energy required to offset the spin alignment decreases leading to decrease in the T_C.

The Mössbauer study was also carried out, analysis being done using NMOSFIT computer program based on Meerwal's program [33]. Typical spectra are shown in Figure 9. The results have been reported [27].

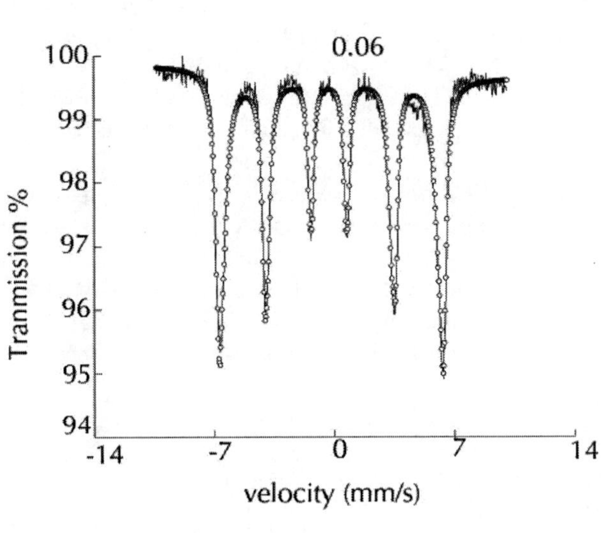

FIGURE 9 Typical analyzed Mössbauer spectra.

The values of isomer shift of the Fe^{3+} ion at tetrahedral A site S at octahedral B site are given in Table 2, accurate up to 0.03 mm/s. They are found to be essentially independent of the Ni substitution.

TABLE 2 Mossbauer parameters of $Li_{0.4-0.5x}Zn_{0.2}Ni_xFe_{2.4-0.5x}O_4$.

Concentration (x)	Isomer shift (IS) (mm/s)		Quadrupole splitting (Q) (mm/s)		Hyperfine field (H)(T)	
	IS_B	IS_A	Q_B	Q_A	H_B	H_A
0.02	0.22	0.20	0.02	0.02	40.8	39.4
0.04	0.22	0.21	0.04	0.04	40.8	39.1
0.06	0.20	0.19	0.01	0.01	40.7	38.7
0.08	0.20	0.19	0.01	0.01	40.7	38.2
0.1	0.22	0.19	0.02	0.01	40.8	38.1

This may be attributed to the fact that the substitution of Ni^{2+} ion has negligible influence on the s-electron charge distribution of Fe^{3+} ion. However, the isomer shift for Fe^{3+} ion at the tetrahedral A site was found to be slightly less positive than that for Fe^{3+} ion at the octahedral B site. This has been explained in an earlier reported work, on the basis of the difference in the Fe^{3+}-O^{2-} inter nuclear separations for the A and B site ions. The A site ion may have a smaller bond separation which leads to a larger overlapping of orbital of Fe^{3+} ion and O^{2-} ion at the A site with the result that the Fe^{3+} ion at A site experienced a larger covalence leading to smaller isomer shift [34, 35].

The quadrupole splitting for all the ferrite samples was observed to be negligible within the experimental error (Table 2). This indicates that overall cubic symmetry is preserved. The variation of internal magnetic field at the tetrahedral site (H_A) and octahedral site (H_B) with Ni^{2+} ion concentration can be seen in Table 2. It is observed that H_A slightly decreases with increasing Ni^{2+} ion concentration while H_B remains almost constant. The variation of magnetic field can be understood on the basis of Neel's molecular field model and the cationic distribution.

In a spinel ferrite each Fe^{3+} ion at A site has twelve B site ion as nearest neighbors, while a B site Fe^{3+} ion has six A site and six B site ion as nearest neighbors. Now according to Neel's molecular field model, the AB superexchange interaction predominates the intra sublattice AA and BB interactions. It is also known that in lithium ferrites Zn^{2+} ion has preference for A site while Ni^{2+} for B site [2, 36, 37]. When both this ions are substituted simultaneously, the cationic distribution can be written as:

$$(Zn_zFe_{1-z})_A[Li_{0.5-0.5z-0.5x}Fe_{1.5+0.5z-0.5x}Ni_x]_BO_4$$

The Zn has been kept fixed at 0.2 and the amount of Ni^{2+} ion, x is varied. The cationic distribution therefore, becomes:

$$(Zn_{0.2}Fe_{0.8})_A\,[Li_{0.4-0.5x}Fe_{1.6-0.5x}Ni_x]_BO_4$$

In the proposed cationic distribution $Zn^{2+}$and Li^{1+} ions being non magnetic do not contribute to the magnetic interaction of the sublattice and hence to the nuclear magnetic field. The net internal magnetic field is mainly due to the interactions between the Fe^{3+} ions at A and B sites and the Ni^{2+} ion at the B site. As can be seen from the cation distribution with increased substitution the magnetic neighbors of Fe^{3+} ion at B site remains unchanged, thus accounting for constant value of H_B. The magnetic environment of Fe^{3+} ion at A site becomes more non-magnetic as Ni^{2+} ions with magnetic moment $2\mu_B$ replaces Fe^{3+} ions with magnetic moment $5\mu_B$. This results in the observed slight decrease in hyperfine field at A site.

However, H_A is slightly smaller than H_B. The lower value of A site field may be attributed to the more covalent nature of $Fe^{3+}O^{2-}$ bond for the A site [38, 39] which leads to a greater degree of spin delocalization at the A site, and hence to a smaller value of H_A. This agrees with the isomer shift observations.

1.7 CONCLUSION

The citrate precursor synthesis method is able to produce materials at the nanoscale. It has several advantage of taking short duration of time, low temperature synthesis method with enhanced properties. It will overcome certain drawbacks which arise due to high temperature and long duration of sintering and hence need arise to lower down the sintering temperature. These are all achieved in the citrate precursor synthesis method.

KEYWORDS

- **Citrate precursor synthesis method**
- **Curie temperature**
- **Homogenization**
- **Mössbauer spectroscopy**
- **Nanomaterials**

REFERENCES

1. Fox, A. G., Miller, S. E., and Weiss, M. T. Behavior and applications of ferrites in the microwave region. *Bell Syst. Tech. J.,* (1955).
2. Smit, J. and Wijn, H. P. *J. Ferrites*, Phillips technical library, Eindhoven (1959).
3. Baba, P. D., Argentina, G. M., Courtney, W. E., Dionne, G. F., and Temme, D. H. *IEEE Trans. Mag.*, **8**(1), 83 (1972).
4. Kim, H. T. and Im, H. B. *IEEE Trans. Mag.*, **18**(6), 1541 (1982).
5. Sankaranarayanan, V. K., Pankhurst, Q. A., Pedickson, D., and Johnson, C. E. *J. Magn. Magn. Mater.*, **130**, 288 (1994).
6. Gajbhiye, N. S., Bhattacharya, U., and Darshane, V. S. *Thermoch. Acta.*, **264**, 219 (1995).
7. Pechini, M. U. S. Patent., **3**, 330, 697 (1967).
8. Galceran, M., Pujol, M. C., Aguilo, M., and Diaz, F. *J. Sol-Gel Sci. Techn.*, **42,** 79 (2007).
9. Cannas, C., Falqui, A., Musinu, A., Peddis, D., and Piccaluga, G. *J. Nanoparticle Res.*, **8**, 255 (2006).
10. Gorter, E. W. *Philips Res. Repts.* Einthoven., **9**, 295 (1954).
11. Hafner, S. *Schweiz. Min, petrogr. Mitt*, **40**, 207 (1960).

12. Blasse, G. *J. Inorg. Nucl. Chem.*, **25**, 136 (1963).
13. Bragg, W. H. *Nature*, **95**, 561 (1915).
14. Nishikawa, S. Structure of some crystal of the spinel groups. *Proc. Math. Phys. Soc.*, Tokyo, **8**, 199 (1915).
15. Alex, G. *Modern Ferrites Technology*. Van Nostrand Reinhold, New York (1990).
16. Simonet, W. and Hermosin, A. *IEEE Trans. Mag*., **14**(5), 903 (1978).
17. Hsu, J. H., Ko, W. S., Shen, H. D., and Chen, C. H. *IEEE Trans. Mag*., **30**, 4875 (1994).
18. Ibetombi, S., Sumitra, Ph., and Prakash, C. *J. Alloys. Comp*., **475**, 328 (2009).
19. Ibetombi, S. *Synthesis of Li-Zn-Co ferrites and Li-Zn-Ni ferrites by citrate precursor method and their structural, Mössbauer, magnetic and electrical characterization*. PhD thesis, Manipur University, India (2010).
20. Kramers, H. A. *Physica*., **1**, 182 (1934).
21. Neel, L. *Ann. Phys*., **3**, 137 (1948).
22. Anderson, P. W. *Phys. Rev*., **797**, 350 (1950).
23. Yafet, Y. and Kittel, G. *Phys. Rev*., **87**, 290 (1952).
24. Lotgering, F. K. *Phillips Res. Rep*., **11**, 190 (1956).
25. Weiss, P. *J. Phys*., **6,** 661 (1907).
26. Gilleo, M. A. *J. Phys. Chem. Solids*., **13**, 33 (1960).
27. Ibetombi, S. Sumitra, Ph., and Prakash, C. *J. Magn. Magn. Mater*., **321**, 2779 (2009).
28. Sattar, A. A., El Sayed, H. M., Agami, W. R., and Ghani, A. A. *Amer., J. Appl. Sci*., **4**, 89 (2007).
29. Naughton, B. T. and David, R. C. *J. Amer. Ceram. Soc*. **90**, 3541 (2007).
30. Soohoo, R. F. *Theory and Application of Ferrites*, Prentice Hall Inc. Englewood Cliffs USA, (1960).
31. Dionne, G. F. *J. Appl. Phys*., **45**, 3621 (1974).
32. Ibetombi, S., Sumitra, Ph., Sarma, H. N. K., and Chandra, P. AIP Conference Proceedings, **1003**, 136 (2008).
33. Meerwall, E. V. *Comp. Phys. Commun*., **9**, 117 (1975).
34. Goodenough, J. B. and Loeb, A. L. *Phys. Rev*., **98**, 391 (1955).
35. Hudson, A. and Whitfield, H. *J. Mol. Phys*., **12**, 165 (1967).
36. Kisan, P., Sagar, D. R., and Swarup, P. *J. Less Common Met*., **108**, 345 (1985).
37. Singh, A. K., Goel, T. C., Mendiratta, R. G., Thakur, O. P., and Prakash, C. *J. Appl. Phys*., **92**, 3872 (2002).
38. Fatseas, G. A. and Krishnan, R. *J. Appl. Phys*., **39**(2), 256 (1968).
39. Watson, R. E. and Freeman, A. *J. Phys. Rev*., **123**, 2027 (1961).

2 Shape Controlled Synthesis, Characterization, and Optical Properties of Silver Nanostructures

P. Kumbhakar, A. Maity, R. Sarkar, and A. K. Mitra

CONTENTS

2.1 Introduction 19
2.2 Experimental Details 21
2.2.1 Materials 21
2.2.2 Synthesis 22
2.2.3 Characterization 22
2.3 Theoretical Study of Optical Absorptions in Silver Nanoparticles 22
2.3.1 Optical Absorption in Spherical Metal Nanoparticles 22
2.3.2 Optical Absorption in Rod-like Metal Nanoparticles 23
2.4 Experimental Discussion and Results 23
2.4.1 Sem, Tem, and Fourier Transform Infrared (FTIR) Studies 23
2.4.2 UV-visible Absorption Studies 31
2.4.3 PL Emission Studies 34
2.4.4 Possible Mechanism Growth of Different Nanostructures Synthesized in Cases I And II 36
2.5 Conclusion 38
Keywords 39
Acknowledgment 39
References 39

2.1 INTRODUCTION

There are the novel chemical synthesis of silver nanoparticles (NPs) of different size and shapes, like nanorod, nanospindle, nanocube, nanosphere, nanoellipsoid, and some plate like structures. Here silver nitrate ($AgNO_3$) has been used as the precursor, polyvinyl pyrolidone (PVP) as the stabilizer and benzoic acid as the reducing agent and the temperature of the reaction has been varied from 60 to 120°C at 10°C interval to obtain different nanostructures of silver. It has been found that by taking the higher value of the ratio between $AgNO_3$ and PVP but with the same amount of benzoic acid, different shapes of silver NPs, such as spherical, triangular, pentagonal, hexagonal, and some elongated shapes are produced. The nanostructures of the samples are characterized using scanning electrons and transmission electron microscopes. The UV-visible spectra of all the samples are collected and the absorption spectra of the first set of synthesized samples exhibit two absorption peaks, one in visible region (transverse) and another in near infrared region (longitudinal) but longitudinal band is absent for the second set of samples that is with higher $AgNO_3$ and PVP ratio. It is found that as the temperature increases; the longitudinal peak wavelength is blue shifted. So both temperature and the $AgNO_3$ and PVP ratio have dominant roles in producing different morphologies, size and shape of the silver NPs. The photoluminescence (PL) emissions in the UV-visible region have been obtained from the second set of samples and it is found that the intensity of the most intense PL peak at 345 nm reduces with the increase in the temperature of synthesis.

The synthesis of metal NPs become the center of attraction in research and technology, because they exhibit several chemical and physical properties having great potential in biological imaging of tumor cell and photo thermal therapy of cancer [1-3], catalysis [4], photonics [5, 6], surface enhanced raman spectroscopy (SERS) [7], surface enhanced fluorescence (SEF) [8] and so on. In each of these applications, the size as well as the shape of the NPs has a dominant role and so controlling the size and shape of the metal NPs has a great impact on technological application and scientific advancements. It becomes a challenge to material scientists to control the shape of metal NPs of different sizes. Recently, metal NPs of different shapes like, sphere [9, 10], cube [11, 12], triangle [13, 14], rod [15, 16], hexagon [17], spindle [18, 19], and wires [20, 21] have been synthesized by many authors.

Each shape has different optical properties that are available from UV to infrared (IR) region, like spherical particles exhibit optical properties in the visible range, rod and cube like particles in the visible (transverse band) and near IR (longitudinal band) regions. Therefore, the synthesis and characterization of properties of silver and gold nanorod or cube have attracted a great attention, as the color of the light they scatter can be tuned from near IR to visible region by simply controlling the aspect ratio. Since, these nanostructures scatter and absorb light in IR region (where absorbance is the minimum in biological tissue and the transmission is the maximum), they can be used in photo thermal therapy and in optical imaging of cancer [1-3, 22] as the absorbed energy produced localized heating due to thermal dissipation. In sensor applications, [23] the absorption and scattering are measured as the function of changing chemical and physical environment of the surface of NPs. There have been some theoretical and experimental work in past on the synthesis and characterization of optical properties of different sizes and shapes of metal NPs [24-26].

There are several reports on the shape evolution of different metal nanostructures from one shape to another [26-50]. The photon induced shape transition of gold nanorod into spherical particle by laser ablation has been reported [30]. In their report they also found Ø (phi) shaped particle as the stable reaction intermediate from rod to sphere transition and the first transition or change in shape starts from the middle of the nanorod which is melted by laser energy [30]. The rod to sphere and sphere to prolate shape conversion by femtosecond laser pulse is also studied by other authors [27-29]. A series of shape conversion of silver nanoplates in one system in which silver nanoprisms can be transformed to nanodisk and *vice versa* by adding more citrate solution in a light induced reaction [31]. Recently, solvothermal techniques have been used to synthesize metal NPs and it has been shown that small triangular nanoplates can be transformed into nanobelt and nanofilms by controlling temperature, reaction time, and concentration of the reactants [32]. A hydrothermal [33] report on controlled growth of silver NPs has reported on different shape by controlling capping agent to precursor ratio, temperature, and so on.

However, here we have presented a novel technique to synthesize silver NPs of various shapes and sizes through a chemical route by simply controlling the temperature or the ratio of the reactants of the chemical reaction. We have carried out the work in two cases. In the first case (case-I), we choose a particular ratio of $AgNO_3$ PVP and some amount of benzoic acid, which acts as the reducing agent and then in the second case (case-II) we increased the value of the ratio of $AgNO_3$ PVP but with the same amount of benzoic acid. The uniqueness of this effort is that we have found rod, spindle, cube, sphere, ellipsoid, and some plate like structure in case-I in different temperatures in a simple run, consecutively with different aspect ratios that is, to get a particular shape we have to give the desired temperature. This is a very simple method in compared to the reported methods [31-33]. Next in the case-II we have increased the precursor amount only and the amount of other compounds are kept fixed and interestingly, hexagonal, pentagonal, triangular plates, and some elongated structures are synthesized, for which the growth mechanism is completely different. The possible mechanisms of growth of different silver nanostructures are presented. UV-visible absorption characteristics of the samples prepared in case-I show that two absorptions peaks are present one due to transverse mode of oscillations and another due to the longitudinal modes of oscillations. Whereas the absorption characteristics of the samples prepared in case-II shows that there is only one peak due to transverse mode of oscillations in all the samples. The PL emissions from the samples prepared in case-II, *viz.*, have been collected and it is found the samples exhibit PL emission in the UV-visible wavelength regions. It is also found that the intensity of PL emission at 354 nm decreases gradually as the synthesis temperature increases and this occurs due to the increase in the average size of the particles with the increase in the temperature of the synthesis of the samples.

2.2 EXPERIMENTAL DETAILS

2.2.1 Materials

The Silver nitrate ($AgNO_3$) is used as precursor, poly(*N*-vinyl-2-pyrolidone), (PVP) is used as capping agent, benzoic acid is used as reducing agent, and methyl alcohol (CH_3OH) is used as dispersive media. All the chemicals used in this study are analytical grade reagents (Merck) and are used as received.

2.2.2 Synthesis

The synthesis is performed within a beaker in two cases as described earlier. In the first case (case-I) 0.0025 mole of $AgNO_3$ is mixed with 10 ml methanol with simultaneous mixing of benzoic acid solution (0.5 gm in 10 ml methanol) and PVP solution (1 gm in 10 ml methanol), and stirred vigorously with magnetic stirrer at room temperature. After 20 min, the light yellow solution is kept at heater and temperature was increased from 60 to 120°C with 10°C interval. During heating, after every 10°C raise in the temperature, the samples are collected in different test tubes and labeled accordingly. The color of the samples gradually changes as yellow, deep yellow, brownish, and reddish. The color of all the samples after few days becomes greenish gray. In the second case (case-II) we have increased the ratio between $AgNO_3$ and PVP by taking 0.025 mol $AgNO_3$ and by keeping other parameters same. The color of the freshly prepared samples changes from light yellow to deep yellow as the temperature increases and it becomes reddish brown after few days.

2.2.3 Characterization

The shape and the microstructures of the samples have been determined by using scanning electron microscope (SEM) (Hitachi, Model No. S-3000 N) and transmission electron microscope (TEM) (JEOL JEM 2100). Optical absorption characteristics of the samples are obtained by using UV-visible spectrophotometer (Hitachi-3010) and PL emissions measurements have been carried out by using Spectrofluorimeter (Perkin Elmer, LS 55). For UV-visible and PL measurements, samples are dispersed in methanol and kept in a quartz cuvette of path length 10 mm.

2.3 THEORETICAL STUDY OF OPTICAL ABSORPTIONS IN SILVER NANOPARTICLES

2.3.1 Optical Absorption in Spherical Metal Nanoparticles

The metallic NPs having sizes much smaller than wavelength of absorbing light acts as dipole in the presence of the electric field of an incoming radiation and the total extinction coefficient of small metal NPs are given by the Mie theory [36]. In quasistatic approximation, extinction coefficient κ for N particles of volume V is given by:

$$k = \frac{18NV\varepsilon_m^{3/2}}{\lambda}\frac{\varepsilon_2}{[\varepsilon_1 + 2\varepsilon_m]^2 + \varepsilon_2^2} \quad (1)$$

where λ is wavelength of absorbing radiation and ε_m is the dielectric constant of the surrounding medium, which is taken as constant with the variation of wavelength. The real and imaginary parts of metal dielectric constant $\epsilon(\omega)$ $[\epsilon(\omega) = \epsilon_1(\omega) + i\epsilon_2(\omega)]$, are ϵ_1 and ϵ_2 and ω is the angular frequency. It is to be noted that extinction coefficient is maximum when $\varepsilon_1 = -2\varepsilon_m$ and $\epsilon_2(\omega)$ is either small or weakly dependent [34] on ω. The bandwidth mainly depends on $\epsilon_2(\omega)$. Now from Equation (1) plasmon absorption is independent on particle diameter (<20 nm) within dipole approximation. But, experimentally a size effect is found when plasmon bandwidth increases for decrease of particle diameter D. As the size increases scattering becomes predominant and in the end it dominates extinction at large particle size.

2.3.2 Optical Absorption in Rod-like Metal Nanoparticles

The optical response is significantly changed when the shape of the particle is changed. For silver nanorod absorption spectrum splits into two distinct bands, corresponding to oscillation in transverse and longitudinal directions of the nanorod. The transverse mode shows resonance at near 420 nm which is superposed with plasmon band of spherical NPs. The resonance frequency of longitudinal mode is strongly depends on aspect ratio R (R = length of rod/width of rod). The optical absorption spectrum of silver nanorod with aspect ratio a can be expressed by using an extension of the Mie theory [36]. According to Gans [51] the extinction coefficient of N particles of volume V is given by:

$$k = \frac{2\pi N V \varepsilon_m^{3/2}}{3\lambda} \sum_j \left(\frac{(\frac{1}{P_j^2})\varepsilon_2}{(\varepsilon_1 + \frac{1-P_j}{P_j}\varepsilon_m)^2 + \varepsilon_2^2} \right) \tag{2}$$

Here, $\epsilon_m(\omega)$ is the dielectric constant of surrounding medium which is capping agent PVP in our case. Throughout whole calculation the value of ϵ_m has been taken constant. The P_j values are depolarization factor for the three axes A, B, and C of the nanorod with (A > B = C) and for nanodisk (A = B > C).
where

$$P_A = \frac{1-e^2}{e^2}\left[\frac{1}{2e}\ln\left(\frac{1+e}{1-e}\right) - 1\right] \tag{3}$$

$$P_B = P_C = \frac{1-P_A}{2} \tag{4}$$

$$e = \sqrt{1-\left(\frac{B}{A}\right)^2} = \sqrt{1-\frac{1}{R^2}} \tag{5}$$

2.4 EXPERIMENTAL DISCUSSION AND RESULTS

2.4.1 SEM, TEM, and Fourier Transform Infrared (FTIR) Studies

The SEM images of the samples prepared in case-I at different reaction temperatures are shown in Figures 1(a)–1(g). The images shown in Figures 1(a)–1(g) have been taken after 2 weeks of synthesis of the samples. From the Figure 1(a) it is seen that for samples prepared at 60°C temperature, the shape of the majority of the NPs are nanorods having average aspect ratio of ~7 and few of them are spherical in shape. It is seen from Figure 1(b) that particle shape of the prepared NPs at 70°C temperature remains rod-like, however with slightly lower average aspect ratio of ~6, and it is seen from Figure 1(c) that majority of the NPs which are prepared at 80°C temperature are spindle shaped with dimensions of 494 × 281 nm that is with average aspect ratio of ~1.7. It is found from Figure 1(d) that at 90°C temperature cube like particles having average particle size of 360 nm with sharp corner are synthesized. At 100°C temperature, the mixture of spherical and cube like particles having average particle size of 340 nm are obtained as can be seen from the Figure 1(e). At 110°C temperature spherical and ellipsoidal particles are found as can be seen from Figure 1(f). At 120°C temperature, the particles shape becomes plate like structures and few spherical particles are also present as shown in Figure 1(g). Thus, from the analyses of SEM images, it is found that there is a shape transition from longer rod to sphere and reaction intermediates are shorter nanorods of lower aspect ratio, spindle, cube, and plate like structures.

Figures 2(a)–2(g) show the high resolution transmission electron microscopy (HRTEM) images of the NPs prepared in the second case, that in case-II. It is seen from Figures 2(a)–2(g) that the shapes and morphologies of the synthesized NPs in this case are completely different from the former ones. In this case, as the temperature is increased from 60 to 120°C different shapes of silver NPs, such as some irregular plates (aggregated) with stacking fault, some pentagonal shaped along with elongated and small spherical particles (majority), hexagonal plates with some small spherical particles, hexagonal plates along with few triangular plates and small spherical particles, and some irregular shaped particles are synthesized. The particle size distributions for different samples are calculated from the respective HRTEM images and those are shown in Figures 3(a)–3(g). It is found that the average sizes of the particles are 10, 12, 11, 14, 16, 36, and 20 nm, respectively for the samples synthesized at 60, 70, 80, 90, 100, 110, and 120°C temperatures.

The temperature and ratio between precursor and capping agent (because in this case precursor amount is increased at least 10 times so the reaction rate becomes slow) have great effect on this shape modification which has also been reported by some other researcher [48]. Thus, we have seen that temperatures as well as the concentration of solution have a dominant role for shape evolution. In the study [30] gold nanorod to sphere transition has been reported by femtosecond laser ablation and the reaction intermediates are Ø shaped particle. In our case, we have get spindle, cube, and so on. Particle as reaction intermediates and finally spherical particles with some rectangular plates and the growth mechanism is completely different from it because initially there is no such different shaped particle. From the UV-visible absorption spectra we have seen that only surface plasmon resonance (SPR) peak is present in as produced sample. So, initially small NPs are produced and nucleated into small nanocluster and produced different types of seed. The nature of seed of different temperature sample is obviously different due to increased temperature treatment and increased reaction rate.

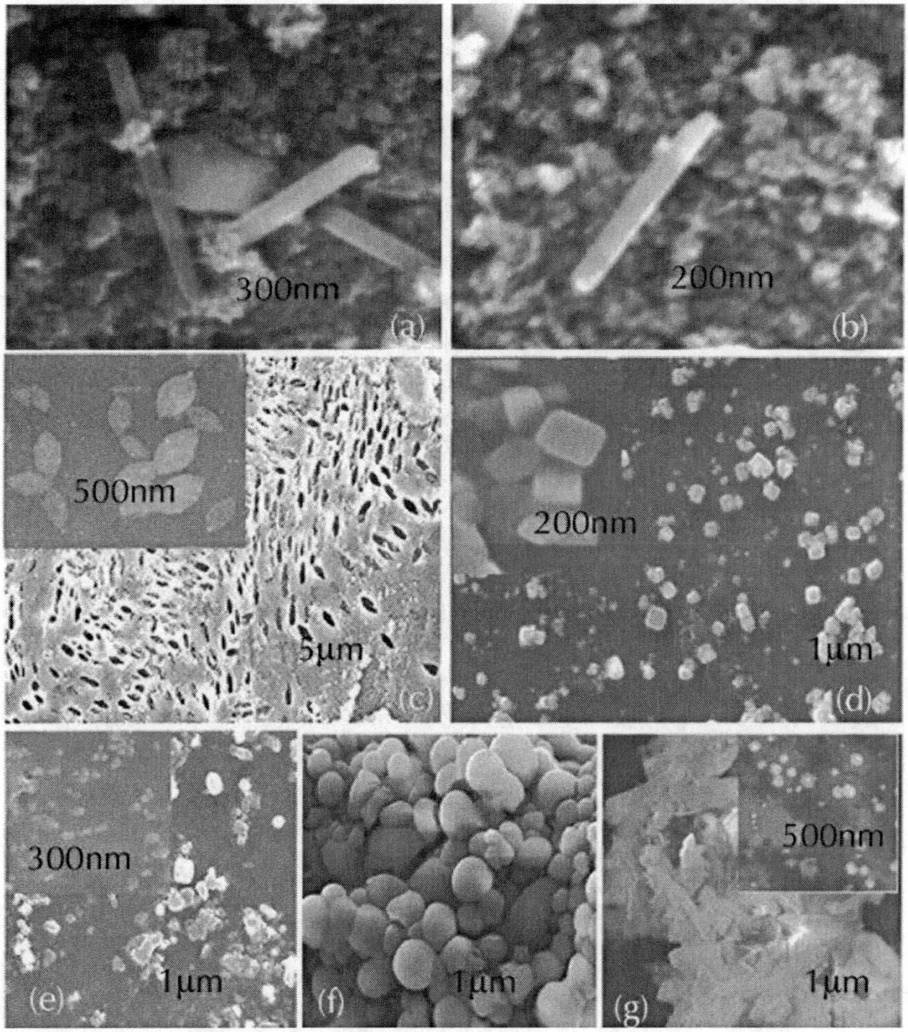

FIGURE 1 The SEM images of the samples synthesized in case-I. Figures marked as (a), (b), (c), (d), (e), (f), and (g) correspond to the synthesis temperature of 60, 70, 80, 90, 100, 110, and 120°C, respectively.

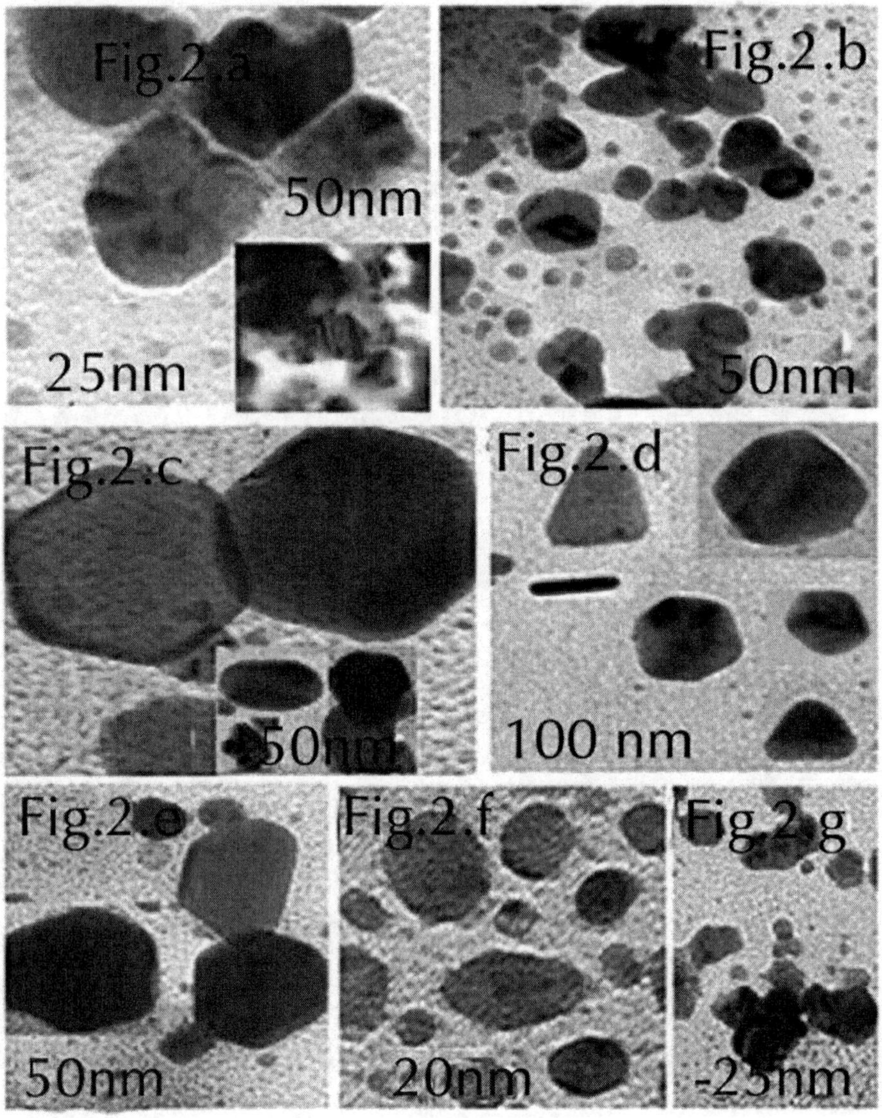

FIGURE 2 The TEM images of the samples synthesized in case-II. Figures marked as (a), (b), (c), (d), (e), (f), and (g) correspond to the synthesis temperature of 60, 70, 80, 90, 100, 110, and 120°C, respectively.

The fourier transform infrared (FTIR) spectra of pure PVP and PVP protected Ag NPs are shown in Figure 4 as curve 1 and 2, respectively, *viz.* for the sample which is prepared in the case-I at 60°C temperature. From the Figure 4 it is seen that a transmission minima appeared at 1,660 cm^{-1} for PVP due to C=O bond and it is slightly blue shifted to 1,652 cm^{-1} for PVP protected silver. The shift of the carbonyl bond frequency is associated with formation of coordination bond between silver atom and oxygen atom of carbonyl group [52].

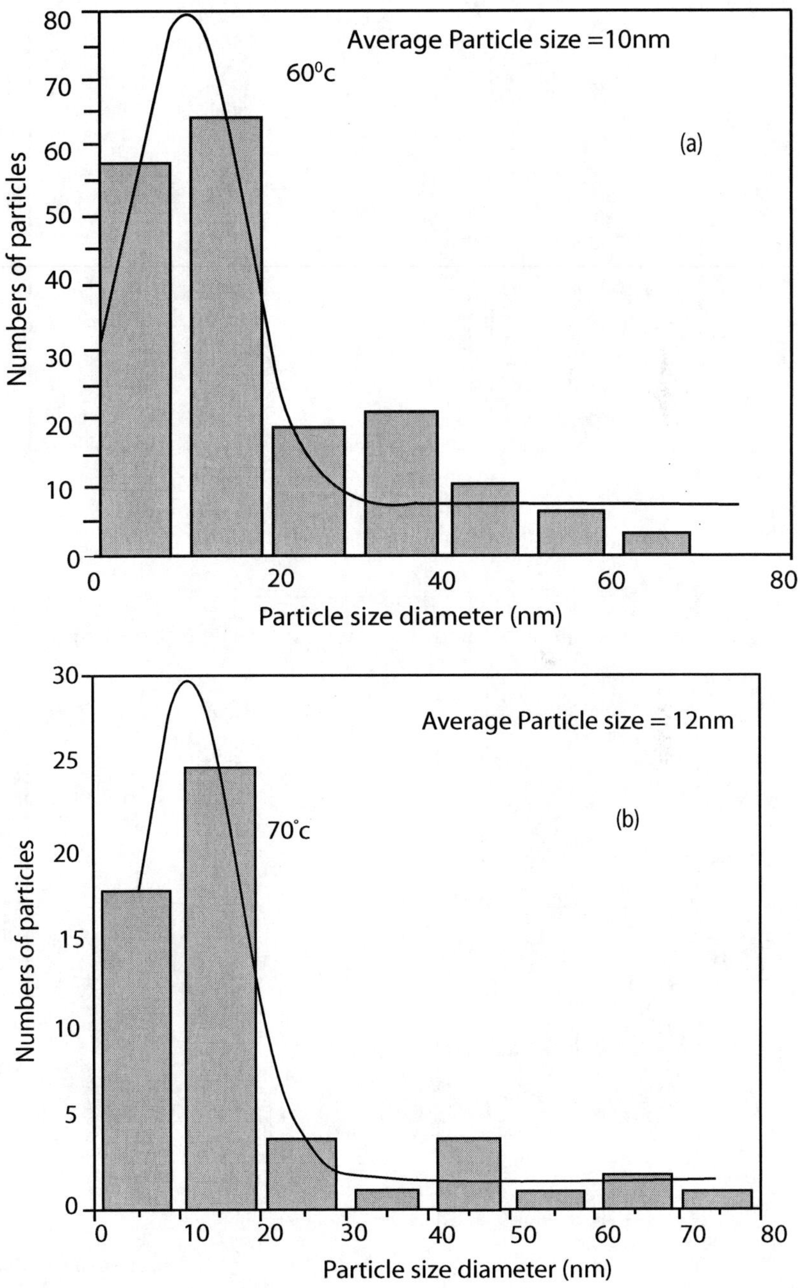

FIGURE 3 *(Continued)*

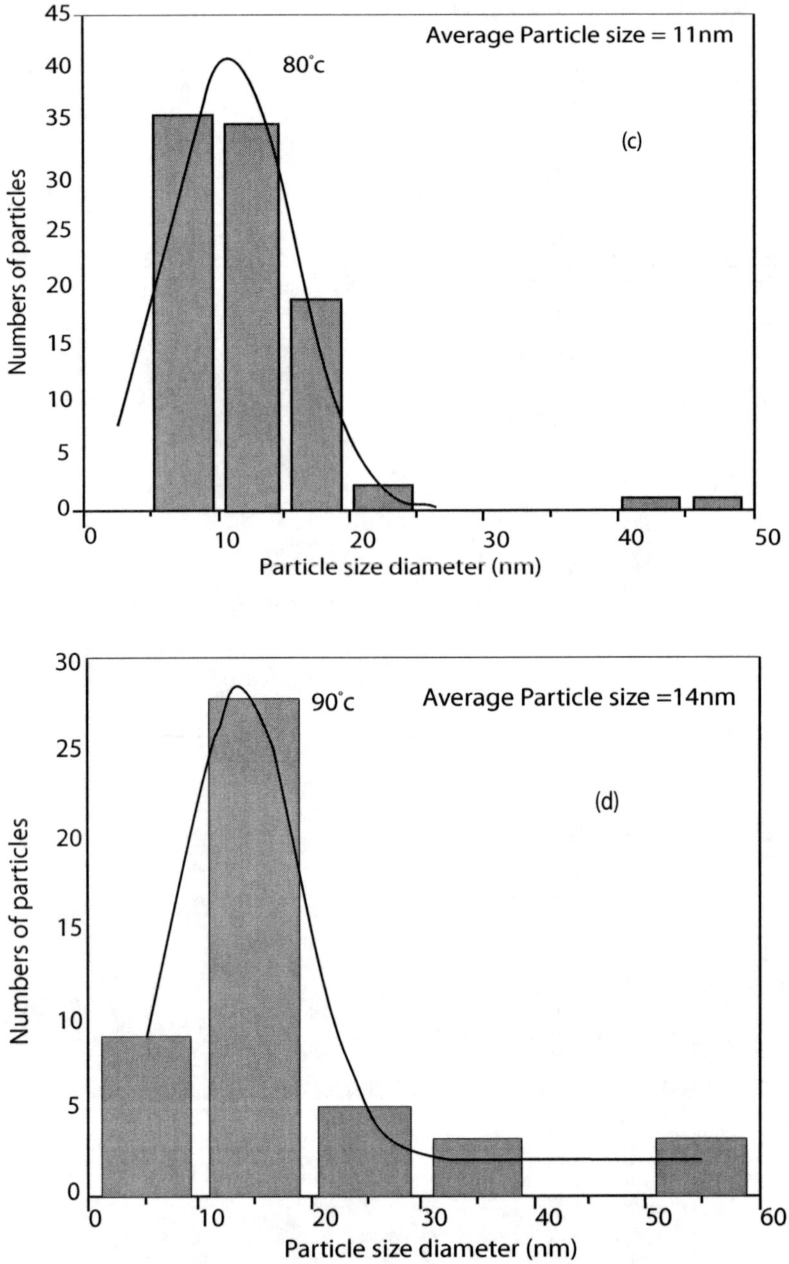

FIGURE 3 *(Continued)*

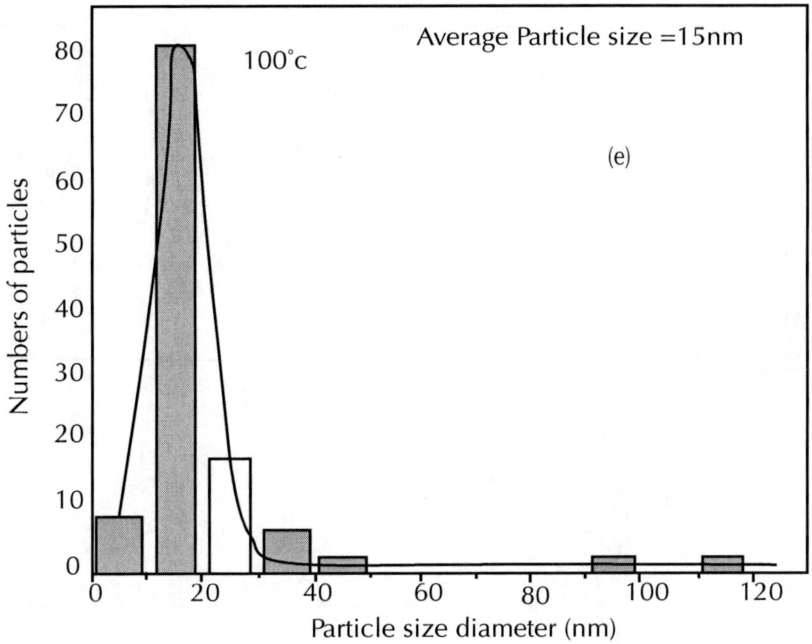

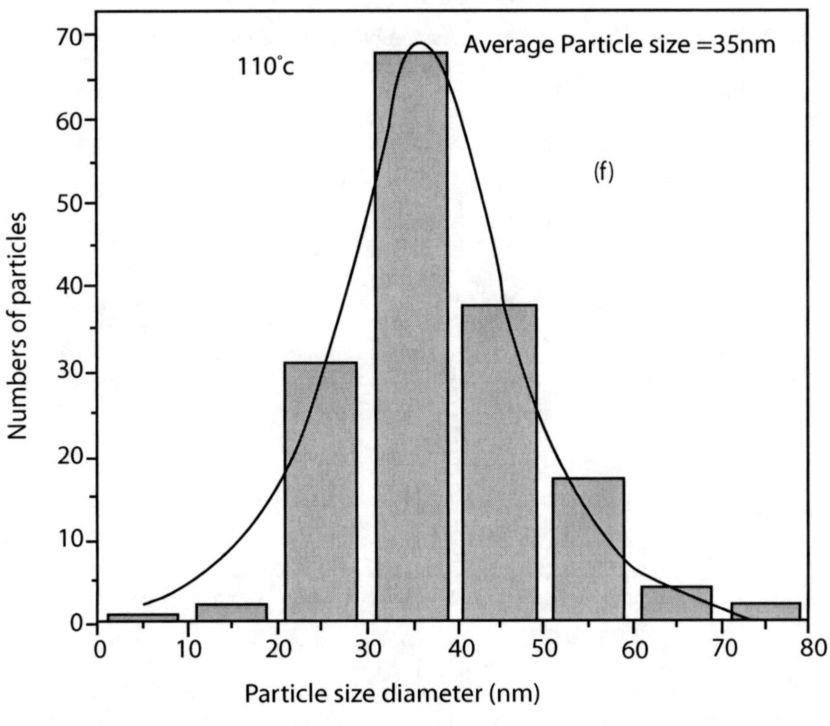

FIGURE 3 *(Continued)*

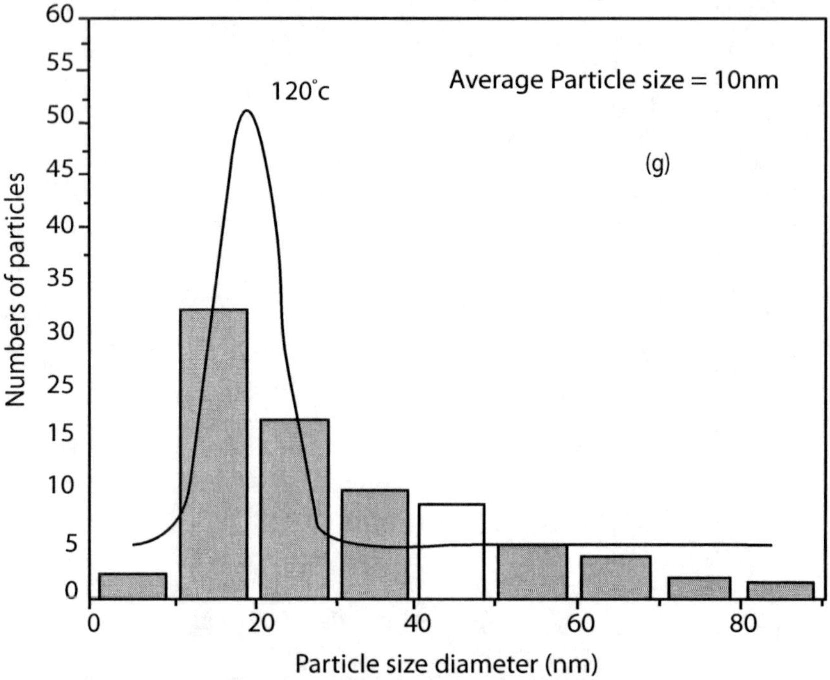

FIGURE 3 Particle size distributions for the samples synthesized in case-II. Figures marked as (a), (b), (c), (d), (e), (f), and (g) correspond to the synthesis temperature of 60, 70, 80, 90, 100, 110, and 120°C, respectively.

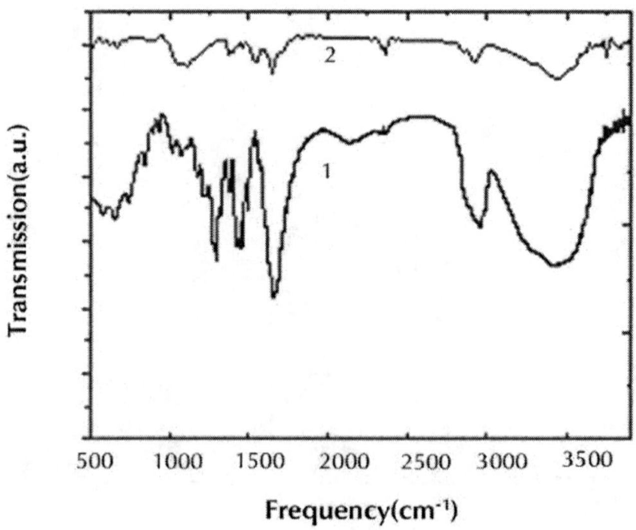

FIGURE 4 The curves marked as 1 and 2 of Figure. 4 show the FTIR spectra of pure PVP and PVP protected Ag NPs, respectively, *viz.* for the sample which is prepared in the case-I and at 60°C temperature.

2.4.2 UV-visible Absorption Studies

The UV-visible absorption spectra of all the samples are collected at room temperature and those are shown in Figures 5(a) and 5(b) for the samples which are prepared in case-I. The absorption characteristics of all the samples have been collected at different dates after synthesis, *viz.* Figure 5(a) and 5(b) show the absorption characteristics of the fresh sample and aged sample after 15 days of synthesis, respectively. From the Figures 5(a) and 5(c) it is found that only SPR peaks are appeared in all the samples in the wavelength range of 425–430 nm in the as produced samples but in the aged samples, other peaks are appearing in the longer wavelength region in addition to the conventional SPR peak of silver NPs. The appearance of the second peak at longer wavelengths is due to the longitudinal plasmon oscillations in the non spherical NPs (rod, spindle, bar, cube etc). As the synthesis temperature increases, there is a negligible change in the position of SPR peak due to transverse mode but there is a considerable change in longitudinal mode in red and infrared region in the aged samples.

It has been observed from the SEM studies that as the temperature of synthesis are increased, the aspect ratio of the synthesized NPs changes considerably and it also matches with UV-visible study, where longitudinal plasmon band is blue shifted with increase in temperature that is longitudinal plasmon band is very sensitive to the change in temperature. The change in the peak position of the aged (after 15 days of synthesis) sample with the variation of the synthesis temperature due to both transverse and longitudinal plasmon oscillations are shown in Figure 6.

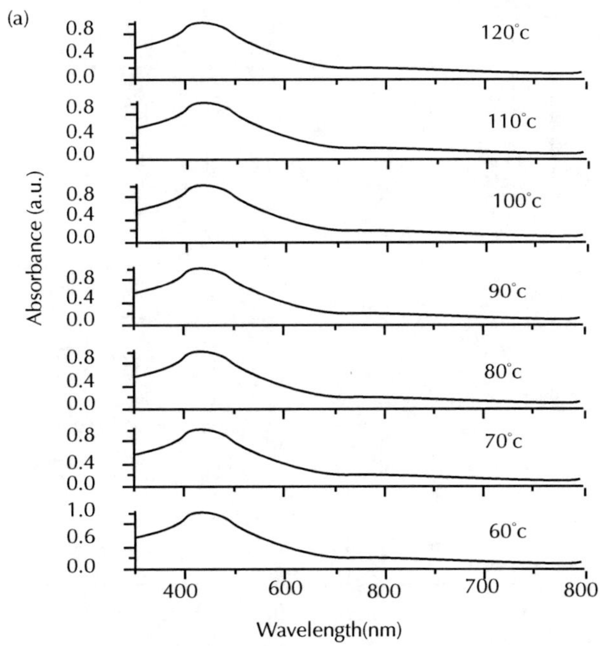

FIGURE 5 *(Continued)*

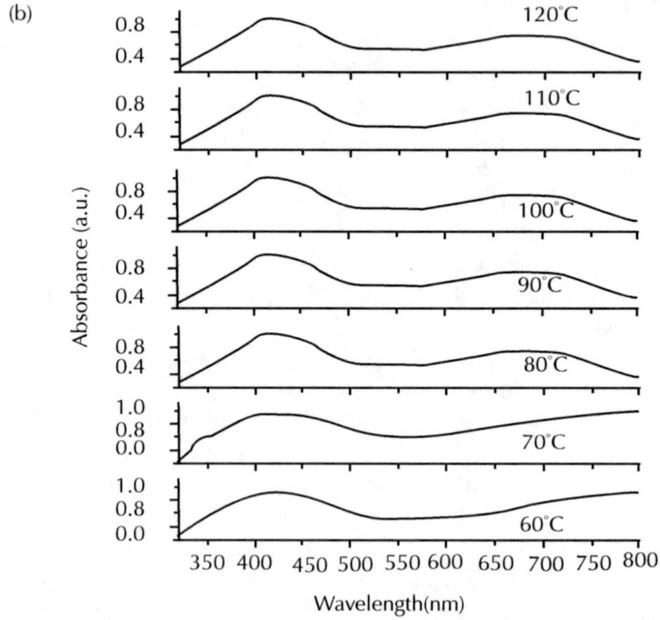

FIGURE 5 The UV-visible absorption spectra of all the samples synthesized at different temperatures in case-I. Figures (a) and (b) show the spectra of as produced and aged sample after 15 days of synthesis.

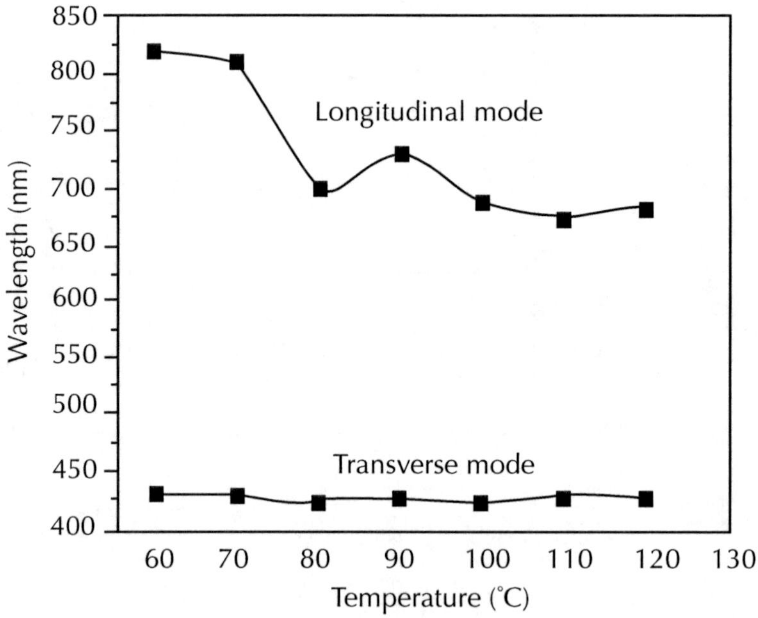

FIGURE 6 The variation of the absorption peak positions due to transverse and longitudinal oscillations with the variation of the synthesis temperature collected at different time (case-I).

From the Figure 6 it is seen that with the aging time there is no significant variation of the SPR peak positions due to transverse mode of oscillations with the variation of the temperature of the synthesis. But there is a blue shift of longitudinal peak positions with the increase in temperature of synthesis which corresponds to the decrease of the aspect ratio of the synthesized NPs. This result commensurate with SEM results where we have found that there is change in aspect ratio of the synthesized nanorods, *viz.* the average aspect ratio of the nanorods synthesized at 60 and 70°C temperatures are 7 and 6, respectively, and the average aspect ratio of the spindle shaped NPs synthesized at 80°C is 1.7. Thus, change in temperature of the reaction results in change in shape and morphology of silver NPs which is confirmed from both SEM and absorption data analyses. It is to be noted that quasi-spherical particles are always present more or less in near about every samples and causes the appearance of SPR resonance peak in 410–430 nm wavelength range. The theoretical calculation of absorption characteristics of rod or elliptical like silver nanostructures shows that as the aspect ratio of the nanorods is decreased the peak absorption due to longitudinal band becomes blue shifted.

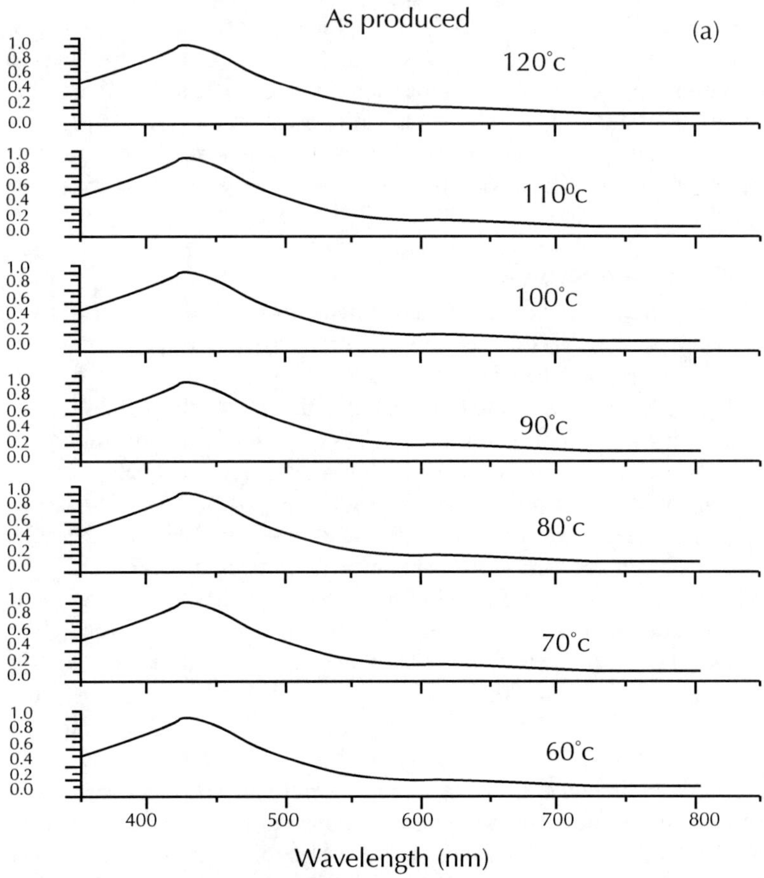

FIGURE 7 The UV-visible absorption spectra of all the samples prepared in case-II are shown in (a) and (b) for as produced and aged sample after 15 days of synthesis, respectively.

We have also measured the UV-visible absorption characteristics of the samples which are synthesized in the second case that is case-II, where the amount of $AgNO_3$ that is used is 10 times larger than that used in case-I. Figures 7(a) and 7(b) show the absorption characteristics of all the prepared fresh and 15 days aged samples synthesized at different temperatures. From the Figures 7(a) and 7(b), it is seen that unlike for the sample prepared in the case-I, in the present samples the longitudinal band is absent but absorption peaks have appeared in all the samples in the visible region due to transverse mode of oscillations. From TEM studies of the samples, as discussed before, a shape transition from a regular shape to irregular one has been observed. We have found pentagonal, hexagonal, spherical, rod-like and triangular shaped NPs in low temperature region (60–100°C) but for higher temperature (110–120°C), we have found irregular shaped NPs. So, the ratio between PVP and $AgNO_3$ is very crucial for different shape formation and its transition to another shape. It is to be noted that the UV-visible absorption characteristics of as produced sample of both cases (Case-I and Case-II) gives SPR peaks in near about same positions but longitudinal peak arises for the case-I samples and more pronounced peaks are observed in the long wavelength region for the aged samples. It is belived that during temperature treatment small metal cluster arising from reduced silver atoms created as seeds initially. The seeds may be singly or multiply twinned or may be single crystal and also different for different temperature treatment. Now, when nulcleation starts each type of seed can still grow into a nanocrystal with several possible shapes and as a result corresponding absorption spectra also changes considerably. The shape evoulation for both case is explained later.

2.4.3 PL Emission Studies

The luminescence efficiency of noble metals are very low [37] as compared to that of semiconductor materials. In semiconductors there is an enough energy band gap to dissipate a phonon when there is radiative transition between valence and conduction band and gives luminescent radiation. But in noble metals (bulk) there is no band gaps and there is a high probability of radiation less or non radiative transition and ultimately gives low intensity luminescence which has been observed before and is attributed to interband absorption edge (330 nm for bulk silver) [38]. The PL emission efficiency (quantum yield) from bulk silver was 10^{-10}. But, when particle size decreases to nanometer scale it has a great surface area than its bulk counterpart and luminescence become possible *via* electron-hole recombination and gives some improved intensity luminescence. For example, PL efficiency in the order of 10^{-4} have recently been found in gold nanorods [39] which is a great enhancement in comparison to a value of PL efficiency of 10^{-10} achieved in a smooth gold films [40]. We know that NPs whose diameter is smaller than the wavelength of light will respond as a dipole in an optical field. Under this dipole approximation the absorption and emission is controlled by the surface excited energy bands or surface active sites. Now, the photoelectron at the surface energy states absorbs light at its plasmon resonance frequency and some amount of light radiates in light near the wavelength of absorption and some amount contributes to heat energy. The collective coherent excitations of the free electron in conduction band produce a strong absorption and scattering light by the particles. This coherent oscillation is named as SPR band. The bandwidth and frequency of SPR band depends on several factors like size, morphology, and optical constants of interacting particle and surrounding medium [41]. The SPR excitation in particles induces strong local electric field that enhances the PL quantum yield of Ag NPs [41]. In most reports, PL emissions from noble metal NPs, *viz.*, in gold [40], silver [41], and copper [42] the emitted PL band is either close to interband transition or it is SPR enhanced.

The visible luminescence of Ag is arises due to excitation of electrons from occupied *d* bands into states above the Fermi level. Subsequent electron–phonon and hole–phonon scattering process leads to an energy loss and finally photo luminescent radiative recombination of an electron from an occupied *sp* band with the hole. The optical properties of silver depend on both interband and intraband transitions between electronic states. The strength of these transitions is determined by the spectral overlap of two factors: (i) the energy dependent joint density of electronic states and (ii) the radiation coupling efficiency defined by interfaces and nanostructure shape resonances. According to Boyd et al. [53], the PL emissions can be classed into these two categories. The first concerns the electronic transitions within silver, and the second describes the optical coupling and enhancement specific to each sample. Visible PL process thus occurs when an electron in the *d* band is excited to an unoccupied state in the conduction band. This leads to creation of a hole in the *d* band, which recombines with electron. Though, the recombination occurs through nonradiative mechanisms, but the hole can also radiatively recombine with electrons and exhibit visible PL emissions.

We have measured PL emission characteristics *viz.*, from the samples synthesized in case-II and pronounced PL emissions are observed in UV-visible regions with the peaks at 345, 384, 486, and 531 nm wavelengths under an excitation wavelength of 220 nm and those are shown in Figure 8(a). The emission at 345 nm is near to the bulk silver emission band at 330 nm [38, 52] and can be assigned to interband radiative transition between *sp* band electrons with hole in the *d* band [38, 41, 52]. The second intense peak appeared at 384 nm is very close to the interband absorption edge of bulk silver (3.2 eV or 388 nm). The red shift of this band relative to the PL band from bulk silver (330 nm) may be caused by the coupling of the incoming exciting photons and outgoing emitted photons with SPR [41]. Due to this coupling, surface plasma and the states of photon attract each other and ultimately PL peak is shifted towards the SPR resonance side as the excitation wave length increases. From Figure 8(b) it is seen that the intensity of the most intense PL peak at 345 nm reduces for the samples which are synthesized at higher temperatures. As has been described before that as the synthesis temperature increases from 60 to 120°C, there is also red shift in UV-visible absorption SPR peak due to the increase of the particle sizes as confirmed from TEM data.

Now, as the temperature increases the PL intensity decreases synchronously as the particle size (as obtained from TEM analysis) increases as shown in Figure 8(c). There are two distinct size effects for controlling PL intensity. If the particle size is greater than 30 nm the radiative damping effect causes the decrease in the local electric field inside the particles which decreases the PL spectra because the enhancement of the PL (compared to bulk) due to the coupling with the SPR excitation is damped by the radiative losses. So, PL intensity decreases. Now, if the particle size is less than the mean free path of the electron within the particle volume, the scattering from the surface and interface damping causes another damping of the SPR excitation and naturally leads to a decrease in the PL intensity.

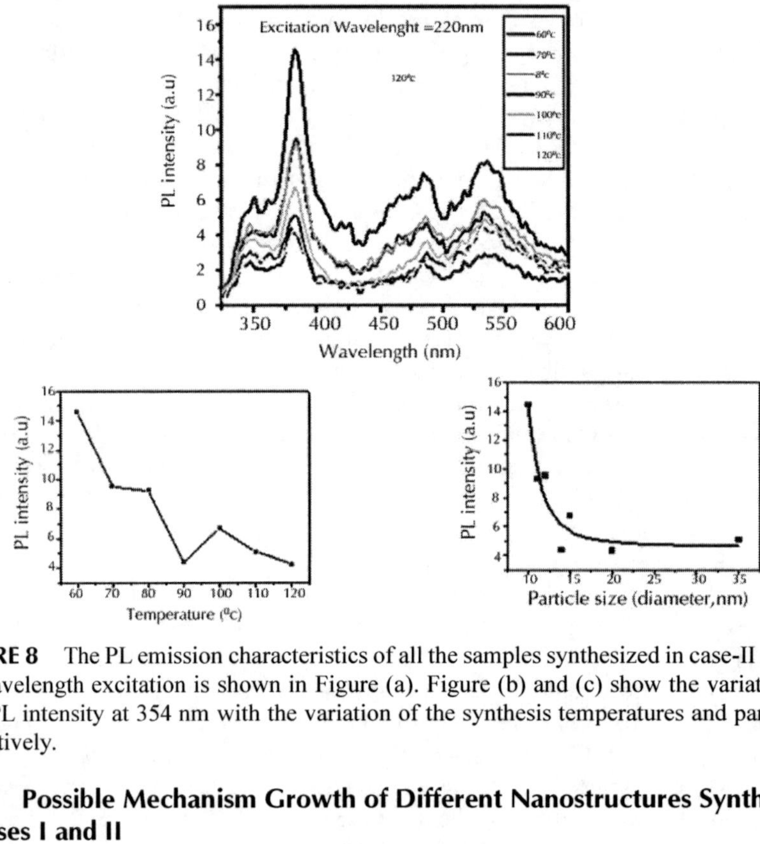

FIGURE 8 The PL emission characteristics of all the samples synthesized in case-II under 220 nm wavelength excitation is shown in Figure (a). Figure (b) and (c) show the variation of the peak PL intensity at 354 nm with the variation of the synthesis temperatures and particle size, respectively.

2.4.4 Possible Mechanism Growth of Different Nanostructures Synthesized in Cases I and II

Case-I

During the non equilibrium growth of NPs the final shapes are controlled by many factors that contribute to the kinetics of growth. Several factors like the size, structure, composition of the seed, nature, and concentration of metal precursor and reducing agent, molar ratio of precursor to capping agent, the selective adsorption of additives to different crystal facets, and so on, are all clearly capable of influencing the kinetics of particle growth and, hence, the particle morphology. At first we consider, the case-I, where we have found at lower temperature rod and spindle like NPs. The growth mechanism can be described by thermodynamic control (Figure 9(a)) where single crystal and multiple twinned crystals are produced initially as "seeds" which have certain stable structure and well defined crystallinity [47]. In the solvothermal solution synthesis, seeds are created by "nucleation" of small cluster of reduced atoms. The shape of the seeds is primarily determined by surface energy which should be minimized. The order of surface energy at different crystallographic plane is {111}< {100}< {110}. The outer surfaces of the seeds are expected to be enclosed with minimized surface energy surfaces {111} and {100} and shape may be quasi-spherical or cubo-octahedral. Now in the solvothermal solution phase a twin defect may arise inside the seed. Due to the presence of twinning, strain energy is increased and it is minimized by expanding the {111} lowest surface energy and minimum surface area. In short, seed may be single crystal, single twinned or multiple twinned structures. Now experimentally crystallinity of seed may be controlled by reduction rate which is again critical to the growth of seeds. When reduction rate is extremely fast most seeds become single crystal because sizes are increased rapidly. But when reduction rate is slowed down, the multiple or single twinned seeds may appear because seeds can get enough time to maintain small size due to slow addition rate of atom [11].

In our case at the low temperatures, the reduction rate is slow and multiple twinned seeds arise but when the temperature is elevated, the reduction rate is increased and single crystal seeds are produced. After the creation of different types of seeds at different temperatures, each type can still grow with several possible shapes which are depicted in Figure 9(a). In lower temperature (60–70°C), multiple twinned seeds grow into decahedron which finally gives the twinned fivefold "nanorod" shaped structure and this twining in nanorod again further creates spindle like structure. In the high temperature region single crystal seeds becomes cubic or cubo-octahedron shapes depending upon the growth rate ratio between <100> and <111> directions. So, we have found in higher temperatures, several cubical and cubo-octahedral shapes due to single crystal seeds as illustrated in Figure 9(a). At larger temperatures, saturation sets in and depending upon the growth rate ratio nearly spherical and ellipsoidal particles are formed. In the case of homogeneous nucleation, the equilibrium shape of a crystal reaches when surface energy for a given volume is the minimum. If the surface energy is isotropic as for the case with liquid, the equilibrium shape will be spherical as the sphere has the minimum surface area for a given volume. In the case of crystalline solids the surface energy is anisotropic. Therefore, total surface energy is not only determined by the surface area but also by the nature of the crystal planes. Therefore, the particular shape is energy minimized when facets surrounding the crystal give minimum surface area for a given volume, which results in a "polyhedron and spherical shape". For a face centered cubic (*fcc*) crystal structure, according to surface energy at different crystallographic facets is given in the order $\gamma\ \{111\} < \gamma\ \{100\} < \gamma\ \{110\}$. So, according to this condition it is more prone to obtain a tetrahedral or an octahedral shape enclosed by {111} planes but these shapes are not the minimum area shapes at all. But if the case is like truncated octahedron (enclosed by eight {111} facets and six {100} facets which is high energy facets) it looks like nearly spherical shape which has again reduces surface area and free energy. The regular polyhedron shapes can be obtained at lower temperature, where the surface energy anisotropy is the maximum. But as the temperature increases the surface energy anisotropy decreases and more rounded parts appear in the equilibrium shape [11, 47].

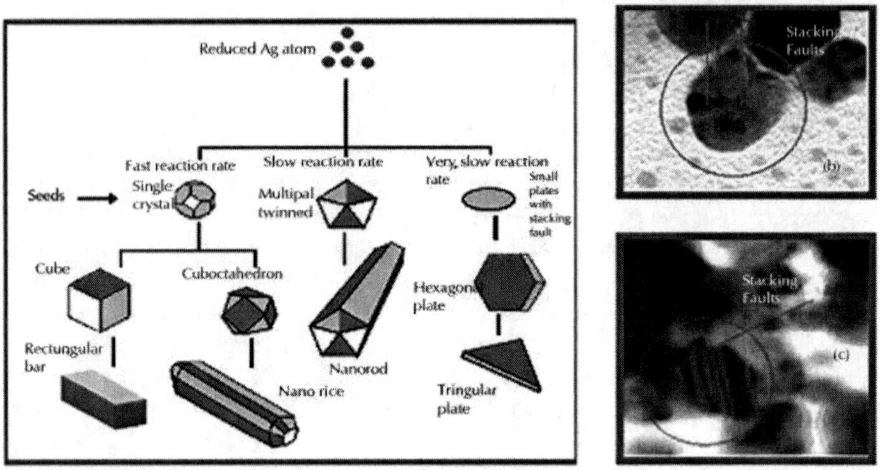

FIGURE 9 (a) A schematic demonstration of the shape evaluation of silver nanostructures from different seed crystals. Figures (b) and (c) demonstrate the stacking faults as observed in the TEM images of the sample synthesized at 60°C temperature in case-II.

Case-II

In the case-II, we have increase the $AgNO_3$ amount and kept the amount of reducing agent and capping agent same. Naturally, the reduction rate will be slower than the former one. The concentration of reduced atom should be so lower that it can not aggregate into thermodynamically favored crystalline seeds. As a result, kinetic control dominates and stacking faults are introduced in the seeds to induce the formation of Ag thin nanoplate. From TEM study we have found at lower temperature such circular thin nanoplate with stacking faults initially (Figures 9(b) and 9(c)). The presence of a single twin plane in the seed leads to the formation of triangular prisms, whereas the presence of two parallel twin planes forms hexagonal nanoplates. Single-twinned seeds obtain hexagonal shape due to the sixfold symmetry of the *fcc* system [47]. Thus as the temperature increases more silver atoms reduced and circular seed plates becomes hexagonal, pentagonal or triangular nanoplate that is also observed by TEM in 70–100°C. The relative amount of hexagon and pentagon is greater than the triangular nanoplates. This is due to slow growth rate of seed particles. In higher temperature (110–120°C) the anisotropic growth predominates and it becomes irregular elongated particles. This behavior is also found by some author [50] where they found cubo-octahedron, triangular prism, hexagonal, and pentagonal in lower temperature heat treatment but it becomes irregular anisotropic elongated shapes in higher temperatures.

2.5 CONCLUSION

We have presented here a simple chemical route to synthesize silver NPs of different shapes, sizes, and aspect ratios. It is found that to synthesize a particular shape of NPs (like nanorods, nanocrystals etc.) we have to control simply the temperature of the reaction. The absorption by nanorods in the near infrared region can be applicable to biological imaging and targeting of cancer cell and also in IR detections. Here, PVP do a major role by attaching it in particular crystallographic plane of silver helps to grow different morphologies of silver NPs. The increased temperature, concentration of solution, and stabilizing agent PVP obviously control the whole growth mechanism by twinning and stacking fault mechanism. The longitudinal plasmon band is very sensitive to the change in temperature that is responsible to change in aspect ratio. The presented methodologies may be used in future to synthesize other *fcc* metal nanostructures of varied shapes and sizes. These metal nanostructures with controlled shapes might be useful in different applications, such as in catalysis, in optical imaging as well as in photothermal cancer therapy.

KEYWORDS

- **Hybrid energy**
- **Longitudinal and transverse absorption**
- **Photoluminescence**
- **Shape controlled synthesis**
- **Silver nanostructures**

ACKNOWLEDGMENT

Authors are grateful to DST, Govt. of India for the partial financial support (Grant No. SR/FTP/PS-67/2008). Authors are also grateful to Dr. A. Patra, Dept. of Chemistry, NIT Durgapur for providing the FTIR facility.

REFERENCES

1. Huang, X., El-Sayed, I. H., Qian, W., and El-Sayed, M. A. *J. Am. Chem. Soc.*, **128**, 2115–2120 (2006).
2. Li, Z., Huang, P., Zhang, X., Lin, J., Yang, S., Liu, B., Gao, F., Xi, P., Ren, Q., and Cui, D. *Mol. Phar.*, **7**, 94–104 (2010).
3. Lee, S. E., Sasaki, D. Y., Perroud, T. D., Yoo, D., Patel, K. D., Lee, L. P. *J. Am.Chem. Soc.*, **131**, 14066–14074 (2009).
4. Bawendi, M. G., Steigerwald, M. L., and Brus, L. E. *Annu. Rev. Phys. Chem.*, **41**,477–496 (1990).
5. Wang, W. and Asher, S. A. *J. Am. Chem. Soc.*, **123**, 12528–12535 (2001).
6. Wang, Y. and Toshima, N. *J. Phy. Chem. B*, **101**, 5301–5306 (1997).
7. Hayne, C. L. and Van. Duyne, R. P. *J. Phy. Chem. B*, **107**, 7426–7433 (2003).
8. Bharadwaj, P., Anger, P., and Novotony, L. *Nanotechnology*, **18**, 044017 (2007).
9. Meng, C., Gang Feng, Y., Wang, X., Cheng Li, T., Zhang, J. Y., and Qian, D. J.*Langmuir*, **23**, 5296–5304 (2007).
10. Filippo, E., Serra, A., and Manno, D. *Sensors and Actuators B*, **138**, 625–630 (2009).
11. Sun, Y. and Xia, Y. *Science*, **298**, 2176–2179 (2002).
12. Willey, B. J., Chen, Y., McLellan, J. M., Xiong, Y., Li, Z. Y., Ginger, D., and Xia, Y. *Nano Lett.*, **7**, 1032–1036 (2007).
13. Jin, R., Cao, Y. C., Hao, E. G., Metraux, S., Schatz G. C., and Mirkin, C. A. *Nature*, **425**, 487–490 (2003).

14. Torres, V., Papa, M., Crespo, D., and Moreno, J. M. *Microelectronic Engineering*, **84**, 1665–1668 (2007).
15. Link, S., Mohamed, M. B., and El-Sayed, M. A. *J. Phys. Chem B*, **103**, 3073–3077 (1999).
16. Kang, S. K., Chah, S., Yun, C. Y., and Yi, J. *Korean J. Chem. Eng.*, **20**, 1145–1148 (2003).
17. Kvitek, L. and Prucek, R. *Journal of Material Science*, **22**, 2461–2473 (2005).
18. Li, Z., Gu, A., and Zhou, Q. *Cryst. Res. Technol.*, **44**, 841–844 (2009).
19. Li, J., Lin, Y., and Zhao, B. *J. Nanoparticle Research*, **4**, 345–349 (2002).
20. Graff, A., Wagner, D., Ditlbacher, H., and Kreibig, U. *Eur. Phys. J. D.*, **34**, 263–269 (2005).
21. Sarkar, R., Kumbhakar, P., Mitra, A. K., and Ganeev, R. A. *Curr. Appl. Phys.* **10,** 853–857 (2010).
22. Jain, P. K., Huang, X., El-Sayed, I. H., and El-Sayed, M. A. *Acc. Chem. Res.*, **41**, 1578–1586 (2008).
23. Basu, S. and Basu, P. K. *Journal of Sensors* **790476**, 1–9 (Article id 861968) (2009).
24. Link, S., Mohammed, M. B., and El-Sayed, M. A. *J. Phys. Chem. B*, **103**, 3073–3077 (1999).
25. Kilin, D. S., Prezhdo, O. V., and Xia, Y. *Chem. Phys. Lett.*, **458**, 113–116 (2008).
26. Lee, K. S. and El-Sayed, M. A. *J. Phys. Chem. B*, **110**, 19220–19225 (2006).
27. Link, S. and El-Sayed, M. A. *J. Phys. Chem. B*, **103**, 8410–8426 (1999).
28. Teng, B., Xu, S., An, J., Zhao, B., and Xu, W. *J. Phys. Chem. C*, **113**, 7025–7030 (2009).
29. Stalmashonak, A., Unal, A., Graener, H., and Seifert. G. *J. Phys. Chem. C*, **113**, 12028–12032 (2009).
30. Chang, S. S., Shih, C. W., Chen, C. D., Lai, W. C., and Wang, C. R. C. *Langmuir*, **15**, 701–709 (1999).
31. Tang, B., Xu, S., An, J., Zhao, B., and Xu, W. *J. Phys. Chem. C*, **113**, 7025–7030 (2009).
32. Li, Q., Laye, K. J., Lee, K. B., Kim, H. T., Lee, J., Myung, N. V., and Chaoa, Y. H. *J. of Colloid and Interface Science*, **342**, 8–17 (2010).
33. Zou, J., Xu, Y., Hou, B., Wu, D., and Sun, Y. *China Particuology*, **5**, 206–212 (2007).
34. Uwe, K. and Michael, V. *Optical Properties of Metal Clusters*, Springer, Berlin (1995).
35. Link, S. and El-Sayed, M. A. *J. Phys. Chem. B*, **103**, 8410–8426 (1999).
36. Mie plot version 3.4 is available at www.philipalven.com.
37. Cleveland, C. L., Landman, U., Schaaff, T. G., Shafigullin, M. N., Stephens, P. W., and Whetten, R. L. *Phys. Rev. Lett.*, **79**, 1873–1876 (1997).
38. Apell, P., Monreal, R., and Lundqvist, S., *Phys. Scr.*, **38**, 174 (1988).
39. Mohammed, M. B., Volkov, V., Link, S., and El-Sayed, M. A. *Chem. Phys. Lett.*, **317**, 517–523 (2000).
40. Dulkeith, E., Niedereichholz, T., and Klar, T. A., Feldman, *Phys. Rev. B*,**70**, 205424 (2004).
41. Yeshchenko, O. A., Dmitruk, I. M., Alexeenko, A. A., Losytskyy, M. Y., Kotko, A. V., Pinchuk, A. O. *Physical Review B*, **79**, 235438 (**2009**).
42. Mooradian, A. *Phys. Rev. Lett.*, **22**, 185–187 (1969).
43. Zhang, A., Zhang, J., and Feng, Y., *Journal of Luminescence*, **128**, 1635–1640 (2008).
44. Zhiliang, J., Zhongwei, F., Tingsheng, L., Fang, L., Fluxin, Z., Jiyun, X., and Xiaghui, Y., *China Ser. B*, **44**, 175–181 (2001).
45. Pastrnack, R. F. and Collins, P. J. *Science*, **269**, 935–939 (1995).
46. Kim, S. H., Choi, B. S., Kang, K., Choi, Y. S., and Yang, S. I. *Journal of Alloys and Compounds*, **433**, 261–264 (2007).
47. Sau, T. K. and Rogach, A. L. *Adv. Matter.*, **21**, 1–24 (2009).
48. Wang, Y. Q., Liang, W. S., and Gen, C. Y. *J. Nanopart Res*, **12,** 655–661 (2010).
49. Yang, S., Wang, Y., Wang, Q., Zhang, R., Yang, Z., Guo, Y., and Ding, B. *Cryst. Growth Des.*, **7**, 2258–2261 (2007).
50. Wang, Y. Q. and Liang, W. S. *J. Nanopart Res.*, **12**, 655–666 (2010).
51. Gans, R. *Ann. Physik*, **47**, 270 (1915).
52. Hong, H. K., Park, C. K., and Gong, M. S. *Bull. Korean. Chem. Soc.*, **31**, 1252–1256 (2010).
53. Boyd, G. T., Yu, Z. H., and Shen, Y. R. *Phys. Rev. B*, **33**, 7923–7935 (1986).

3 Nanocomposites of Poly(Methyl Methacrylate) and Brookite Titania Nanorods

Niranjan Patra, Alberto Barone, Marco Salerno, Gianvito Caputo, and Athanassia Athanassiou

CONTENTS

3.1 Introduction 41
3.2 Sample Preparation 42
3.3 Sample Characterization Methods 42
3.4 Discussion and Results 43
3.5 Conclusion 49
Keywords 49
References 49

3.1 INTRODUCTION

A polymer nanocomposite was produced by using poly(methyl methacrylate) (PMMA) as the matrix and crystalline Brookite TiO_2 nanorods (NRs) as the filler, in loading range between 5 and 30% in weight. The colloidal NRs were synthesized through low temperatures hydrolysis of titanium tetraisopropoxide, and showed a prolate shape with length to diameter aspect ratio around 20. The PMMA/Brookite composites were characterized through atomic force microscopy (AFM) and transmission electron microscopy (TEM) for the morphology, and through differential scanning calorimetry (DSC) and nanoindentation for the thermal and mechanical properties, respectively. It was found that, with respect to the bare PMMA, the glass transition temperature is increased of about 10°C for all the composites, whereas the reduced modulus and hardness are substantially increased only for 10 wt% loading. The TEM analysis showed evidence of NRs aggregation on increase loading.

The TiO_2-based nanostructure materials have emerged in the past decades as a platform on which a variety of appealing physical-chemical properties coexist with biocompatibility [1]. Presently investigated applications include photocatalytic systems relying on controlled spatial organization of titania polymorphs [2], and light responsive coatings with simultaneous antireflective, antibacterial, self-cleaning, and antifogging behavior [3].

In the most research work has dealt so far with the tetragonal Anatase and Rutile phases, due to the relative ease with which these polymorphs can be attained [1]. However, recent investigations have highlighted that TiO_2 in the orthorhombic Brookite crystal structure can exhibit superior electrochemical [4], catalytic, and photocatalytic [5] performances.

In this work, thin film (<1 μm thickness) nanocomposites of Brookite TiO_2 NRs dispersed in PMMA have been prepared by solvent spin coating. The morphology of the top film surface has been checked by AFM, whereas the internal NRs distribution of the films has been checked by TEM. The functional thermal and mechanical properties of the nanocomposites have been investigated by DSC and nanoindentation, respectively.

3.2 SAMPLE PREPARATION

Titanium (IV) chloride ($TiCl_4$, 99.999%), titanium (IV) isopropoxide ($Ti(OiPr)_4$, 99.999%), oleic acid ($C_{17}H_{33}CO_2H$, 90%), oleyl amine ($C_{17}H_{33}NH_2$, 70%), trimethylamine *N*-oxide dihydrate ($(CH_3)_3NO{\cdot}2H_2O$, 98%), and 1-octadecene ($C_{18}H_{36}$, 90%), as well as all solvents, were purchased from Aldrich, and were used as received. Water was bi-distilled (Millipore Q).

The synthesis was carried out under air free conditions using a standard Schlenk line setup, as described [6]. After the synthesis, extraction/purification procedures of the nanocrystals were carried out under ambient atmosphere, to remove precursor and surfactant/octadecene residuals. The purified nanocrystals were finally dispersed in chloroform, providing a milky suspension due to the nanocrystals size, which was stable in time. According to the used procedure, Brookite TiO_2 NRs are prepared, which has been confirmed by X-ray diffraction on solid samples drop casted from the prepared solutions.

The hybrid solutions with Brookite TiO_2 NRs dispersed in a liquid polymer phase were prepared by adding solutions of NRs to PMMA, with different relative NRs to PMMA mass concentration Φ (shortly termed 'loading' in the following) of 5, 10, 20, and 30 wt%. The nanocomposite thin films for AFM and nanoindentation measurements were prepared by spin coating the NRs–PMMA solutions onto properly cleaned glass substrates at 1,000 rpm for 60 s, (using a Sawatec SM-180-BT spinner, Germany). The thickness of the films was in all cases ~0.7 μm, as measured by a profilometer (XP-2, AMBIOS Technology, USA), which is sufficiently high to provide reliable nanoindentation measurements.

3.3 SAMPLE CHARACTERIZATION METHODS

The top surface of the nanocomposites films was investigated with MFP-3D AFM (Asylum Research, USA), working in ambient air in Tapping mode with gold coated silicon probes (NSG10, NT-MDT, Russia) with a resonant frequency of ~250 kHz.

The low resolution TEM images were recorded with Jeol JSM 1011 microscope operated at an accelerating voltage of 100 kV. The samples for this analysis were prepared by spin coating 1:10 diluted nanocomposite solution onto carbon coated Cu grids, which allowed acquire ~200 nm thick films, appropriate for TEM analysis. Since, the goal in this analysis is to check the distribution of NRs in the nanocomposite solution, no solution with 0% NRs loading was considered. On the contrary, the starting solution with NRs only, without polymer matrix, was used.

The DSC was carried out on a Pyris Diamond SII (Perkin-Elmer, USA), heating from 50 to 200°C with a rate of 10°C/min in nitrogen atmosphere (flow rate 20 ml/min). The sample weights were of 7-8 mg in all the measurements. The instrument was calibrated using Indium and Zinc as the standard materials.

The nanoindentation experiments were performed by using a Nanotest instrument (Micro Materials Ltd., UK) equipped with Berkovich pyramidal indenter, with nominal radius of curvature of the tip ~50 nm. The equipment was calibrated by iterative indentation cycles into fused silica (with loads ranging from 0.5 to 200 mN). The indentations were performed in a cabinet with constant temperature of 23°C, in mechanical and electrical low noise conditions and under load control. The maximum load range (0.12-0.18 mN) was chosen such that the indentation size and the tip effect at low loads were minimized, and the influence of the glass substrate at higher loads was negligible. For each load, indentations were repeated ten times on different regions of the film surface, with 0.01 mN/s loading and unloading rate and 60 s dwell period at peak load.

3.4 DISCUSSION AND RESULTS

After optical microscopy inspection at low magnification, higher magnification analysis of the film top surface was performed by means of AFM. For each NRs-PMMA loading Φ, both 10 and 3 μm scan size images were taken, in several regions of the films. In Figure 1(a) and (b) representative AFM images at the higher magnification are shown.

The nanocomposite films with increasing NRs loading exhibit increasing surface roughness. The red regions show areas where the height was above the maximum value in the color bar by the side, taken as the height distribution mean plus two standard deviations. In Figure 1(c) the summarizing plot of root mean square (RMS) roughness *versus* loading Φ shows a linear increase, in the considered Φ range. This result is qualitatively similar to the studied case of Anatase TiO_2 NRs [7]. However, in the present case the roughness measured at the highest loading Φ = 30% is between one and two orders of magnitude higher (~15 nm instead of ~2 nm for 3 μm scan and ~100 nm instead of ~3 nm for 10 μm scan, respectively). In particular, the significantly higher increase for the 10 μm scan images shows that the roughness is better sampled in this case on such a lateral size scale. Both this large lateral scale of the roughness texture and the high RMS values are hints of NRs aggregates forming close to the film surface.

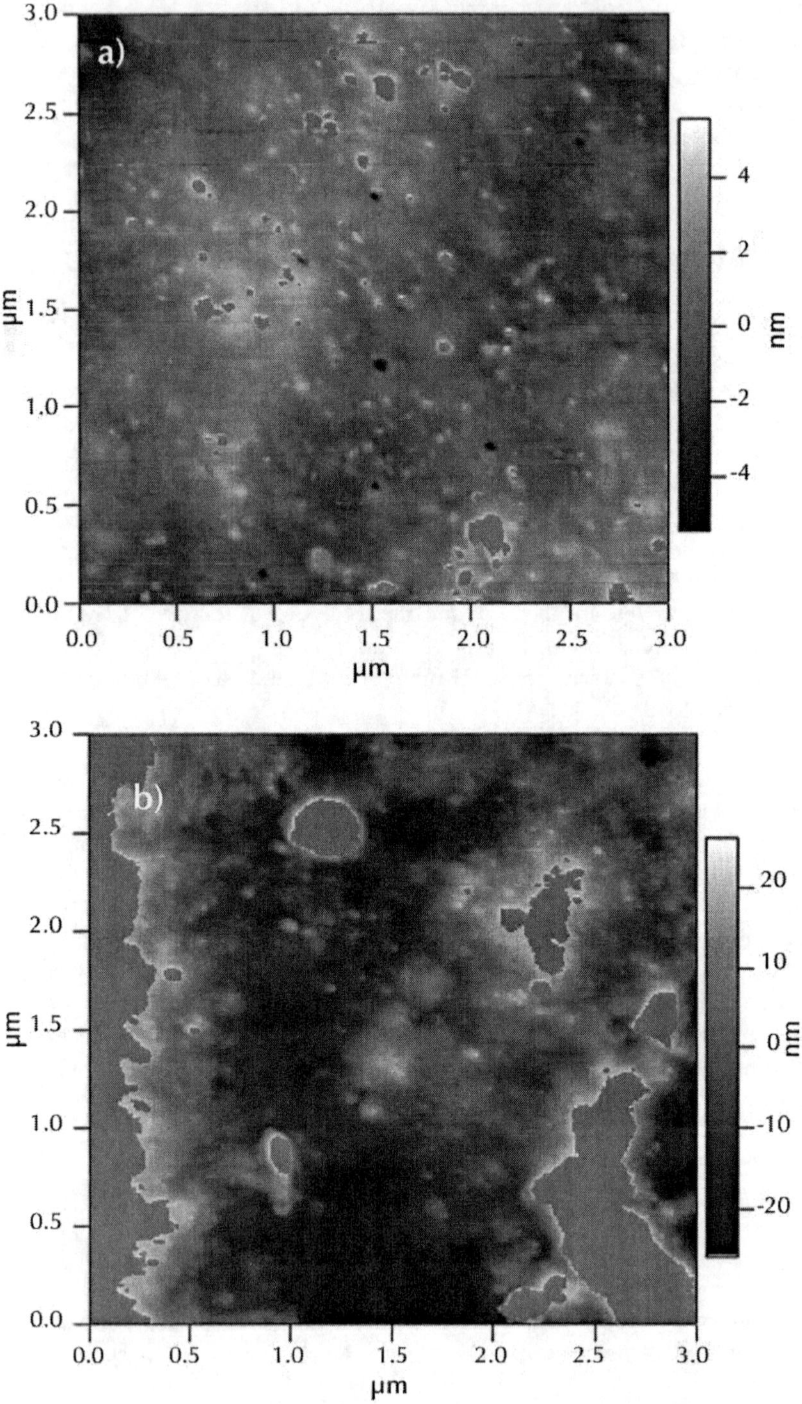

FIGURE 1 *(Continued)*

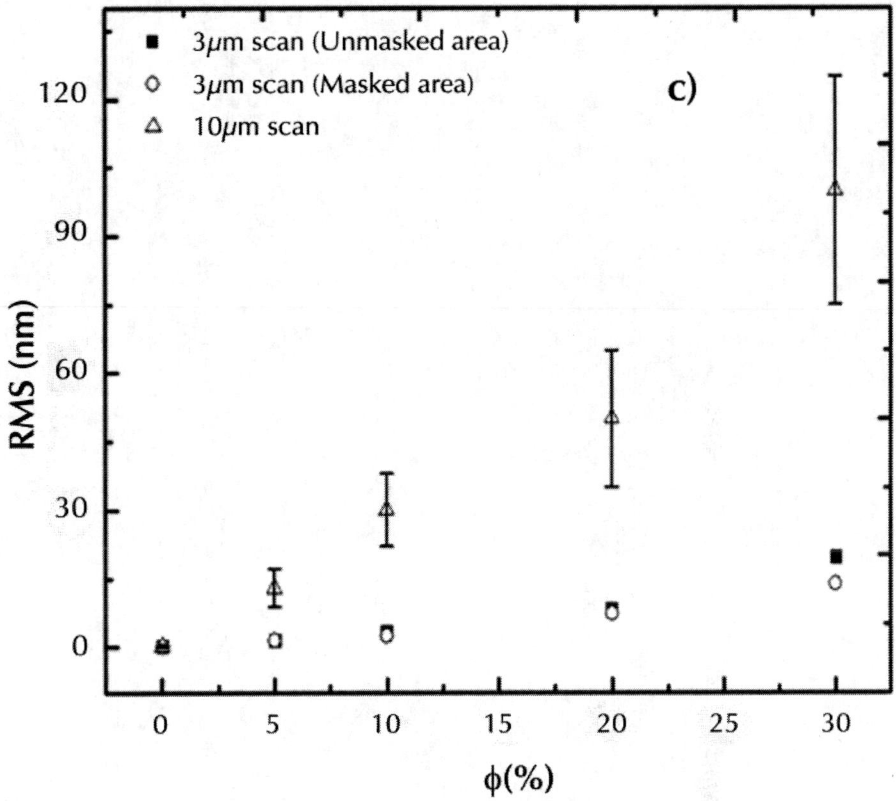

FIGURE 1 High resolution of AFM topographical maps (3 μm scan size), showing nanocomposite films with different NRs loading Φ in the starting solution of (a) 5%, (b) 20%, and (c) plots of RMS roughness extracted from the AFM images.

The imaging nanocomposite thin films with TEM allowed us to better assess the possible occurrence of NRs aggregates under the surface. The results are shown in Figure 2.

Figure 2 shows the TEM micrographs of (a) bare TiO_2 Brookite nanocrystals and (b) PMMA-TiO_2 NRs nanocomposites. In Figure 2(a) it appears clearly that the NRs have rather high elongation. Since, the diameter is ~5 nm and the average length is ~100 nm, the typical aspect ratio is ~20. This is much higher than the investigated Anatase TiO_2 NRs [7], which was ~8 (diameter of 3-4 nm and length of 25-30 nm). The probably as a consequence of this high aspect ratio, aggregates are formed in the nanocomposites (Figure 2(b)), due to the high probability of the NRs to meet and stick to each other already from the solution phase, despite the surfactant capping. Even higher agglomeration may occur in nanocomposites with higher loading (Φ >5%).

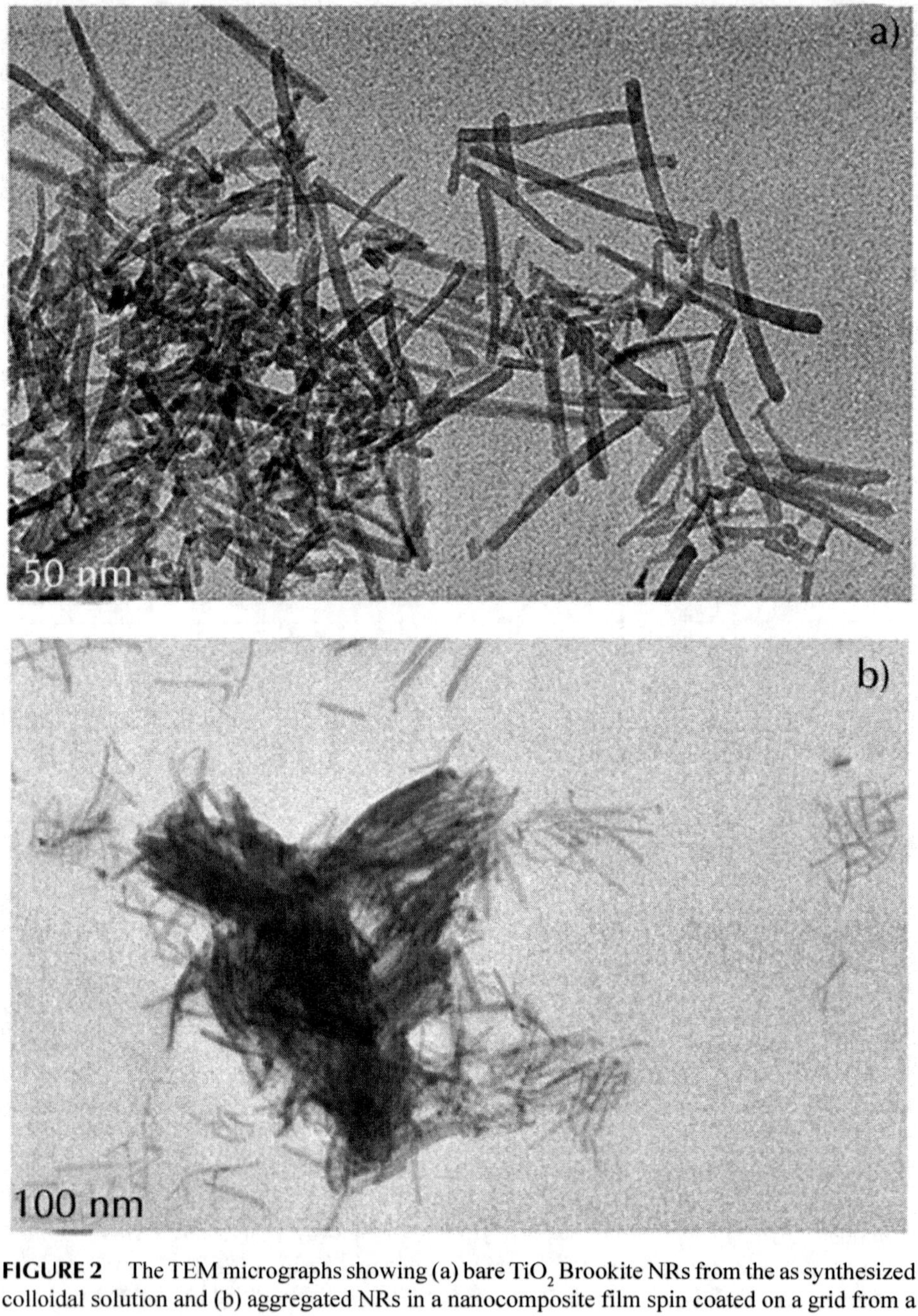

FIGURE 2 The TEM micrographs showing (a) bare TiO_2 Brookite NRs from the as synthesized colloidal solution and (b) aggregated NRs in a nanocomposite film spin coated on a grid from a 1:10 diluted $\Phi = 5\%$ solution.

The DSC measurements allowed us to collect curves of heat flow *versus* temperature in the nanocomposite films prepared from solutions with different Φ, which are shown in Figure 3(a). The glass transition temperature T_g of the respective nanocomposites is identified by the flex point in these curves, and has been plot in Figure 3(b). The change in specific heat capacity at constant pressure ΔC_p *versus* the inorganic filler ($\Delta C_p = 0$) has also been plot in Figure 3(b), as a useful control parameter for the experiment. This parameter should be as constant as possible, and indeed only minor decrease occurs for the samples with NRs loading Φ = 10%. In Figure 3(b) it also appears that the T_g increases suddenly for Φ = 5%, and comes to an almost constant value for all the higher loadings. This behavior is different from the observed Anatase NRs [7], for which a rather linear increase was observed over the whole 5-30% Φ range. It seems that a threshold effect occurs in Brookite NRs nanocomposites, already at a loading as low as Φ = 5%. This is probably due to the high aspect ratio of the Brookite NRs, which get easily agglomerated for loading higher than that limit.

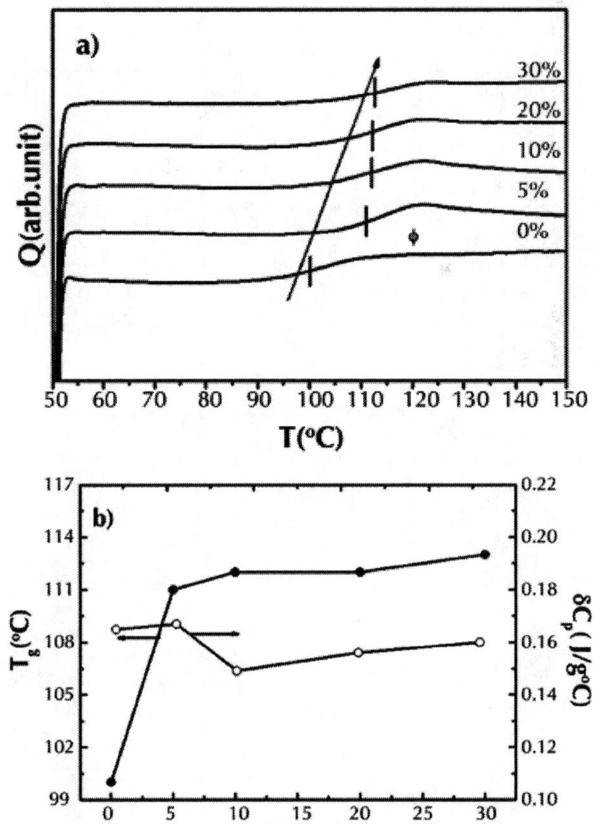

FIGURE 3 The DSC thermal traces of nanocomposites prepared from solutions with different Brookite NRs loading Φ = 0, 5, 10, 20, and 30%.

The indentation experiments were carried out to determine the hardness (H) and the reduced modulus (E_r) of the nanocomposite materials, whose results are shown in Figure 4. For the roughest films with $\Phi = 30\%$ (see Figure 1(c)) it was not possible to obtain clear and repeatable measurements, probably due to the uncertain determination of the contact point in our load penetration curves. In Figure 4 the statistical deviations of the data points have been represented with a box diagram, where the inner square symbol and horizontal line represent the mean and the median, respectively. The box boundaries are the lower and upper quartile, and the outer vertical bars extend to largest and smallest values within ±1.5 interquartile range from box boundaries, (data points outside this range are plotted separately as crosses). Considering the mean values, one can see that compared to TiO_2 Anatase NRs as the filler [7] only minor changes in both H and E_r are observed in the present case, on increasing Φ. In particular, for H the apparent increase occurring between $\Phi = 0$ and $\Phi = 10\%$ is followed by a consistent drop for $\Phi = 20\%$. Furthermore, the spreads of the measurements are comparatively large, which is attributed to the increase in sample in homogeneity also measured as film surface roughness.

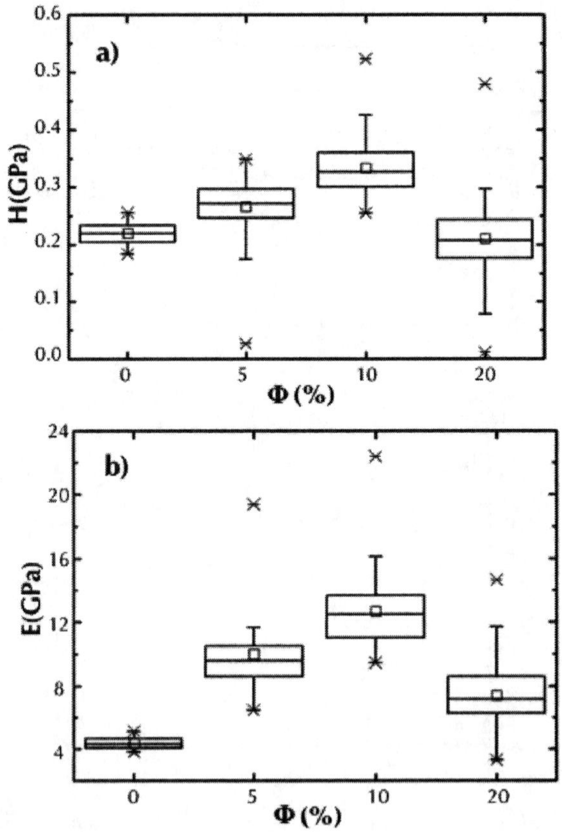

FIGURE 4 Statistical box diagrams of (a) hardness H and (b) reduced modulus E_r, measured at different Brookite NRs loading $\Phi = 0$, 5, 10, and 20%.

The significance of the observed differences on Φ was statistically analyzed by analysis of variance (ANOVA). For H only the point at Φ = 10% was significantly higher than the others (at a level of significance $p < 0.05$). Probably for Φ = 20% the aggregated NRs have the effect of a weak point rather than a reinforcing agent, and on pressing the film they push away the polymer and sink in without opposing much resistance, due to the lack of an extended networking. The differences occurring on Φ are more marked for E_r, for which ANOVA showed that all the nanocomposite data points (Φ = 5, 10, and 20%) are significantly higher ($p<0.05$) than the value for bare PMMA (Φ = 0). Furthermore, again the difference is stronger for Φ = 10%, ($p<0.01$). Obviously, the mechanical effects of NRs aggregation occur for higher Φ than the thermal effect of T_g.

3.5 CONCLUSION

The thin films of nanocomposites of PMMA and Brookite nanocrystals were prepared, and the resulting morphology, thermal, and mechanical properties were characterized. It was found that these long NRs (aspect ratio ~20) tend to agglomerate even at a comparatively low loading of 5% wt, which also reflects in a highly increased surface roughness (at least one order of magnitude higher) if compared to similar nanocomposites made with Anatase nanocrystals (of aspect ratio ~8). The glass transition temperature increased of approximately the same amount (~+10°C) for all the composite loadings. The hardness increased significantly only at 10% loading, whereas the reduced modulus was clearly increased at all loadings, with a maximum also at 10%.

KEYWORDS

- **Brookite composites**
- **Differential scanning calorimetry**
- **Nanocomposites films**
- **Nanorods**
- **Transmission electron microscopy**

REFERENCES

1. Chen, X. and Mao, S. S. Titanium dioxide nanomaterials: Synthesis, properties, modifications, and applications. *Chem. Rev.*, **107**, 2891–2959 (2007).
2. Kawahara, T., Konishi, Y., Tada, H., Tohge, N., Nishii, J., and Ito, S. A patterned TiO_2(anatase)/ TiO_2 (rutile) bilayer-type photocatalyst: Effect of the anatase/rutile junction on the photocatalytic activity. *Angew. Chem. Int. Comm.*, **41**, 2811–2813 (2002).
3. Fujishima, A. and Zhang, X. Titanium dioxide photocatalysis: Present situation and future approaches. *C. R. Chim.*, **9**, 750–760 (2006).
4. Koelsch, M., Cassaignon, S., Guillemoles, J. F., and Jolivet, J. P. Comparison of optical and electrochemical properties of anatase and brookite TiO_2 synthesized by the sol–gel method. *Thin Solid Films*, **403–404**, 312–319 (2002).
5. Shibata, T., Irie, H., Ohmori, M., Nakajima, A., Watanabe, T., and Hashimoto, K. Comparison of photochemical properties of brookite and anatase TiO_2 films. *Phys. Chem. Chem. Phys.*, **6**, 1359–1362 (2004).

6. Buonsanti, R., Grillo, V., Carlino, E., Giannini, C., Kipp, T., Cingolani, R., and Cozzoli, P. D. Nonhydrolytic synthesis of high-quality anisotropically shaped brookite TiO_2 nanocrystals. *J. Am. Chem. Soc.*, **130**, 11223–11233 (2008).
7. Patra, N., Barone, A. C., Salerno, M., Caputo, G., Cozzoli, P. D., and Athanassiou, A. Thermal and mechanical characterization of PMMA-TiO_2 nanocomposites. *Advanced materials Research*, **67**, 209–214 (2009).

4 Biomorphic Scaffolds for the Regeneration of Load-bearing Bones

Simone Sprio, Andrea Ruffini, and Anna Tampieri

CONTENTS

4.1 Introduction 51
4.2 Mechanical Biomimesis of Bone Scaffolds........ 54
4.3 Biomorphic Transformations to Create New Bone Tissue 55
4.4 Synthesis of Biomorphic Bone Scaffolds Starting by Ligneous Templates 57
4.4.1 Pyrolysis 59
4.4.2 Carburization 60
4.4.3 Oxidation 65
4.4.4 Carbonation 67
4.4.5 Phosphatization........ 69
4.5 Preliminary in vivo Assessment of Biomorphic HA Scaffolds 74
4.6 Final Remarks........ 76
Keywords 77
Acknowledgment 77
References........ 77

4.1 INTRODUCTION

The bone tissue diseases are among the most disabling pathologies and affect an increasing number of people worldwide. Particularly, the diseases that most seriously impact on the people life and activity are those involving long bone portions subjected to mechanical load. The impact of such problems is particularly relevant among the aged people (i.e. due to osteoporosis) but in the last decade the number of relatively young patients is continuously increasing also due to modern life style which include intense sport activity, tendency to obesity, and so on. In this case the pain and disability also impact on the psychological state, leading to increasing anxiety, depression, and altered perception of their social role. Nowadays, this feeling is shared also by aged people, due to the expectance of a longer active life. For this reason the social and economic impact of the pathologies related to bone tissue is continuously increasing, estimated in about 20% per year; in response to this pressure medicine is required to offer solutions able to really improve the quality of life of patients at every age and in this respect a wide variety of biomedical devices for bone healing has been developed and launched on the market in the last 40 years.

Such devices were initially designed as mere substitutes of the missing part of bones, especially in load-bearing applications, whereas in the last 15 years a new regenerative approach has led to the development of biomimetic and bioresorbable porous scaffolds, able to take part to the bone regenerative and remodeling processes, so to allow the formation and development of mature and well organized bone tissue and to be progressively resorbed by osteoclasts. Such scaffolds are based on hydroxyapatite (HA) and other bioactive calcium phosphates, closely mimicking the chemical composition of bone. This result allows the complete restoration of the original bone functionality and the recovery of a satisfactory active life and reduced hospitalization time. The regenerative approach became more and more complex in the last years: recently, the development of biohybrid HA/collagen composites [1] have allowed the early repair and regeneration of osteocartilaginous regions [2, 3], taking advantage of the close reproduction of the chemico-physical and morphological features of the newly formed bone. The perfect biomimesis of biohybrid HA/collagen composites depends on the synthesis route employed, which closely reproduces the organic matrix-mediated biomineralization process, occurring *in vivo* [4] when collagen nanofibers spontaneously self-assemble in the extracellular space and act as templates for mineralization (Figure 1). The biologically inspired process of collagen mineralization also allows to incorporate ions suitable for the enhancement of the processes of bone regeneration (Mg, Si) [5].

Although the regenerative approach to bone disease is already effective in a wide range of clinical cases, there is a substantial lack of satisfactory results in load-bearing applications. In fact, in this field the currently employed solutions consist in bioinert prostheses (based on alumina, titanium) characterized by high mechanical strength and toughness; such devices are often coated with a bioactive HA layer to improve cell adhesion and osteointegration [6]. Such prostheses act as a mere support of the biomechanical loads and they do not take actively part to the regenerative processes; in consequence they do not resorbs neither in the long-term. They are conceived for duration of decades, notwithstanding they often meet with failure and must be replaced, especially in the youngest patients, implying repeated highly invasive and painful surgery. The reasons for failure of bioinert prostheses are various: firstly, their scarce osteoconductivity hampers deep cell colonization and a complete osteointegration in the surrounding tissue. Moreover, their excessive mechanical strength, toughness and rigidity, compared to those of the surrounding bone, yield a mismatch in the stress fields at the bone implant interface upon mechanical stimulation, which in turn provokes the necrosis and resorption of the bone in contact with the prosthesis upon repeated biomechanical stimulation; consequently, it results the progressive detachment of the prosthesis in the long term (about 10–15 years).

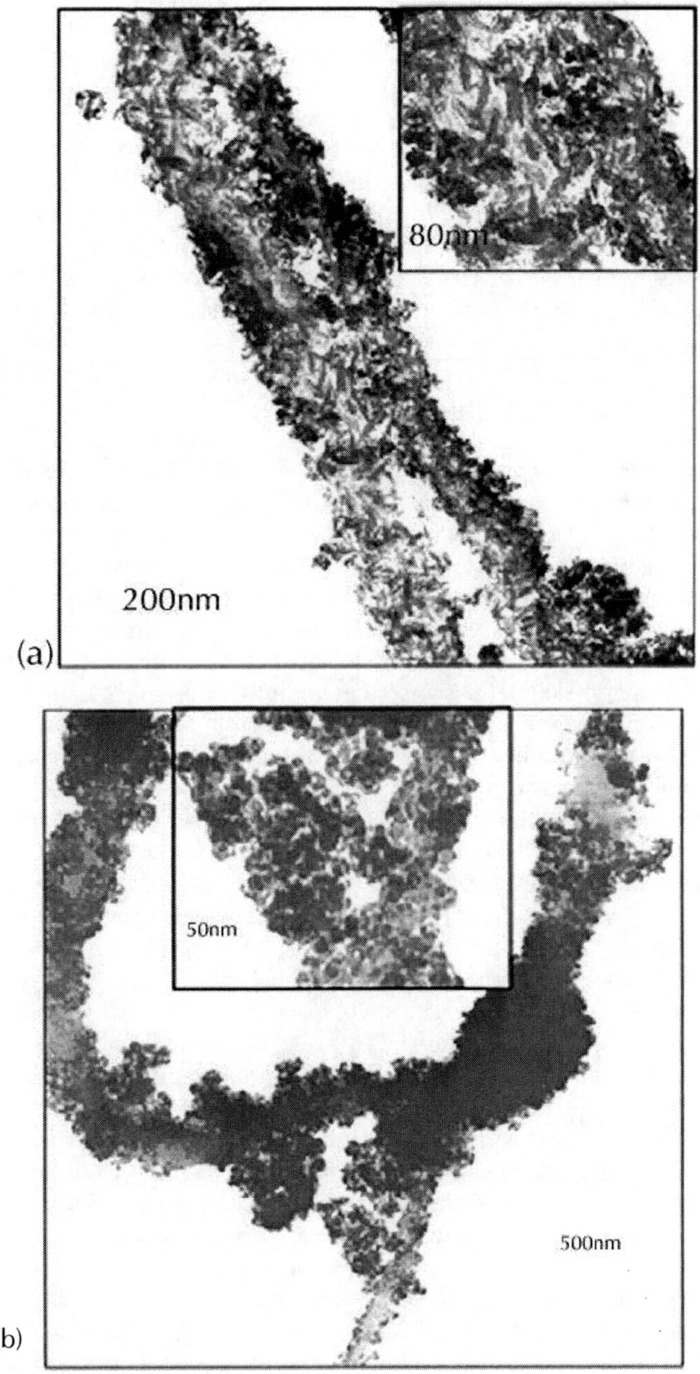

FIGURE 1 TEM micrographs of biohybrid HA/collagen composites mineralized (70:30 wt %) with Mg-Si-HA (a) and Mg-CO_3-HA (b).

4.2 MECHANICAL BIOMIMESIS OF BONE SCAFFOLDS

The major problems to be faced for a regenerative approach to load-bearing bone diseases is the achievement of porous scaffolds endowed with both chemical and mechanical biomimesis. Chemically, the scaffold must be bioactive, osteoinductive, osteoconductive, and (at least partially) resorbable; the election materials are calcium phosphates and in particular, ionically substituted HA [7-11]. Mechanically, the scaffold must withstand the early biomechanical loads when implanted *in vivo* so to activate the regeneration process and rapidly induce the formation and spatial organization of new mature bone; anyway, as above explained, the bioactive scaffolds should exhibit a mechanical behavior not too exceeding that of native bone, to avoid stress mismatch. Ceramic composites were developed by combining HA with bioactive phases exhibiting improved mechanical strength [12-14] but the concern related to a proper morphology of the scaffold is far to be solved.

In fact, recent considerations highlighted that cytoskeleton (the internal scaffold of cells) is not based on simple interactions among individual molecules; rather it is based on a globally integrated architecture, known as tensegrity [15], which represents a basis to coordinate each part of the cytoskeleton with itself as a whole. This concept provides a holistic general description on how cells and tissues are organized in tri-dimensional architectures and allows expressing a relationship between mechanical stimulation and remodeling [16]. Particularly, this concept is applied to the musculo-skeletal tissue so that the information driving the cell activity is processed through a mechano-chemical transduction of the mechanical loads. On the basis of this mechanism, the compressive and tensile forces macroscopically acting on bone propagate down to the cell level and expose cells to shearing forces through the variation of the interstitial fluid flow through the bone canaliculi [17]. In turn, these flow variations generate electric potentials deforming the cell membranes; such deformations act as stimuli for bone cells (presumably, osteocytes) which in consequence release suitable biochemical signals driving the tissue regeneration activity.

Hence, the processes of regeneration and self-healing of the bone tissue depend on a proper transmission of the mechanical loads from the macroscopic down to the cell scale, this transmission is allowed by the peculiar morphological organization of the long bones, which are hierarchically structured on different scale size (Figure 2). At the lowest level, nanometer sized crystals of carbonate apatite are embedded in and surround the fibrous protein collagen; these mineralized fibers lie bundled together and are attached to each other to form lamellae at a width of ~2 μm. The lamellae have various patterns; a very common one is a secondary osteon, in which concentric lamellae form cylindrical structures, ~200 μm in diameter, surrounding a central blood vessel, the Haversian channel. Compact bone, solid to the naked eye, is modified in places to form trabecular bone, which consists of many struts, with the spaces between the struts filled with marrow. The struts are not just randomly arranged, but are related to the direction of the macroscopic mechanical loads on the bone.

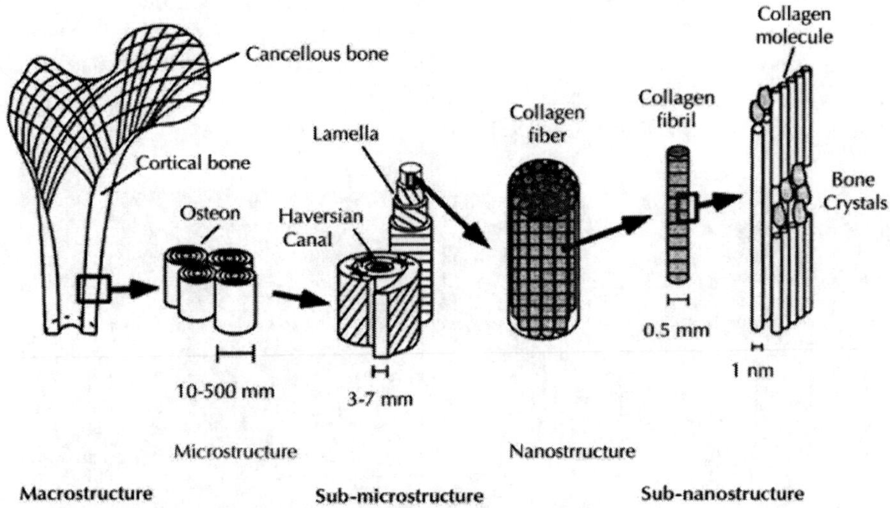

FIGURE 2 Hierarchical structural organization of bone.

The insuperable mechanical and biological properties of the bone thus depend on the complex interaction taking place across all levels of organization and this is the reason why presently, acceptable solutions for a regenerative approach to load-bearing bones still do not exist. In fact, no fabrication technique exists, as far as complex, to reproduce even roughly the bone structure, organized down to the submicron level, and in absence of such organization, the mechano-chemical transduction and the activation of regenerative processes at the cell level is not possible.

Nature instead exhibits plenty of biological structures characterized by hierarchically organized morphologies optimized in millions year of evolution, able to couple, such as bone, an extreme lightness to astonishing functional properties. For this reason, the development of hierarchical structured devices based on natural templates could pave the way for realizing prosthetic devices which could get closer to the extraordinary performance of human tissues.

4.3 BIOMORPHIC TRANSFORMATIONS TO CREATE NEW BONE TISSUE

The general concept of tensegrity, based on the tensional integrity of the elements constituting a defined system, is at the basis of the spatial and functional configuration of most, if not all, of the biologic structures existing in nature [15, 18], as well as of how information and signals are processed and exchanged at a molecular level. Generally biologic structures are organized on the basis of minimum energy principles, so to have the best compromise between mass, volume and functionality; for this reason the vascular system in living tissues is organized so to exchange the fluids necessary for the organ activity with the highest efficiency. All tissues have a structural organization based on this paradigm, otherwise a proper functioning would be compromised, thus a proper approach to the regeneration of tissues must be based on such concept and on this basis scaffolds endowed with a suitable morphology must be obtained (organomorphicity) so to drive progenitor cells to selected differentiation and proper regenerative behavior [19-21]. In long bones, such concept is expressed by their hierarchic structural organization which allows an efficient distribution of the mechanical load and the capacity to self-repair, but also an efficient vascularization which starts from the Haversian channels (see Figure 3).

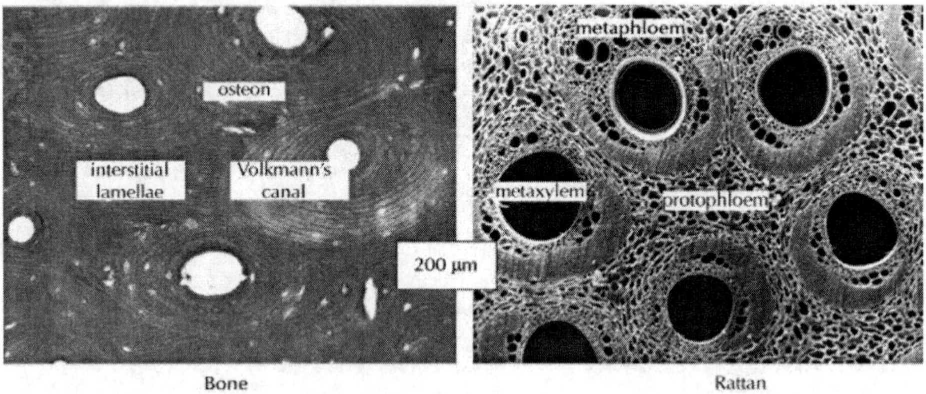

FIGURE 3 Comparison between a section of compact bone tissue and rattan wood.

Considering that it is impossible to model the inorganic structure of bone in the micron scale with the current manufacturing technologies [22, 23], it is instead possible to exploit biologic structures, upon transformation into inorganic scaffolds, whose chemical composition, similar to that of bone, and the maintenance of the hierarchical structure can allow cells to activate the regeneration and remodeling processes through a progressive digestion of the scaffold by osteoclasts and formation of mature bone, driven by the organized morphology of the biomorphic template.

Among the morphologies existing in nature, vegetables, particularly ligneous structures, exhibit a morphology and organization similar to that of bone, which confer impressive biomechanical properties: for example, some tree branches can withstand very heavy loads and others can be stretched and bent into any shape [24, 25]. Wood can be regarded as a cellular material at the scale of 100 µm to cm. At the cell level the mechanical properties are governed by the diameter and shape of the cell cross-section, as well as the thickness of the cell wall. In particular, the ratio of cell-wall thickness to cell diameter is directly related to the apparent density of wood which, in turn is an important determinant of the performance of lightweight structures [26]. The unique hierarchical architecture of the cellular micro-structure confers to wood a remarkable combination of high strength, stiffness, and toughness at low density. [27, 28] The alternation of fiber bundles and channel like porous areas makes the wood an elective material to be used as template in starting the preparation of a new bone substitute characterized by a biomimetic hierarchical structure.

On this basis, selected wood structures could reproduce different bone portions characterized by different porosity and pore distribution, such as cortical and spongious bone. Such structure can adequately drive cells to the complete regeneration of bones, in particular long bones, provided that (i) the size of the macroscopic channels is wide enough to allow the cell colonization and proliferation and (ii) its chemical composition is very close to that of bone, that is a nanostructure nearly amorphous HA partially substituted with carbonate [29, 30], thus able to be progressively resorbed by osteoclasts.

4.4 SYNTHESIS OF BIOMORPHIC BONE SCAFFOLDS STARTING BY LIGNEOUS TEMPLATES

In this respect, native or semi-processed wood and plants may be successfully used as templates for generating ceramic materials through transformation processes, involving pyrolysis and chemical reactions (with solid, liquid or gaseous phases); these materials, known as biomorphic, can exhibit a hierarchic structure very close to the one of the original wood. The nanosize of the building blocks of biological tissues is one of the bases of their self-organization ability and can be replaced by synthetic materials in the perspective of the synthesis of microstructured architectures with controlled organization on multiple length scale [31, 32].

In the past, biomorphic ceramic structures derived by the conversion of wood have been obtained in a variety of compositions, through a sequence of infiltration reactions aimed at coating the inner surfaces of the wood tissue with oxides precursors, which gave rise to consolidated oxide ceramics (Al_2O_3, ZrO_2, TiO_2, MnO) [33-37]. Alternatively, carbon templates obtained by pyrolysis of natural wood precursors have been infiltrated and directly transformed into carbide phases (SiC, TiC, ZrC) [38-41]. Pyrolysis process allows eliminating all the organic fraction of wood, leaving a porous skeleton in carbon, reproducing the cellular structure of the specific wood; pyrolysed templates can then be transformed in inorganic scaffolds by suitable chemical reactions. Particularly, Si/SiC wood derived structures obtained by infiltration of molten silicon were optimized to be employed as bioinert bone scaffolds [42,43]. However, such ceramic structures derived by the conversion of wood exhibit a biomimesis limited to inner morphology and organized structure, but their chemical composition and surface bioactivity are far from the biomimetic concept. A potential improvement can be achieved by specific surface treatments devoted to increase the biological affinity of the scaffolds [44, 45]. Recently, different woods were transformed into HA by sol-gel methods [46] and multistep thermal-chemical transformation processes [47].

In particular, both the approaches succeeded in the achievement of porous HA scaffolds derived by Rattan. Rattan wood was selected as organic template since its structure very closely resembles that of bone: Figure 3 evidences the system of channel like pores (simulating the Haversian system) in rattan, interconnected with a network of smaller channels (such as the Volkmann system).

The close reproduction of the starting wood structure depends on the nature of the interactions occurring between the reactants and the template. The reactions must take place until the inner parts of the biomorphic template and preserve the nanostructure of its constituting elements.

The chemical biomimesis of the biomorphic scaffold depends on both the phase composition and the crystal size; when the reactions of transformation occur at the molecular level, the structural mismatch resulting by the phase change is reduced and the original structure can be preserved down to submicron level. Figure 4 displays the scheme of the multistep transformation process to convert wood into biomorphic HA. The first four steps of the process are carried out through a gas-solid reaction, while the fifth one is a liquid-solid reaction at room conditions or autoclave.

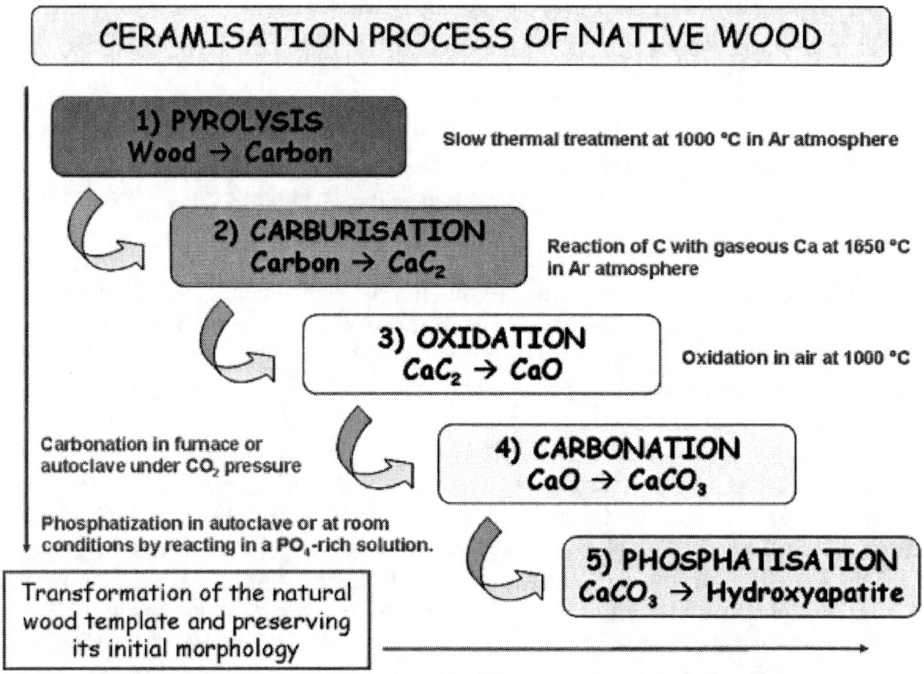

FIGURE 4 Scheme of the multistep transformation process to obtain biomorphic HA.

The criteria for the selection of rattan for bone scaffold development were based on the specifications of regions of spongy bone characterized by suitable porosity (see Table 1). Rattan is characterized by a total porosity of 85% and large pores with diameter 250 ± 40 μm (Figure 5).

TABLE 1 Pore size specifications for bone substitutes.

Characteristics	Vol. (%)	Dimensional range (μm)
Total porosity	50 – 90	–
Micro porosity (intergranular)	5 – 10	0 – 10
Interconnectivity pores	30 – 40	80 – 200
Macro porosity	50 – 65	200 – 600

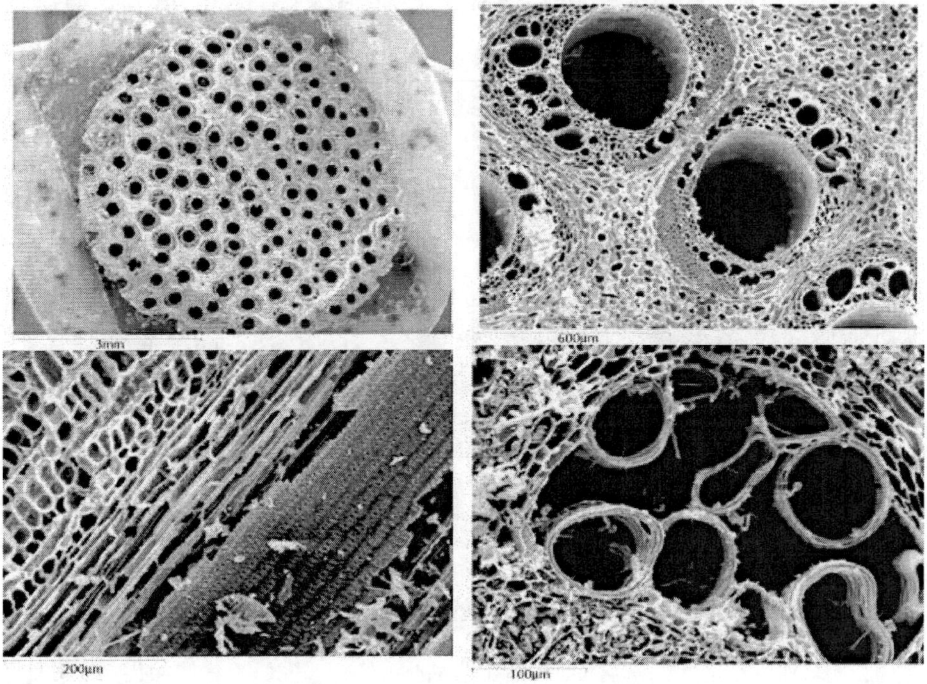

FIGURE 5 SEM images of pyrolyzed rattan wood illustrating its morphology.

4.4.1 Pyrolysis

The pyrolysis of wood is accompanied by the thermal degradation of cellulose, hemicellulose and lignin and a subsequent conversion into a carbon structure. The mechanism of cellulose thermal decomposition is supposed as follows:

- Desorption of adsorbed water (<150°C)
- Dehydration of the cellulose crystal water (150–240°C)
- Chain scissions, breaking C-O and C-C bonds (240–400°C)
- Aromatization (>400°C)

The result of the pyrolysis process is a porous carbon template, characterized by a structure of amorphous graphite. Anisotropic shrinkage in axial, tangential, and radial directions usually results and is associated with the pyrolysis process, according to the starting pore arrangement. Despite the large weight loss during pyrolysis, the carbon template reflected the microstructure and morphology of the native specimen (Figure 6).

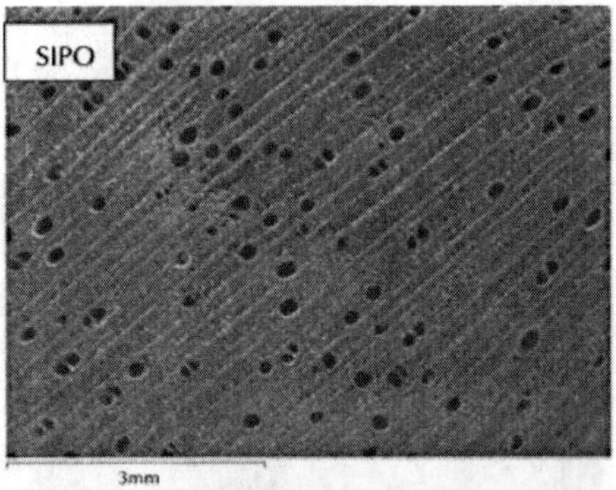

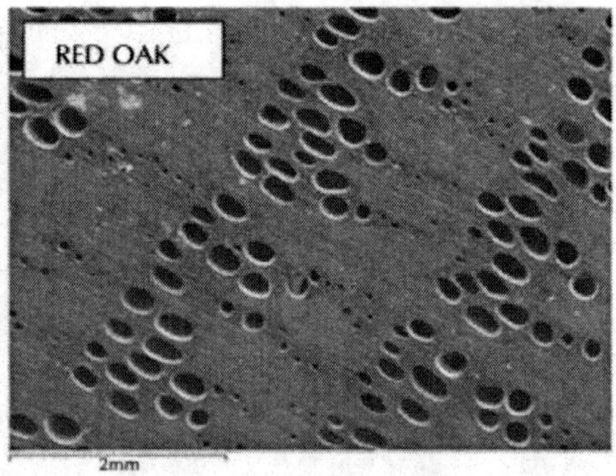

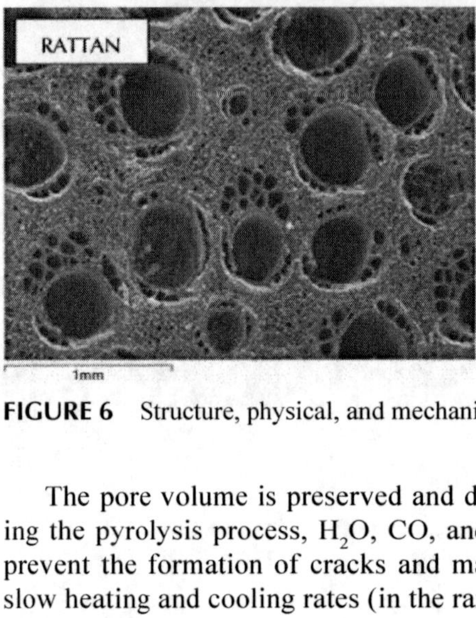

	Porosity (Vol %)	Porosity < 1μm (vol.%)	Density (g/cc)	Mechan Strength (Mpa)
RATTAN	88%	10-20	0.30	11
RED OAK	74%	25-35	0.59	37
SIPO	76%	30-40	0.53	43

FIGURE 6 Structure, physical, and mechanical properties of rattan, sipo, and red oak.

The pore volume is preserved and distributed in a wide range (Figure 7). During the pyrolysis process, H_2O, CO, and CO_2 gases are released; thus, in order to prevent the formation of cracks and macroscopic defects in the carbon template, slow heating and cooling rates (in the range of 1°C/min) must be applied up to final temperature.

4.4.2 Carburization

Multistep transformation process allow to obtain a biomorphic bone scaffold by building the HA molecule step-by-step by adding ions and functional groups at each step. The application of a sequence of steps is necessary to transform carbon into a complex calcium phosphate.

Carburization step provides calcium to the carbon matrix. Calcium must react at high temperature in order to chemically bind with carbon; metallic calcium or calcium hydride can be used as calcium sources and the process must be carried out in inert atmosphere.

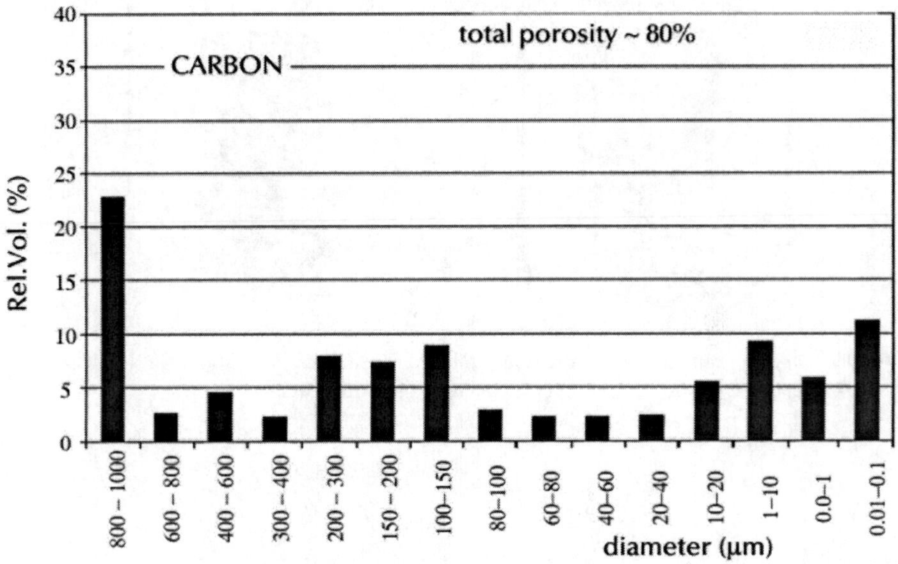

FIGURE 7 Pore size distribution of biomorphic rattan-derived carbon.

The chemical reaction involving the carbon template and calcium to form calcium carbide is:

$$2C + Ca \rightarrow CaC_2$$

In order to achieve the reaction between carbon and calcium at the nanometer level, thus aiding to preserve the local structure, calcium should be in the liquid or vapor state. In the liquid phase method spontaneous infiltration is achieved by immersing the carbon template into the calcium source, so that melted calcium penetrated into the pores and canals by capillarity; in the vapor phase method, the carbon template and calcium are separated and the reaction occur only by solid-gas interaction (Figure 8).

The thermal cycle was set up to allow the slow the melting and subsequent evaporation of the calcium (Figure 9).

In these processes, the most influent parameters on the formation of the products and the final sample structure were found to be the local carbon morphology, the calcium/carbon ratio, the furnace final temperature and dwell time.

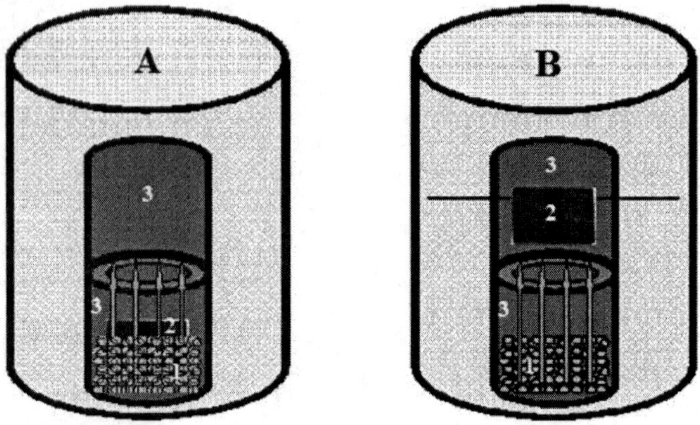

FIGURE 8 Equipment for wood carburization: (a) liquid phase infiltration; (b) vapor phase diffusion. 1: calcium source; 2 carbon template; 3 graphite crucibles.

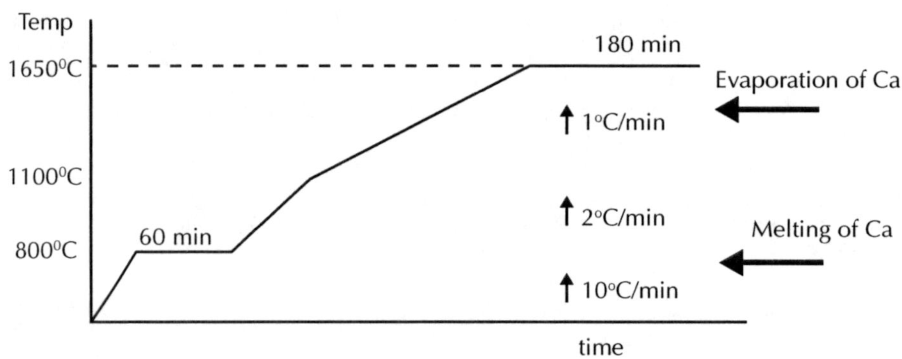

FIGURE 9 Typical thermal cycle of carburization.

The capability of Ca vapor to penetrate into porous carbon template is a determining factor for massive in pore deposition and reaction. Figure 10 shows the onset of CaC_2 nucleation inside the pore of a pine wood-derived carbon template.

Some important differences were observed between liquid and vapor phase infiltration products. In the first ones the reaction calcium carbon occurs mainly superficially (Figure 11), while the vapor phase reaction is more effective, involving more deeply the struts between pores (Figure 12).

FIGURE 10 Nucleation of CaC_2 phase in the pores of wood-derived carbon template.

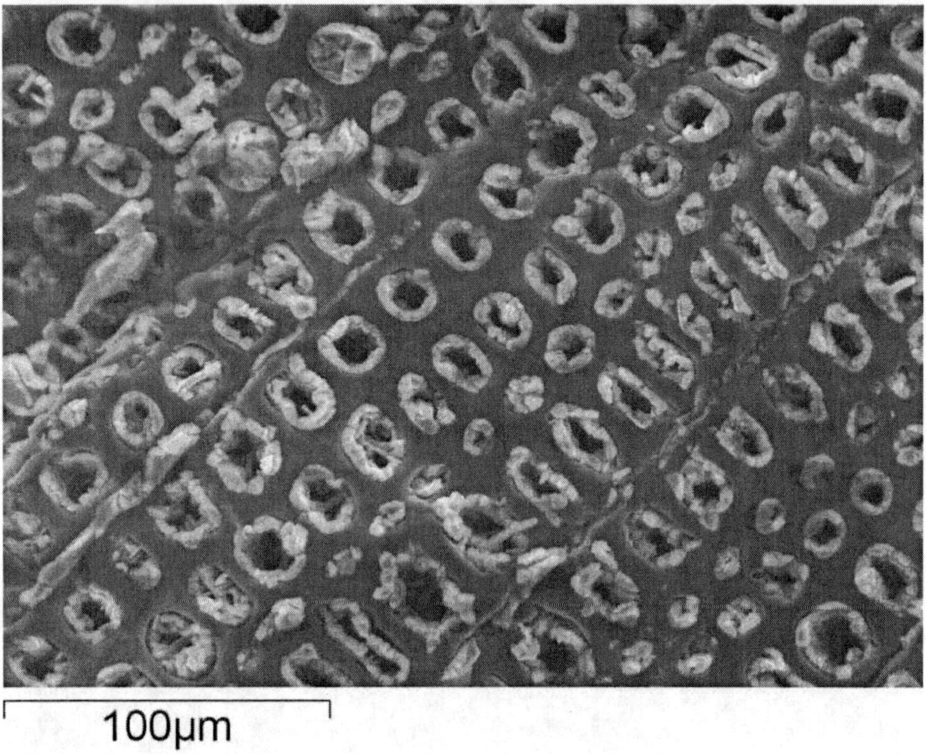

FIGURE 11 The SEM image of wood infiltrated by liquid calcium.

The total pore volume and pore distribution in the structure usually change upon carburization due to the volume expansion caused by the conversion of carbon into CaC_2 ceramic (Figure 13).

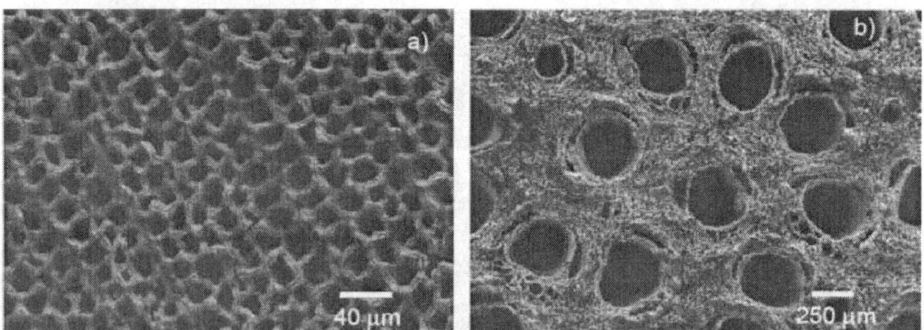

FIGURE 12 The SEM images of biomorphic calcium carbide derived from: (a) pine wood and (b) rattan.

The calcium content in the reaction crucible must be carefully dosed. In fact, the increase of the struts thickness between pores is proportional to the amount of calcium used for the carburization reaction. A molar ratio Ca/C = 1 has given the best results in terms of similar thickness between carbon template and calcium carbide final structure.

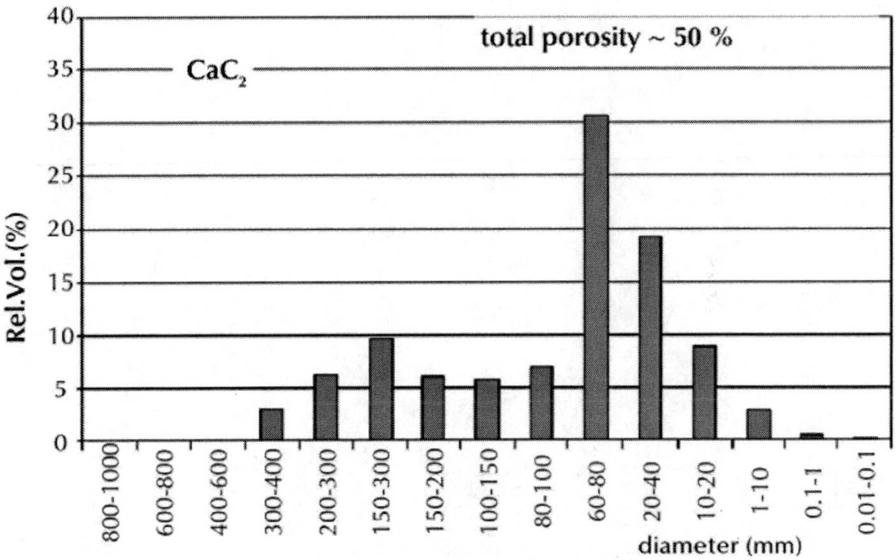

FIGURE 13 Pore size distribution of biomorphic CaC_2.

Figures 6 and 12 highlights that the original microstructure is preserved in detail in pine wood, rattan, and sipo-derived CaC_2. These three woods are characterized by very different porosity, in particular, rattan is very close to spongeous bone and sipo has a smaller pore size and high strength, so to simulate the cortical bone. Two possible steps for vapor phase process: the infiltration of the gaseous calcium into the carbon template and the heterogeneous gas/solid reaction between gaseous calcium and carbon to form CaC_2 on the structure surface, following by the diffusion of gaseous calcium through the CaC_2 layer.

4.4.3 Oxidation

The carbon matrix is to be eliminated during the transformation route towards biomorphic HA; it occurs during the oxidation of the biomorphic CaC_2. This process is not straightforward, since the chemical transformation involves the formation of intermediate calcium hydroxide, following the reaction path:

(1) $CaC_2 + 2H_2O \rightarrow Ca(OH)_2 + C_2H_2$
(2) $Ca(OH)_2 \rightarrow CaO + H_2O$
Total reaction: $CaC_2 + H_2O \rightarrow CaO + C_2H_2$

The two reactions occur at the same time, at temperature above 480°C (temperature of decomposition of $Ca(OH)_2$) and involve the presence of moisture; however, the formation of $Ca(OH)_2$ is detrimental for the maintenance of the original structure, due to the strong volume increase. A careful set up of the process is thus needed: XRD in hot attachment is a valid tool to detect phase composition *versus.* temperature. Figure 14 highlights the sequence of phase changes involved in the oxidation of biomorphic CaC_2.

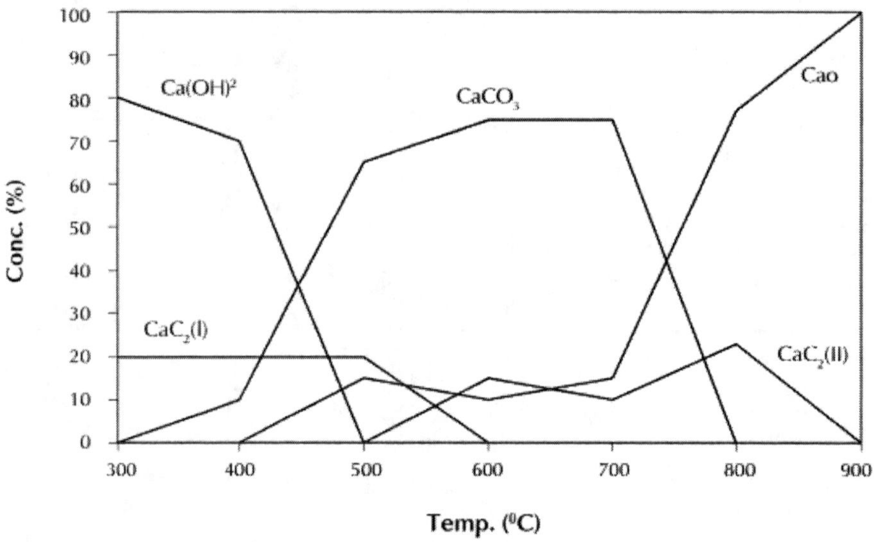

FIGURE 14 Schematic drawing of the volume concentration of the crystalline phases detected by XRD in hot attachment on biomorphic CaC_2.

The clear understanding of the complex phase transitions taking place during heat treatment in air allowed to design a thermal cycle where the hydration of CaC_2 is minimized by fast heating up to 500°C (i.e. higher than the temperature of formation of $Ca(OH)_2$ and then a slow heating up to T > 900°C to favor the reaction 2). Moreover, the slow heating rate reduced the structural deformation of the original CaC_2 porous template.

Upon transformation into biomorphic CaO, the pore size distribution usually changes as well as the total pore volume (Figure 15), also due to the high temperature treatment. It is found that the final porosity strongly depends on the oxidation temperature, so that it has to be carefully chosen, depending on the type of wood used.

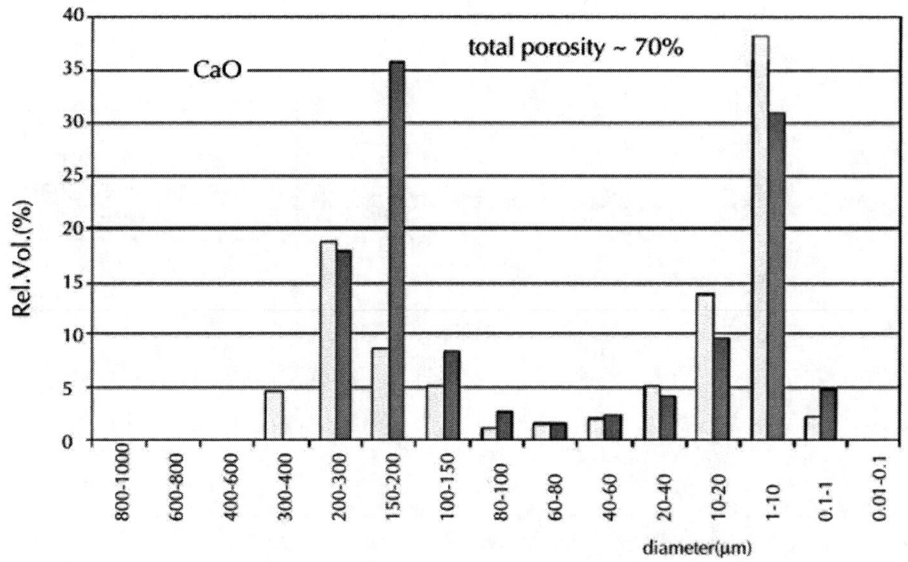

FIGURE 15 Pore size distribution of biomorphic CaO (light: T of oxidation = 1,100°C; dark: T of oxidation = 900°C).

4.4.4 Carbonation

Calcium oxide is very unstable in air, especially in presence of moisture. In order to synthesize biomorphic HA, an intermediate compound must be obtained. Biomorphic calcium carbonate templates can be obtained from calcium oxide *via* either the flux of gaseous CO_2 inside a furnace chamber at temperature > 700°C or *via* the addition of compressed CO_2 into a high pressure closed autoclave. The chemical reaction in both the processes is the following:

$$CaO + CO_2 \rightarrow CaCO_3$$

The gas-solid CO_2–CaO reaction proceeds through two rate controlling regimes: at the initial stage the reaction rapidly occurs by a heterogeneous surface chemical reaction kinetic. Following this initial stage, as a compact layer of $CaCO_3$ is formed on the surface of CaO, the rate limiting step is the diffusion of CO_2 through the $CaCO_3$ layer (Figure 16).

The recourse to high CO_2 pressure is thus necessary in order to obtain a complete phase transformation, especially when dealing with high density woods, whose struts are thicker and less permeable to CO_2 penetration. For porous woods like rattan, the process takes about 24 hr to obtain a fully carbonate biomorphic scaffold, maintaining the wood morphology (Figure 17).

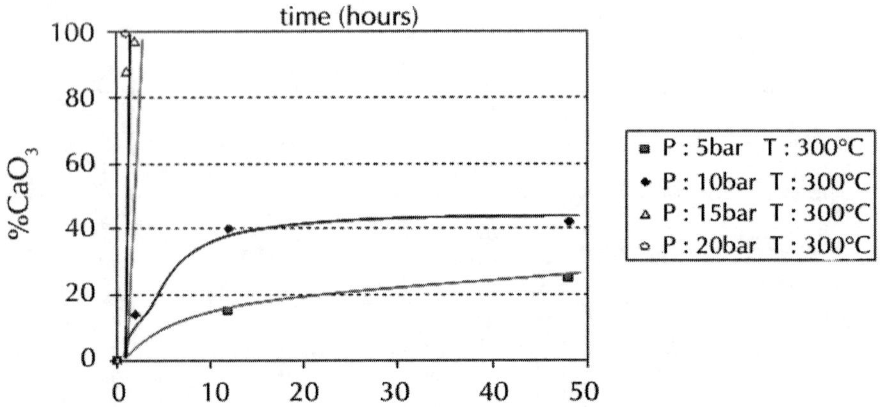

FIGURE 16 Kinetic of carbonation *versus* applied CO_2 pressure.

Figure 18 highlights the wide pore size distribution of biomorphic $CaCO_3$ in spite of the changes occurred upon the uptake of CO_3 molecules.

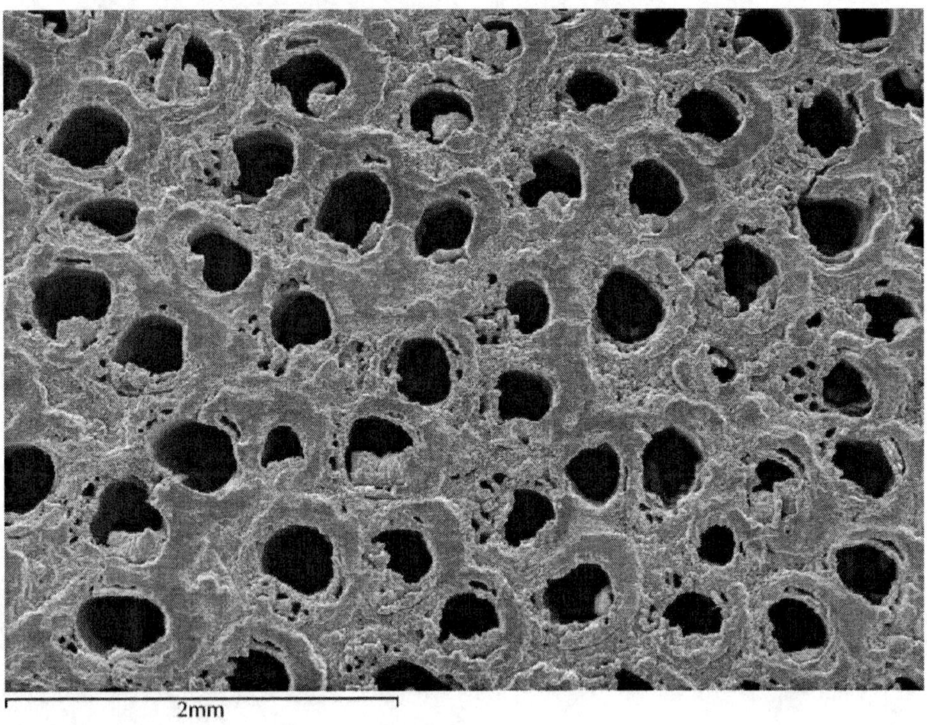

FIGURE 17 Biomorphic $CaCO_3$ derived by rattan.

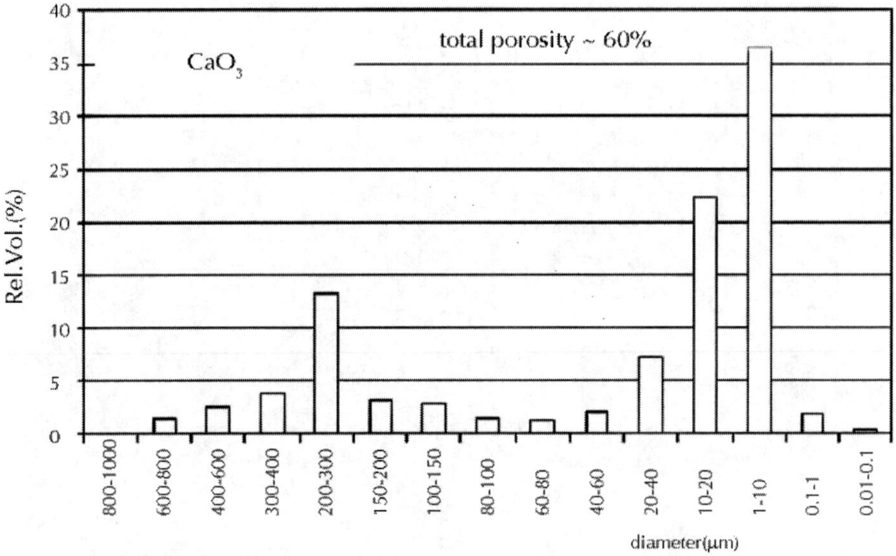

FIGURE 18 Pore size distribution of biomorphic $CaCO_3$.

4.4.5 Phosphatization

The transformation of calcium carbonate into calcium phosphate is well known [48]. Accordingly, the transformation of calcium carbonate templates into biomorphic HA can be performed by soaking in a phosphate solution at mild temperature, for example a possible reaction can be:

$$10CaCO_3 + 6KH_2PO_4 + 2H_2O \rightarrow Ca_{10}(PO_4)_6(OH)_2 + 6KHCO_3 + 4H_2CO_3.$$

The reaction should be carried out at pH values suitable to promote the formation of HA in respect to other phosphate phases (approximate pH range 4–9.5). The phosphatization occurs *via* a mechanism of dissolution reprecipitation, where the cubic crystals of $CaCO_3$ are progressively replaced by elongated grains of HA [49, 50] (Figure 19). The reaction mechanism is topotactic, locally selective, where the dissolution of the $CaCO_3$ occurs, following the precipitation of HA. SEM images (Figure 19 (a–d)) show the sequence of this process where crystals of calcite (cubic shape) are slowly dissolved and transformed into HA by phosphate solution.

The transformation process allowed to obtain biomorphic HA characterized by a very high chemical biomimesis; in fact the apatite phase has a very low crystallinity, like the mineral part of bone (Figure 20). Besides, FTIR spectroscopy reveals a partial carbonation of HA phase (about 3–4 wt %), evidenced by a small band at 880 cm^{-1} and a doublet centered at 1,440–1,550 cm^{-1}, thus reproducing the natural solubility behavior and bioactivity of human bone.

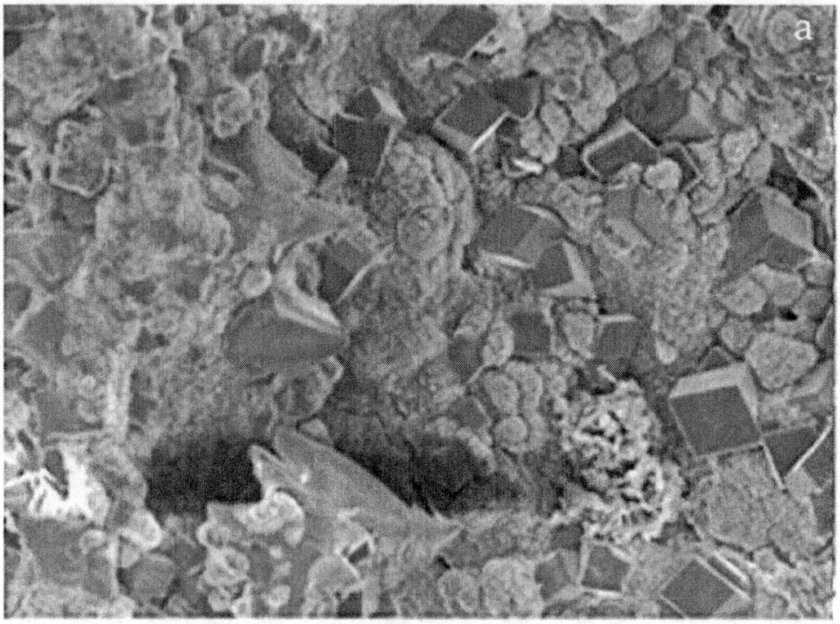

90μm

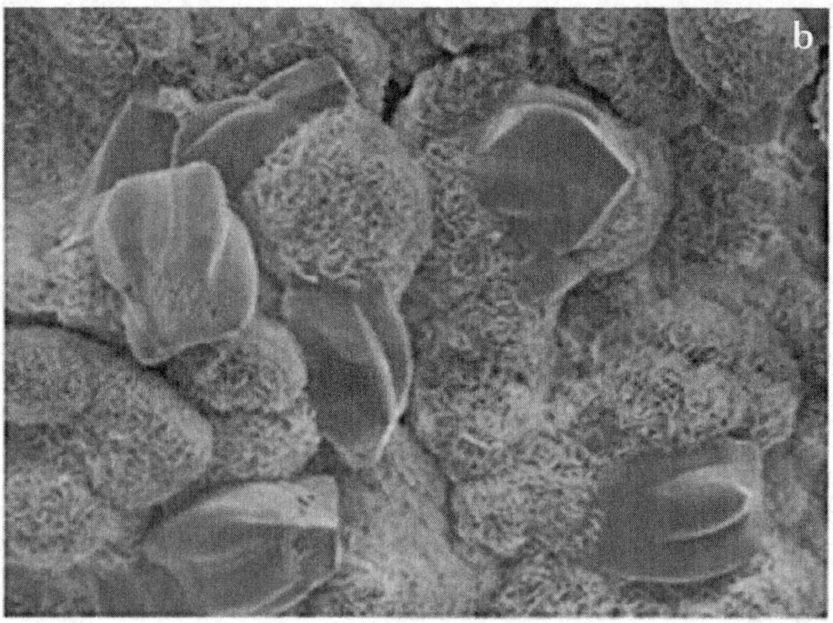

30 mm

FIGURE 19 *(Continued)*

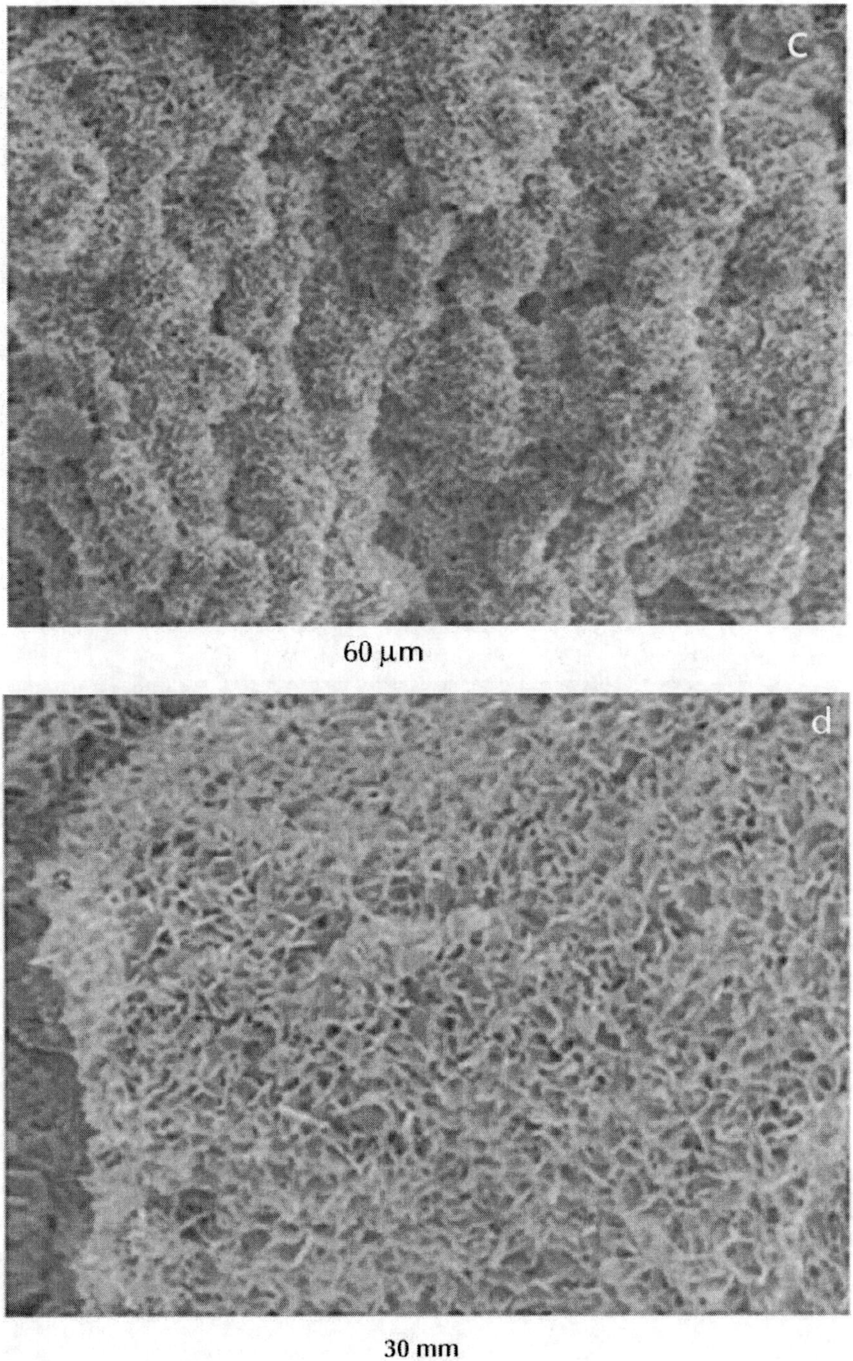

FIGURE 19 (a) Cubic $CaCO_3$; (b) Dissolution of $CaCO_3$; (c) HA formation; and (d) HA final structure.

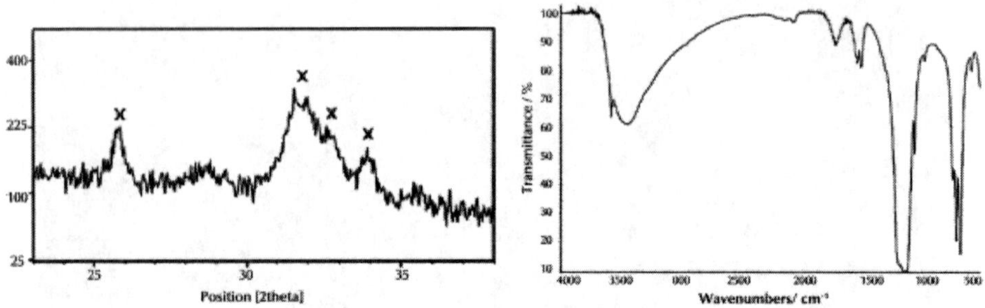

FIGURE 20 XRD and FTIR spectra of biomorphic HA.

Due to the selected hydrothermal pH condition (pH ≈ 7.4) and calcium concentration, the most thermodynamically stable phase is HA and in addition calcium ions cannot reprecipitate as calcium carbonate, according to the solubility product value of HA, which is much higher than that of calcite. The SEM images of the biomorphic carbonated HA (Figure 21), confirm that the starting microstructure is very well preserved in detail.

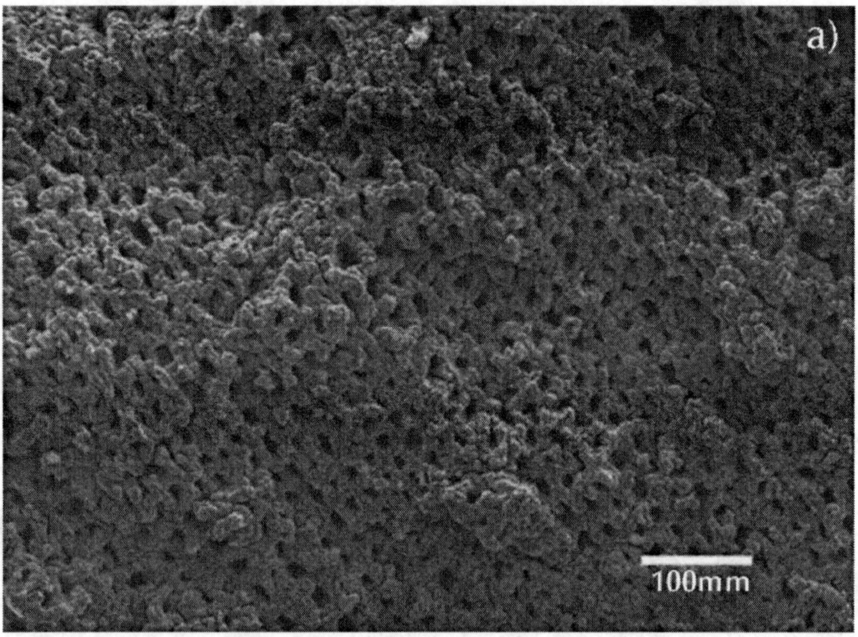

FIGURE 21 *(Continued)*

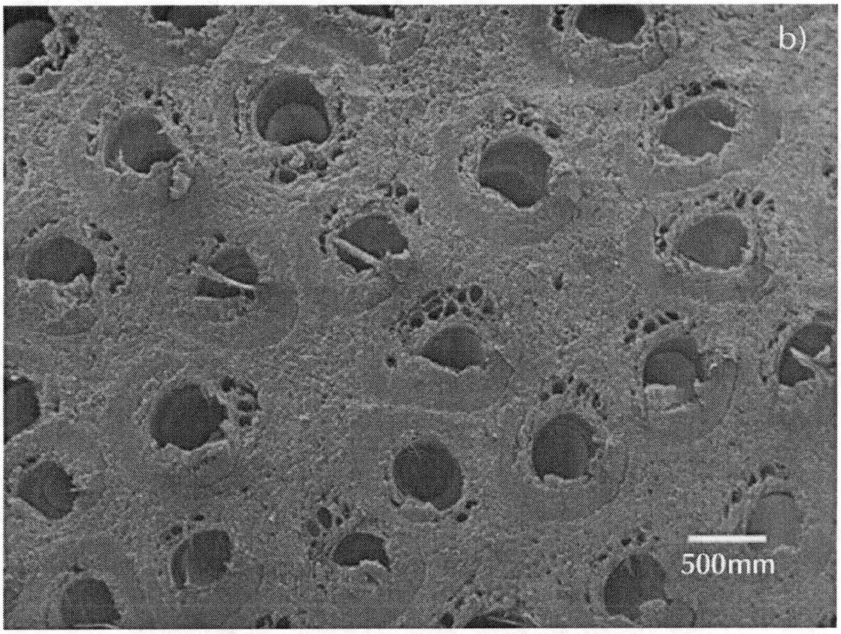

FIGURE 21 The SEM images of biomorphic HA. (a) pinewood and (b) rattan.

The biomorphic HA maintains a high total porosity (~75%) and a good compatibility with total porosity of spongeous bone (between 50 and 90%); it also exhibits a wide pore size distribution (Figure 22).

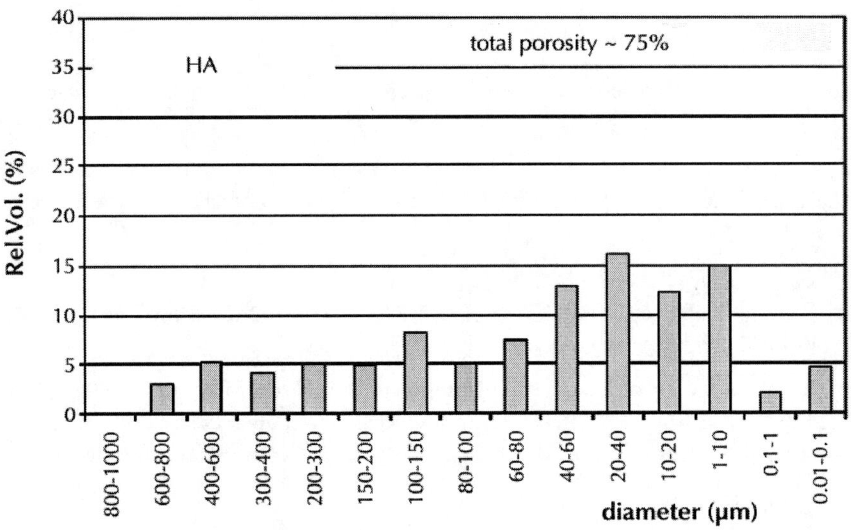

FIGURE 22 Pore size distribution of biomorphic HA.

Besides, Figure 23 highlights a good agreement of the porosity of biomorphic HA (light grey histogram) and spongeous bone (dark grey areas represent its critical porosity ranges). The biomorphic HA contains all porosity ranges necessary for cell colonization (200–600 μm and 80–200 μm) and fluid exchange (0–10 μm and 80–200 μm).

The unidirectional alignment of channels in biomorphic rattan HA induces strongly anisotropic mechanical strength. Along the channel direction, rattan HA samples exhibit higher strength (3.5–5 MPa), compared to the transverse direction (<1 MPa). These features make rattan HA compatible with the characteristics of chancellors bone.

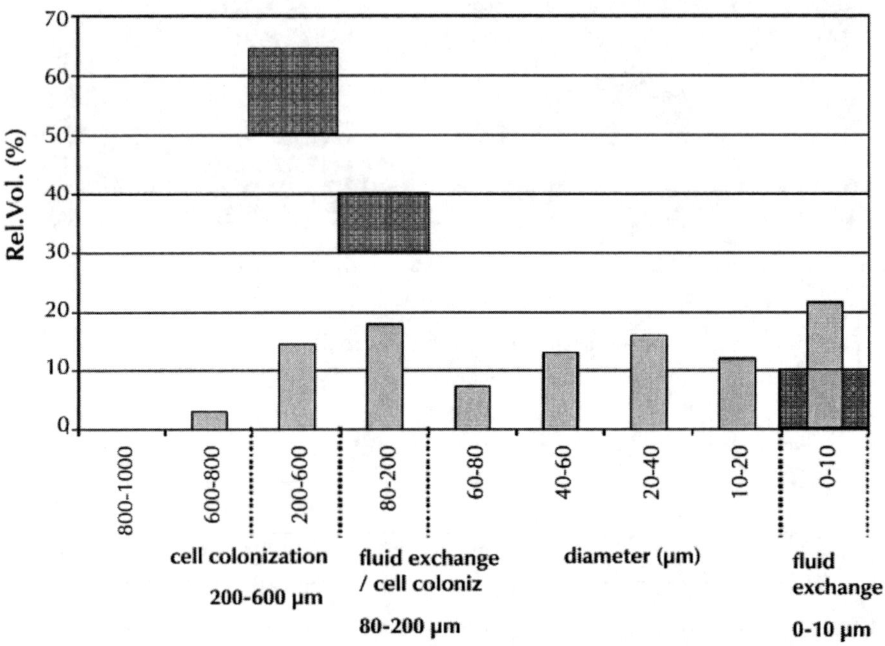

FIGURE 23 Comparison between pore size distribution of biomorphic HA and chancellors bone.

4.5 PRELIMINARY *IN VIVO* ASSESSMENT OF BIOMORPHIC HA SCAFFOLDS

In order to assess the functionality of biomorphic HA scaffolds, *in vivo* tests must be carried out in load-bearing sites. Preliminary tests have been carried out in rabbits, by implanting biomorphic scaffolds in the spongy region of the distal femur (see Figure 24).

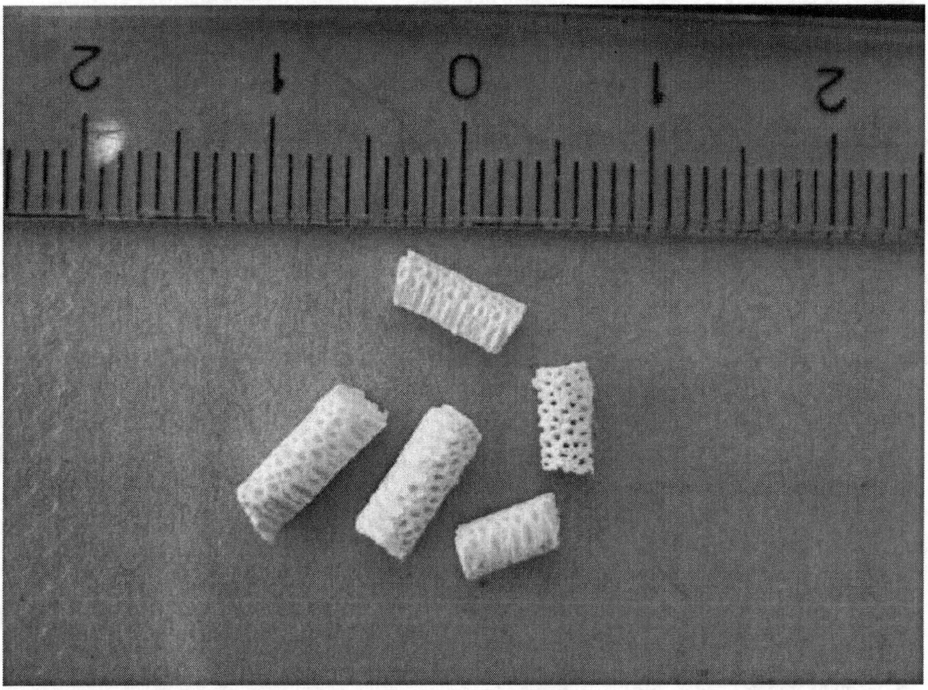

FIGURE 24 Biomorphic HA scaffolds derived by rattan for implantation in rabbits.

After 1 month, the implants were perfectly integrated in the surrounding bone and plenty of new bone formed into the channels of biomorphic HA scaffold (see Figure 25). Thus, biomorphic HA scaffolds can express *in vivo* high bioactivity, osteoinductivity and osteoconductivity.

In order to carry out *in vivo* tests in load-bearing sites, the mechanical strength of scaffolds must be improved. Composite scaffolds can be developed by exploiting the characteristics of dense woods like red oak and sipo, which are much stronger than rattan and can properly simulate the cortical part of bone. Besides, composite biomorphic scaffolds can be obtained, for example by introducing silica and inducing the formation of bioactive calcium silicates, characterized by high mechanical strength, suitable for load-bearing applications [12]. The multistep transformation process is in principle applicable to all types of woods and plants, but it must be adjusted in consideration of the pore organization of the native template. In fact, the different reactions can occur throughout all the template, at the molecular level, provided the diffusion of gaseous phases is allowed (i.e. the by-products of the organic decomposition during pyrolysis, the reactive gases, like CO_2 and O_2); thus the kinetic of the different steps must be tailored accordingly to allow the chemical reactions in the inner part of the template and avoid the formation of cracks and macroscopic defects.

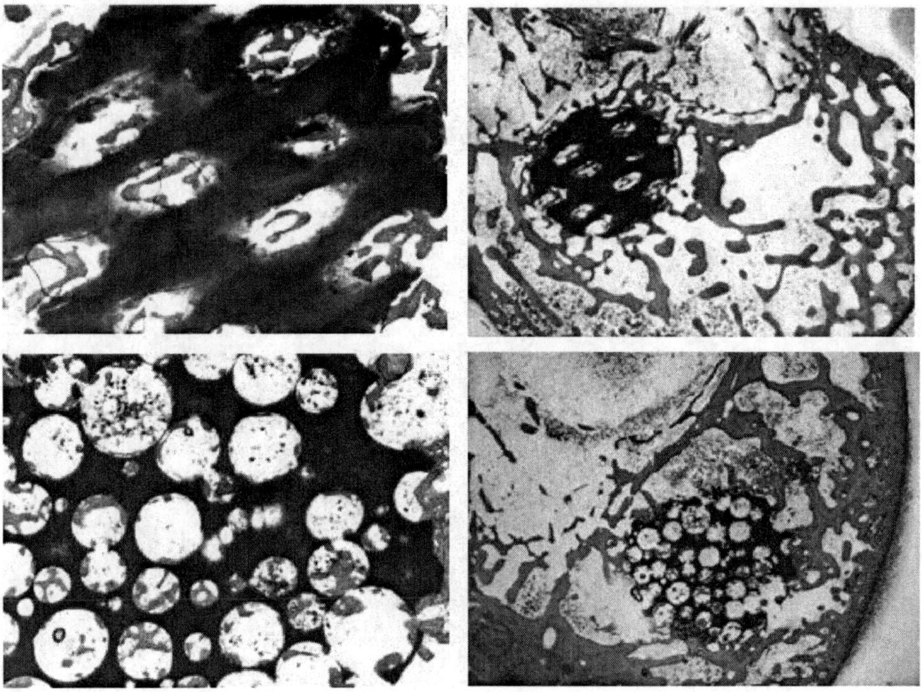

FIGURE 25 Histology of rabbit femur, implanted with biomorphic rattan HA, after 3 months (top) compared with a commercial porous HA scaffold (bottom, ENGI pore, Finceramica Faenza).

4.6 FINAL REMARKS

Biomorphic biomaterials have the unique characteristic of high structural organization and anisotropy, typical of natural structures, which are still impossible to be artificially reproduced in detail. The possibility to transform native structures into biomimetic phases (substituted HA) can push the current boundaries of the state of the art in production of hierarchically structured biomaterials. By combining biology, chemistry, materials science, nanotechnology, and production technologies, smart hierarchical structures existing in nature can be reproduced to develop breakthrough biomaterials that could open the door to a whole new generation of biomedical applications for which no effective solution exists to date.

On a wider scale, the concept of hierarchical structure is useful as a guide for the synthesis of new microstructures responsible of significant improvement in the final material properties; in the long term these materials could be developed and applied in different relevant industrial sectors. An intriguing possibility is that of simultaneously achieving high values of strength and toughness, for which ordinarily there is a trade off, in addition new materials with extreme values of physical properties such as thermal expansion or piezoelectricity can be obtained.

KEYWORDS

- **Bioinert bone**
- **Biological structures**
- **Biomimesis**
- **Carburization**
- **Hydroxyapatite**
- **Rattan wood**

ACKNOWLEDGMENT

The authors gratefully acknowledge the European Commission for the financial support under the FP6 contract number NMP4-CT-2006-033277 (TEM-PLANT).

REFERENCES

1. Tampieri, A., Celotti, G., Landi, E., Sandri, M., Roveri, N., and Falini, G. *J. Biomed. Mater. Res.*, **67A**, 618–625 (2003).
2. Tampieri, A., Sandri, M., Landi, E., Pressato, D., Francioli, S., Quarto, R., and Martin, I. *Biomaterials*, **29**, 3539–3546 (2008).
3. Kon, E., Delcogliano, M., Filardo, G., Fini, M., Giavaresi, G., Francioli, S., Martin, I., Pressato, D., Arcangeli, E., Quarto, R., Sandri, M., and Marcacci, M. *J. Orthop. Res.*, **28**,116–124 (2010).
4. Lowestam, H. A. and Weiner, S. *On biomineralization*. Oxford University Press (1989).
5. Tampieri, A., Sandri, M., Landi, E., Sprio, S., Valentini, F., and Boskey, A. L. *Adv. Appl. Cer.*, **107**, 298–302 (2008).
6. Soballe, K., Hansen, E. S., Brockstedt-Rasmussen, H., and Bunger, C. *J. Bone Joint. Surg. Br.*, **75**, 270–278 (1993).
7. Elliott, J. C. *Structure, chemistry of the apatites and other calcium orthophosphates*. Elsevier, Amsterdam (1994).
8. Sprio, S., Tampieri, A., Landi, E., Sandri, M., Martorana, S., Celotti, G., and Logroscino, G. *Mater. Sci. Eng. C,* **28**, 179–187 (2008).
9. Landi, E., Sprio, S., Sandri, M., Celotti, G., and Tampieri, A. *Acta Biomater*., **4**, 656–663 (2008).
10. Landi, E., Logroscino, G., Proietti, L., Tampieri, A., Sandri, M., and Sprio, S. J. *Mater. Sci: Mater. Med.*, **19**, 239–247 (2008).
11. Landi, E., Uggeri, J., Sprio, S., Tampieri, A., and Guizzardi, S. J. *Biomed. Mater. Res. A*, **94**, 59–70 (2010).
12. Sprio, S., Tampieri, A., Celotti, G., and Landi, E. J. *Mech. Behav. Biomed. Mater.*, **2**, 147–155 (2009).
13. Encinas-Romero, M. A., Aguayo-Salinas, S. C., Santos, J., Castillón-Barraza, F. F. et. al. *Int. J. Appl. Ceram. Tech.*, **5**(4), 401–411 (2008).
14. Heilmann, F., Standard, O. C., Müller, F. A. et. al. *J. Mater. Sci: Mater. Med.*, **18**(9), 1817–1824 (2007).
15. Ingber, D. E. *J. Cell. Sci.*, **104**, 613–627 (1993).
16. Pavalko, F. M., Norvell, S. M., Burr, D. B., Turner, C. H., Duncan, R. L., Bidwell, J. P. *J. Cell. Biochem.*, **88**, 104–112 (2003).
17. Sikavitsas, V. I., Temeno, J. S., Mikos, A. G. *Biomaterials*, **22**, 2581–2593 (2001).
18. Wang, J. H. C., Jia, F., Gilbert, T. W., and Woo, S. L. Y. J. *Biomech.*, **36**, 97–102 (2003).
19. Engler, A. J. et al. *Science*, **324**, 208–212 (2009).
20. Basu, J. et al. *Trends Biotech*, **28**, 526–533 (2010).
21. Toni, R. et al. *Acta Biomed.*, **78**, 129–155 (2007).

22. Ruiz-Hitzky, E. *Chem. Rec.*, 3, 88–100 (2003).
23. Schieker, M., Seitz, H., Drosse, I., Seitz, S., and Mutschler, W. *Eur. J. Trauma*, **2**, 114–124 (2006).
24. Wegst, U. G. K., Ashby, M. F. *Philos. Mag.*, **84**, 2167–2181 (2004).
25. Fratzl, P. and Weinkamer, R. *Prog. Mater. Sci*, **52**, 1263–1334 (2007).
26. Fengel, D. and Wegener, G. *Wood: Chemistry, ultrastructure, reactions, de Gruyter* (1989).
27. Gibson, E. J. *Met. Mater.*, **6**, 333–336 (1992).
28. Lucas, P. W. Darvell, B. W., Lee, P. K., Yuen, T. D. B., and Choong, M. F. *Phil. Trans. Roy. Soc. B*, **348**, 363–372 (1995).
29. LeGeros, R. Z. and Craig, R. G. *J. Bone Min. Res.*, **8**, S583–S596 (1993).
30. Rey, C. Calcium phosphates for medical applications. In *Calcium phosphates in biological and industrial systems*. Z. Amjad (Ed.), Kluwer Academic Publishers, Dordrecht, Netherlands, (1998).
31. Roveri, N. and Palazzo, B. Tissue, Cell and Organ Engineering. In *Nanotechnologies for the Life Sciences*, vol. 9, C. S. S. R. Kumar (Ed.), Wiley-VCH, Weinheim, Germany (2006).
32. Roveri, N., Palazzo, B., and Iafisco, M. *Expert Opin. Drug Deliv.*, **5**(8), 1–17 (2008).
33. Cao, J., Rambo, C. R., and Sieber, H. *J. Por. Mater.*, **11**, 163–172 (2004).
34. Rambo, C. R., Sieber, H. *Adv. Mater.*, **17**(8), 1088–1091 (2005).
35. Singh, M. and Yee, B. M. *J. Eu. Ceram. Soc.*, **24**, 209–217 (2004).
36. Cao, J., Rusina, O., and Sieber, H. *Ceram. Inter.*, **30**, 1971–1974 (2004).
37. Li, X., Fan, T., Liu, Z., Ding, J., Guo, Q., and Zhang, D. *J. Eu. Ceram. Soc.*, **26**, 3657–3664 (2006).
38. Greil, P., Lifka, T., and Kaindl, A. *J. Eu. Ceram. Soc.*, **18**, 1961–1973 (1998).
39. Greil, P., Lifka, T., and Kaindl, A. *J. Eu. Ceram. Soc.*, **18**, 1975–1983 (1998).
40. Binghe, S., Tongxiang, F., Di, Z., and Okabe, T. *Carbon*, **42**, 177–182 (2004).
41. Rambo, C. R., Cao, J., Rusina, O., and Sieber, H. *Carbon*, **43**, 1174–1183 (2005).
42. Parfen'eva, L. S., Orlova, T. S., Kartenko, N. F., Sharenkova, N. V., Smirnov, B. I., Smirnov, I. A. et al. *Phys. Solid St.*, **47**(7), 1216–1220 (2005).
43. Gonzalez, P., Borrajo, J. P., Serra, J., Liste, S., Chiussi, S., Leon, B. et al. *Key Eng. Mater.*, **16**, 1029–1032 (2004).
44. González, P., Serra, J., Liste, S., Chiussi, S., León, B., Pérez-Amora, M. et al. *Biomaterials*, 24, 4827–4832 (2003).
45. Lelli, M., Fortran, I., Foresti, E., Martinez-Fernandez, J., Torres-Raya, C., Varela-Feria, F. M., and Roveri, N. *Adv. Eng. Mat.*, **12**, B348–B355 (2010).
46. Eichenseer, C., Will, J., Rampf, M., Wend, S., and Greil, P. J. *Mater. Sci: Mater. Med.*, 21, 131–137 (2010).
47. Tampieri, A., Sprio, S., Ruffini, A., Celotti, G., Lesci, I. G., and Roveri, N. *J. Mater. Chem.*, **19**(28), 4973–4980 (2009).
48. LeGeros, R. Z., Daculsi, G., and LeGeros, J. P. Bioactive ceramics. In *Musculoskeletal tissue regeneration, Biological materials and methods*. W. S. Pietrzak (Ed.), Humana Press (2008).
49. Ni, M. and Ratner, B. D. Nacre Surface Transformation to Hydroxyapatite in a. Phosphate Buffer Solution. *Biomaterials*, **24**, 4323–4331 (2003).
50. Yoshimura, M., Sujaridworakun, P., Koh, F., Fujiwara, T., Pongkao, D., and Ahniyaz, A. *Mater. Sci. Eng. C*, **24**(4), 521–525 (2004).

5 Porous Nanostructure Ti Alloys for Hard Tissue Implant Applications

J. Jakubowicz and G. Adameks

CONTENTS

5.1 Introduction 79
5.2 Mechanical Alloying of the Tialv and Tizrnb Alloys 81
5.3 Electrochemical Functionalization of the Tialv and Tizrnb Alloys 88
 5.3.1 Electrochemical Anodic Etching 88
 5.3.2 Electrochemical Cathodic Cap Deposition 97
5.4 Corrosion Resistance of the Tialv and Tizrnb Alloys 114
5.5 Conclusion 116
Keywords 116
Acknowledgment 116
References 116

5.1 INTRODUCTION

In this chapter, porous nanocrystalline bioalloys Ti-6Al-4V, Ti-15Zr-4Nb, and Ti-6Zr-4Nb formation is described. The alloys were prepared by mechanical alloying (MA) followed by pressing, sintering, and subsequent anodic electrochemical etching in 1M H_3PO_4 + 2% HF electrolyte at 10 V for time of 30–300 min. We show that, ultra-fine structure improves etching process. The electrolyte penetrates sinters through the empty spaces and the grain boundaries, resulting in effective grains removing and pores-formation. The porosity of the Ti-15Zr-4Nb is larger than the Ti-6Al-4V. The macropore diameter reaches up to 60 μm, while the average micropore size is in the range from 3.5 to 5.46 nm for Ti-15Zr-4Nb and Ti-6Al-4V electrochemically etched alloys, respectively.

The mechanism of the bioactive ceramic CaP layer formation on the porous etched surface was investigated, too. The CaP compounds were cathodically deposited at different potential range (from -0.5 to 10 V), using a mixture of 0.042M $Ca(NO_3)_2$ + 0.025M $(NH_4)_2HPO_4$ + 0.1M HCl. The Ca and P ions penetrate preferentially the pores inside, which results in improved bonding of the bioceramic layer to the metallic background. Change in the depositing potential results in different morphology, porosity, composition, and thickness of the growing CaP layer. We propose the electric field enhancement mechanism of the electrolytic ions flow resulting in CaP growth on the surface irregularities, such as, pores and surrounding hillocks.

The corrosion resistance of the alloys was investigated in Ringer's solution. The electrochemical etching improves the corrosion resistance of the nanocrystalline alloys. The prepared porous nanocrystalline Ti-6Al-4V, Ti-15Zr-4Nb, and Ti-6Zr-4Nb alloys with electrochemically biofunctionalized surface could be a possible candidate for hard tissue implant applications.

The Ti alloys are widely used in medical applications, because of its excellent corrosion resistance, biocompatibility, and mechanical properties [1]. In comparison to the pure Ti, the Ti alloys have better mechanical properties. The elements, such as, Al and V are main alloying additives in the Ti alloys. Unfortunately, V and Al exhibits high cytotoxicity and may induce senile dementia, respectively [2, 3]. The replacing of Al and V by Nb and Zr leads to excellent biocompatibility, because, these elements belong to vital group in the tissue reaction [4]. The most popular Ti-6Al-4V alloy is still in use in the biomedical as well as other technical applications [5, 6]. A new Ti alloys consisting Zr are much more often used in the last time [7]. These alloys have properties promoting them in the hard tissue implant applications, for example low density, high strength, relatively low elastic modulus, high wear resistance, no toxic behavior, corrosion resistance, and biocompatibility. These alloys show long lifetime in human body, which finally extends the time between implant replacement and surgery operations.

The main problem of the metallic materials, for hard tissue implant applications, is the mismatch of the Young's modulus between the Ti type implant (100-120 GPa) and the bone (10-30 GPa), which is unfavorable for bone healing and remodeling [8]. An introduction of pores into the alloy microstructure results in smaller modulus [9]. Decreasing the grain size to a nanoscale could lead to decrease the Young's modulus of the Ti alloys, too. The nanostructure could be generated in a large scale by mechanical methods, for example by MA [10, 11]. Thus, in the alloys for hard tissue implant applications, the best solution seems to be a combination of nanostructure with porosity. Although, pores in whole volume of the implant material can lead to lower mass and modulus, unfortunately the lower mechanical strength is expected, too. So, the electrochemical etching seems to be a very promising method for the surface pores formation. In this case, the bulk background provides good mechanical properties, while the porous surface morphology improves the bone fixing [12]. In the nanocrystalline alloys the large volume of the grain boundaries improves the electrochemical etching [13, 14]. The pore formation (materials removing) is easy and effective, due to the large grains surface area. Additionally, the surfaces etched at positive (*versus* OCP) potentials are oxidized, resulting in thick TiO_x layer, and improving corrosion resistance [15].

On the porous surface bioceramic material could be deposited, which improves the bioactivity of the metallic implant [16]. Usually, bioactive hydroxyapatite (HA) is required on the surface. The HA, which is CaP compound with chemical composition of $Ca_{10}(PO_4)_6(OH)_2$ and Ca/P ratio of 1.67, can easily bond to living tissues, improving biocompatibility of the implant, while the metallic background provide respective mechanical properties. The coating of the implant by CaP compounds results in good osseointegration and short convalescence of the patient after the surgery implantation [17, 18]. Adhesion and proliferation of the human osteoblast cells is accelerated by the topography of the implant. For example, number of the adhered cells to the nanorough TiO_2 nanotubes surface increases by 400% in comparison to microcrystalline Ti [19]. The calcium and phosphorus are components of the bone, and these elements are crucial for good osseointegration. The depositing of the HA layer on the rough and porous implant background improves fixation of the layer, as well as, fixation of the implant with the human body, in comparison to the flat implant surface [20]. The HA can be deposited using electrochemical cathodic deposition method, where the electrolyte consist of Ca and P ions flowing into the negatively charged Ti type surface [21, 22].

We show recent achievements in the formation of nanocrystalline/ultrafine Ti typed alloys prepared by MA with electrochemically biofunctionalized surface, respective for hard tissue implant applications.

5.2 MECHANICAL ALLOYING OF THE TIALV AND TIZRNB ALLOYS

The MA is a process, where the pure elemental powders are mixed and cold welded together, resulting in amorphous or nanocrystalline phases [10, 24]. The powders are hit by balls of the mill, resulting in effective grain size reduction. The weight ratio of the balls to the powders is 10:1. The SPEX 8,000 mixer mill, equipped with hardened steel container with steel balls inside, was applied in work. The protective atmosphere is very important for a high quality materials preparation, so all powders handling was done in Labmaster glove box from MBraun, with argon atmosphere inside and controlled level of oxygen and humidity. The argon atmosphere was in mill container, too. A mixture of respective powders of Ti, Al, V, Zr, and Nb (Figure 1 (a–c) and Figure 2 (a–c)), with stoichiometric ratio, respective to obtain the Ti-6Al-4V, Ti-15Zr-4Nb, and Ti-6Zr-4Nb alloys, was loaded into the container. All powders have initial size below 300 mesh.

After 48 hr of MA the powders are composed from amorphous and/or nanocrystalline phases (Figure 1 (d) and Figure 2 (d, f)). In the mechanically alloyed powders high plastic deformations are generated resulting in high density of dislocation lines and subsequently sub-grains formation, which finally lead to amorphisation [23]. When the powders are trapped between balls, the high energy of the balls hits leads to powders deformation, cold welding, and finally lamellar structure formation (Figure 3) [23].

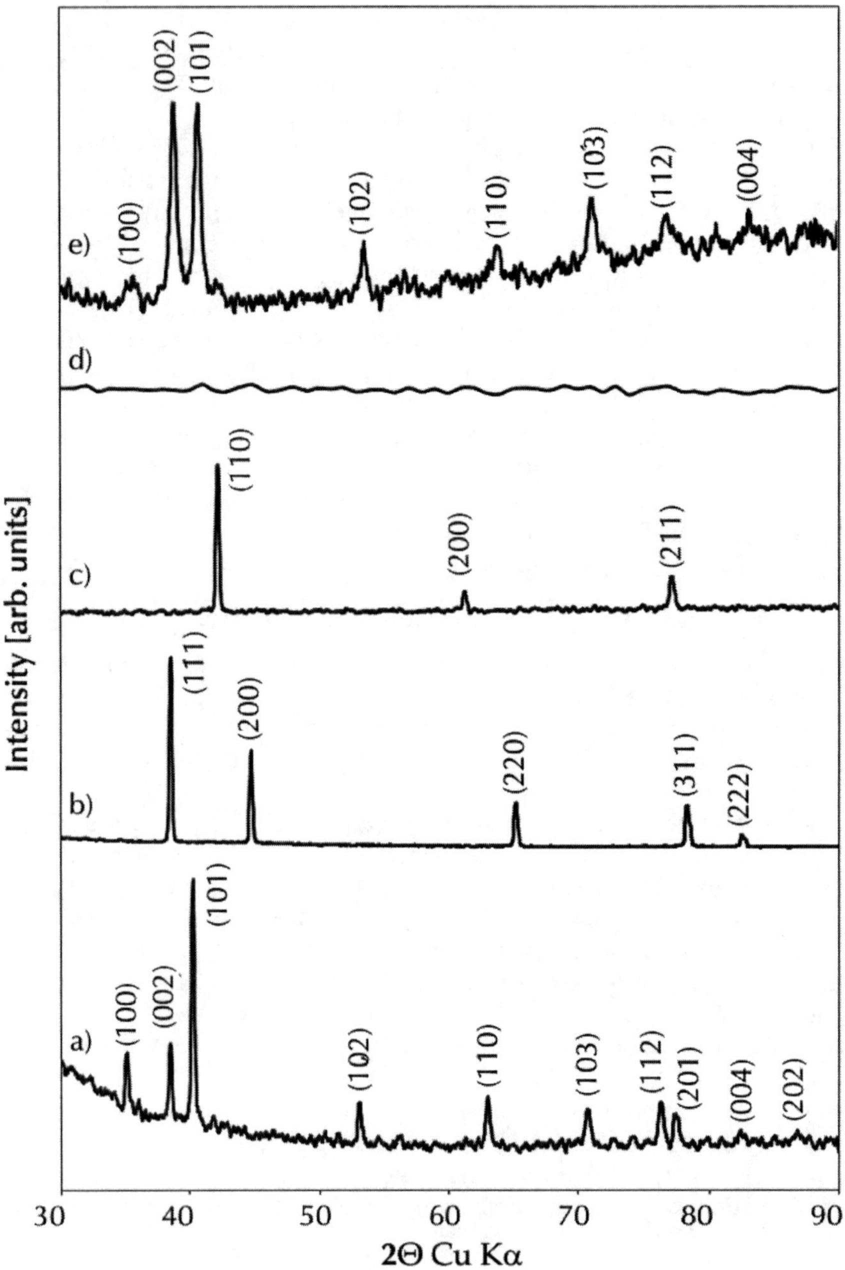

FIGURE 1 The XRD data for pure Ti (a), Al (b), V (c), Ti-6Al-4V after 48 hr MA (d), and after sintering (e).

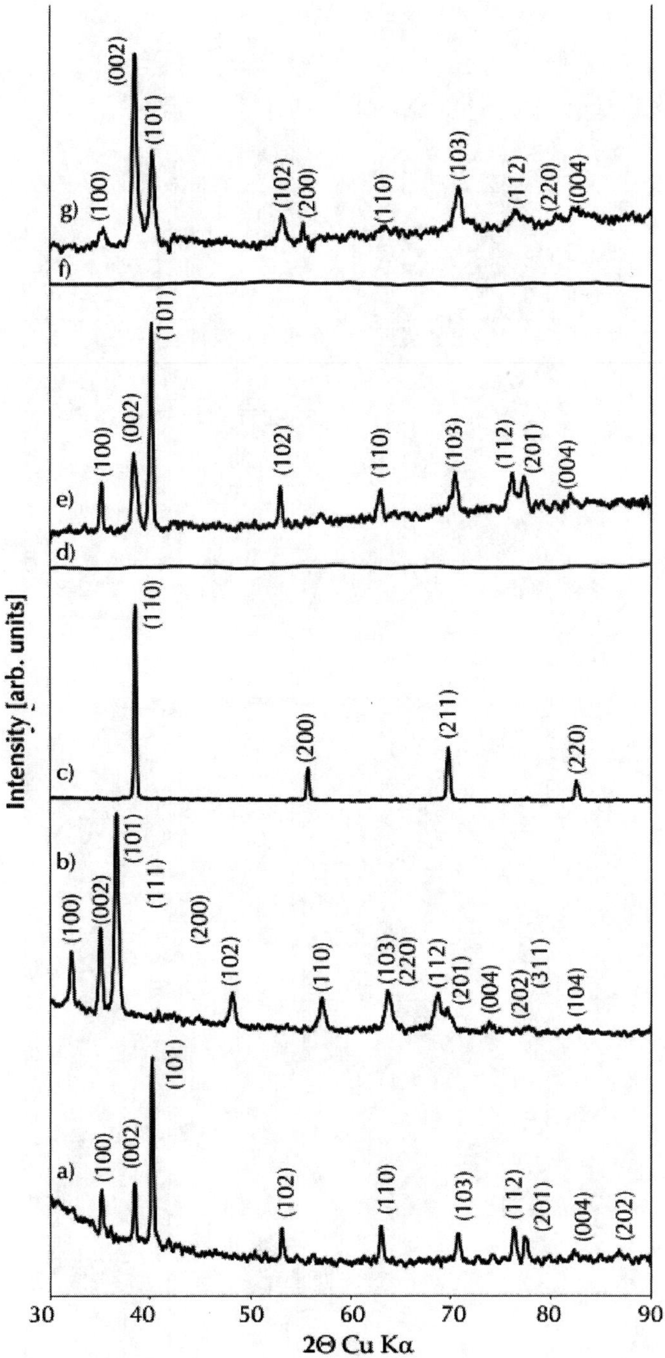

FIGURE 2 The XRD data for pure Ti (a), Zr (b), Nb (c), Ti-15Zr-4Nb alloy after 48 hr MA (d), and after sintering (e), as well as Ti-6Zr-4Nb alloy after 48 hr MA (f) and after sintering (g).

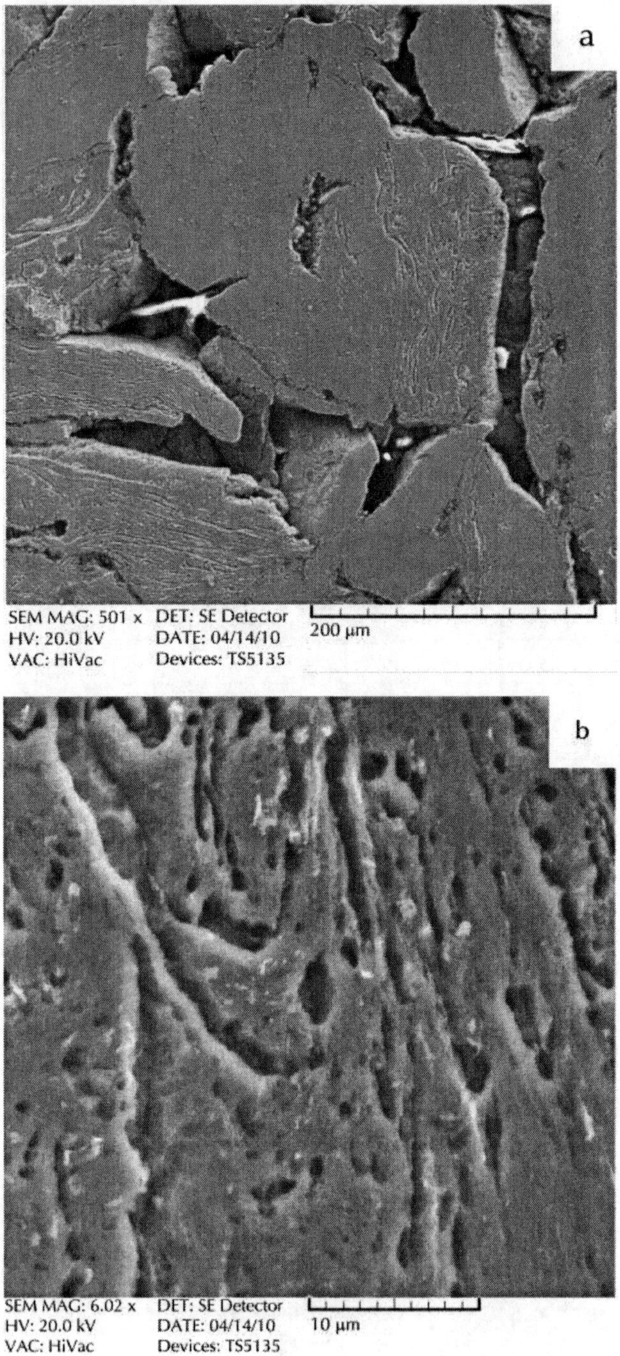

FIGURE 3 Cross section scanning electron microscope (SEM) images of the lamellar (onion type) structure formed in the MA process of the Ti-15Zr-4Nb alloy particles (a, b—different magnifications).

The lamellar or onion type morphology is well visible in large plastically deformed particles (Figure 3 (a)). In these particles, the layers does not strictly adhere one to each other and large volume of small pores occurs (Figure 3 (b)). The continuous is otropic deformations of the particles between balls and walls of the milling container and relatively small stress in the large particles (in comparison to smaller ones) leads to gaps and pores existing between the layers. The long time milling leads finally to amorphous (Figure 4) or nanocrystalline (Figure 5) particles. After MA usually agglomerates are formed, which are composed mainly from the smaller amorphous (Figure 4) or nanocrystalline (Figure 5) areas [23]. Figure 4 shows amorphous agglomerate particle of the mechanically alloyed Ti-6Al-4V alloy. The amorphous structure is unstable and under temperature rising (for example during sintering) crystallization takes place. Due to a high stress and strong plastic deformations observed in the MA process, nanocrystallite areas in the larger powder particles could be formed directly in the process (Figure 5). So, the mixture of amorphous and nanocrystalline areas could coexists together. In these crystalline large particles, which are formed mainly in the Ti-15Zr-4Nb, than in the Ti-6Al-4V, the nanostructure is built from small sub-grains with size below 20 nm (mainly visible in the upper part of the Figure 5). The sub-grains formation is related to high plastic deformations ratio and high density of dislocation lines (visible in the lower part of the Figure 5) formed in the process.

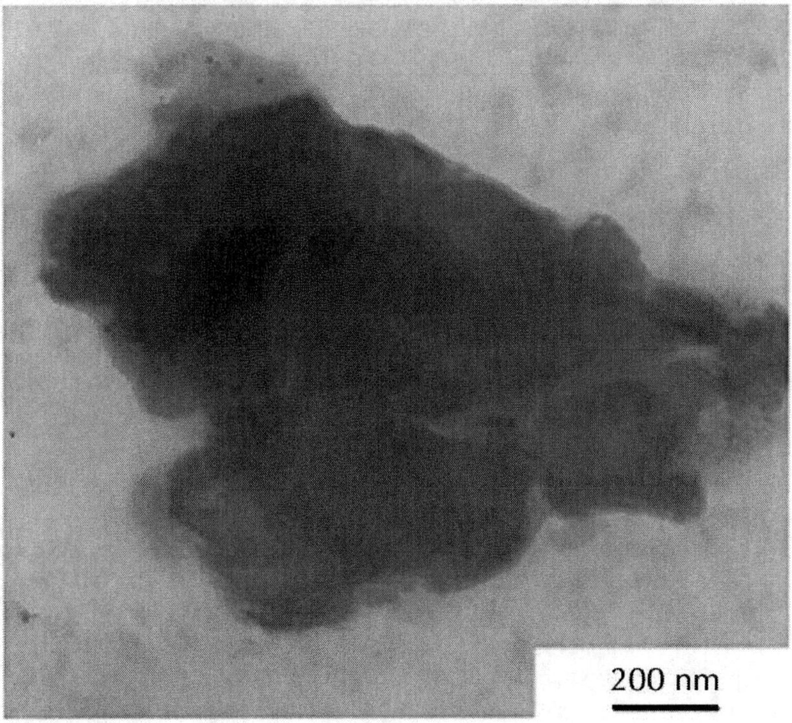

FIGURE 4 The TEM image of the mechanically alloyed amorphous Ti-6Al-4V alloy agglomerate particle.

FIGURE 5 The TEM image of the nanocrystals in the mechanically alloyed large particle of the Ti-15Zr-4Nb alloy.

The subsequent uniaxial pressing at 500 MPa leads to formation of green compacts, which were sintered at 1,000°C for 1 hr. The sinters have crystallographic structure (Figure 1 (e) and Figure 2 (e, g)) corresponding to commercial alloy and density of 3.6 g/cm^3, which is about 80% of the theoretical value for the bulk commercial microcrystalline alloy.

The high temperature sintering leads to increase of the grain size, more in case of the Ti-6Al-4V than the Ti-15Zr-4Nb alloy (Figure 6). For the Ti-6Al-4V and the Ti-15Zr-4Nb alloy the grain size after sintering is in the range 0.2-2 μm and 0.1-1 μm, respectively. In case of the Ti-6Al-4V, the grains have more uniform isotropic shape (Figure 6 (a)). For the Ti-15Zr-4Nb the grains have more anisotropic (lamellar) shape (Figure 6 (b)). Because, a relatively low pressing pressure, after sintering large volume of free spaces remained in the whole volume of the sinters (Figure 6). Hence, sinters density state of about 80% of the theoretical value. The presented pores are useful in the next electrochemical etching stage of the surface biofunctionalization.

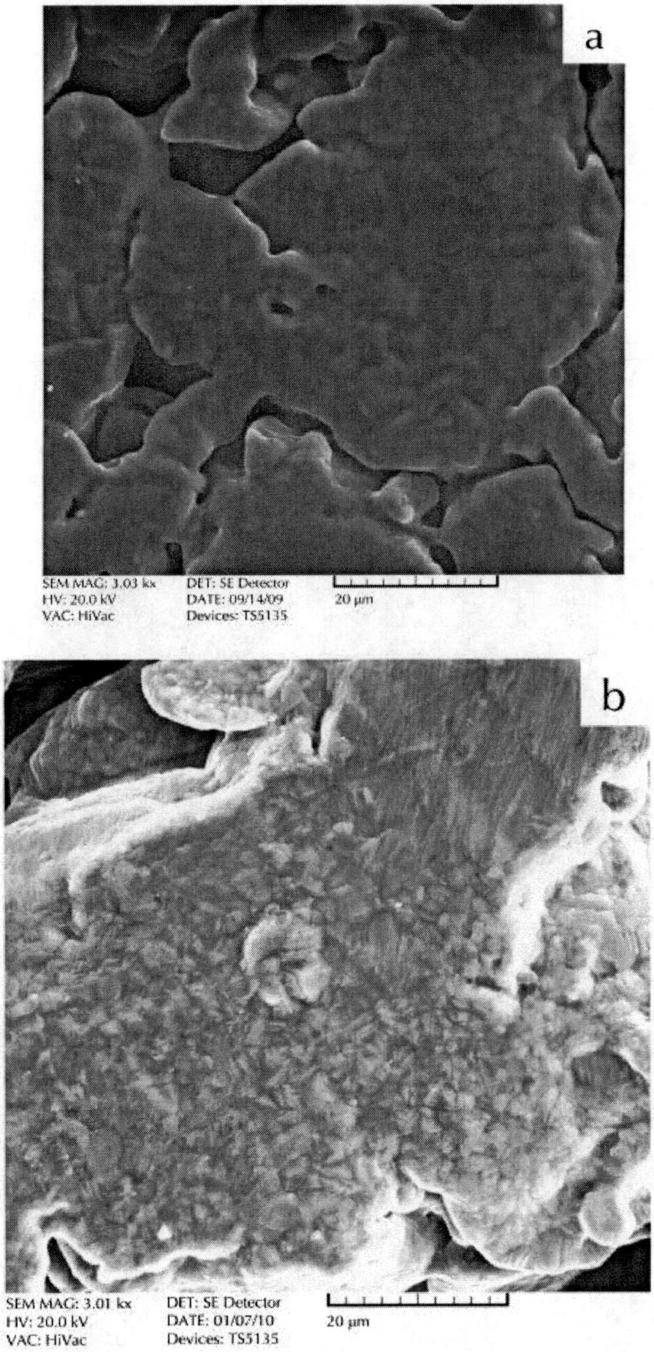

FIGURE 6 The SEM images of the Ti-6Al-4V (a) and the Ti-15Zr-4Nb (b) sinters rough surface.

5.3 ELECTROCHEMICAL FUNCTIONALIZATION OF THE TIALV AND TIZRNB ALLOYS

5.3.1 Electrochemical Anodic Etching

The Ti type nanocrystalline sinters made in MA and powder metallurgy processes have initial porosity useful in decreasing of the Young modulus as well as improving etching process. The electrochemical etching was done using potentiostat connected to respective electrodes (graphite rod—counter electrode, SCE—reference electrode, and Ti alloy sinter—working electrode) in the electrochemical etching cell. The magnetic stirring was applied for the uniform etching, etching products removing of the hydrogen bubbles released from the etched surface. As an etching electrolyte, mixture of the 1M H_3PO_4 + 2% HF in distilled water was applied. The applied etching potential was fixed on 10 V *versus* OCP and the etching time were changed from 30–300 min.

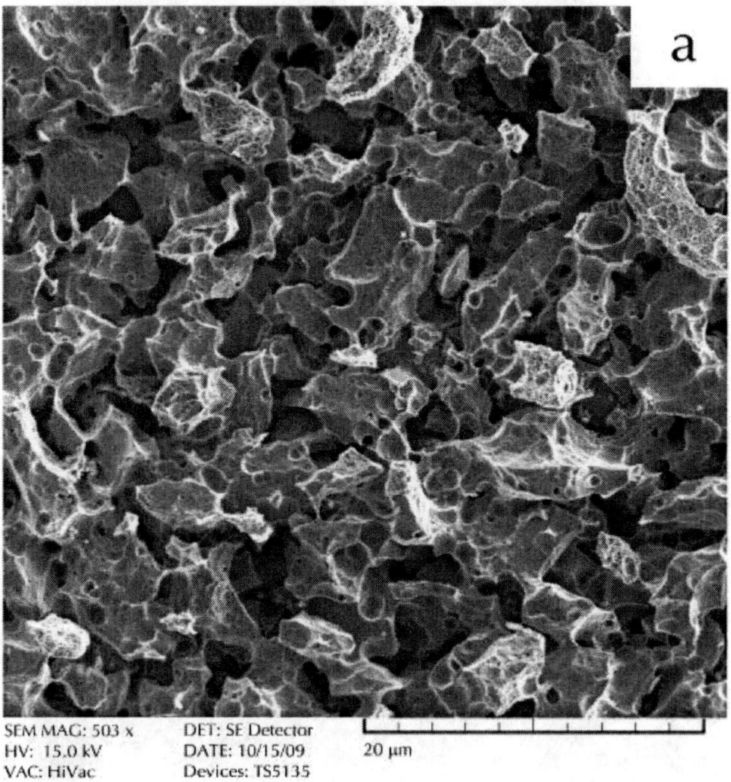

FIGURE 7 *(Continued)*

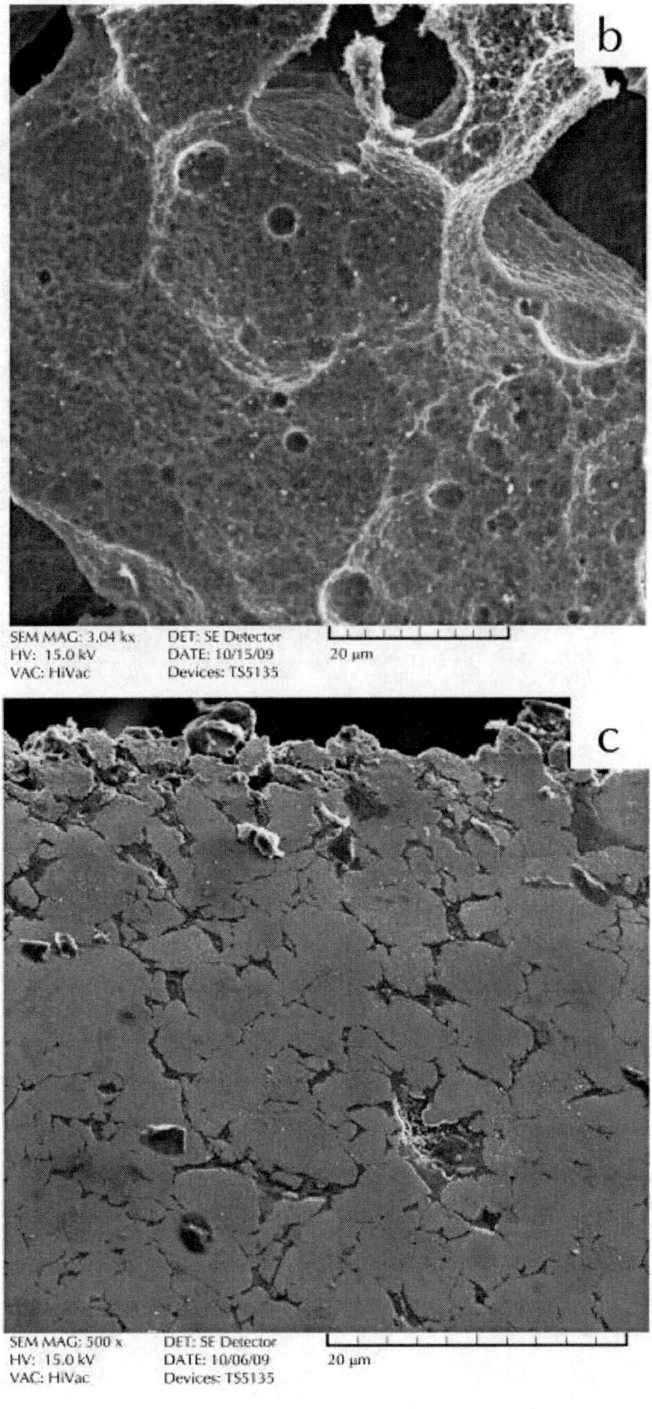

FIGURE 7 *(Continued)*

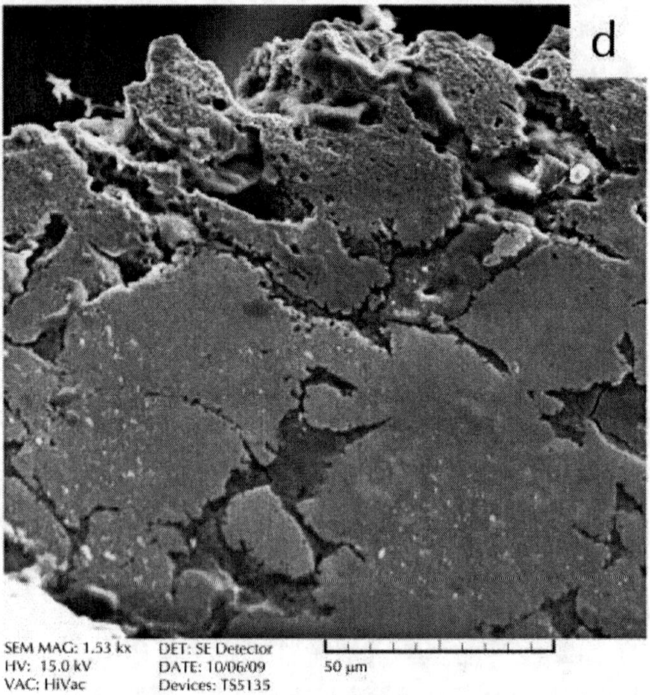

FIGURE 7 The SEM images of the Ti-6Al-4V sintered alloy after electrochemical etching in 1M H_3PO_4 + 2% HF electrolyte at 10 V for 30 min: (a) and (b) surface morphology (different magnifications), (c), and (d) cross section of the surface (different magnifications).

The large volume of the pores and grain boundaries lead to penetration of these features by electrolyte (Figure 7 and Figure 8). The large pores, remained after sintering (Figure 7 (a) and Figure 8 (a)), increases during the surface etching and the small grains are released from the bulk by electrolyte penetration through the grain boundaries (Figure 7 (b) and Figure 8 (b)). The largest pores have diameter up to 60 μm (Figure 7 (a) and Figure 8 (a)), while the smallest have diameter corresponding to the nanograins diameter (Figure 7 (b) and Figure 8 (b)). The etching time in the range of 30–300 min does not significantly affect the pore size. For the effective tissue growth, till now it was commonly accepted, that the pores should be close in size to 100–400 μm (macropores) [24]. Recently, new achievements show that smaller micropores support living cells growth, too [25]. So, the useful range of pores in the implants is very broad, from few nanometers up to 400 μm, supporting growth of different living cells, leading to shortening of the osseointegration time and stronger bonding of the implant to bone. The stronger bonding of the tissue to implant extends lifetime of the implant in the human body.

The initial porosity of the sinters is well visible on the Figure 7 (c) and Figure 8 (c), where the sinters cross section is shown. Depending of the etching time the thickness of the etched layer could be controlled. After 30 min (Figure 7 (c) and (d)) and 300 min (Figure 8 (c) and (d)) etching time, the porous etched layer thickness increases from 50 to 150 μm, respectively. During the etching connected and open channels are formed, resulting in sponge surface morphology (Figure 7 (d) and Figure 8 (d)). The comparable results of surface etching were achieved for the Ti-Zr-Nb alloys.

The chemical composition after etching is shown on Figure 9. Because, during the MA some amount of powders strongly sticks into the milling container walls and materials looses during the etching, the Al content is slightly different from the stoichiometric 6% in the Ti-6Al-4V alloy. The more important is that during the etching, phosphorus is introduced into the surface (from the phosphoric acid). The phosphorus is one of the bone components, so its presence on the surface is highly recommended for osseointegration improvement or HA growth. The rest of the composition (Cl, Fe, and Mg) is the contamination introduced during the process.

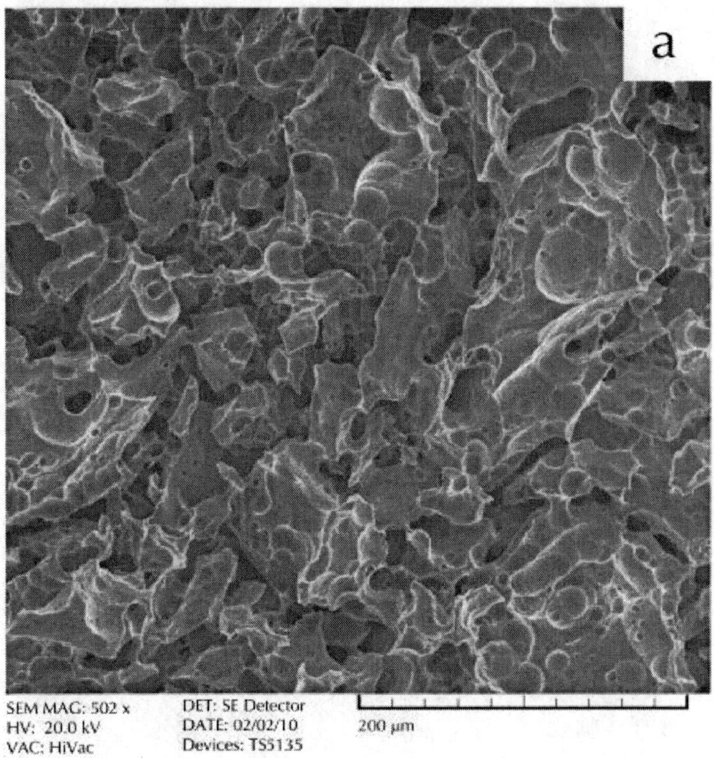

FIGURE 8 *(Continued)*

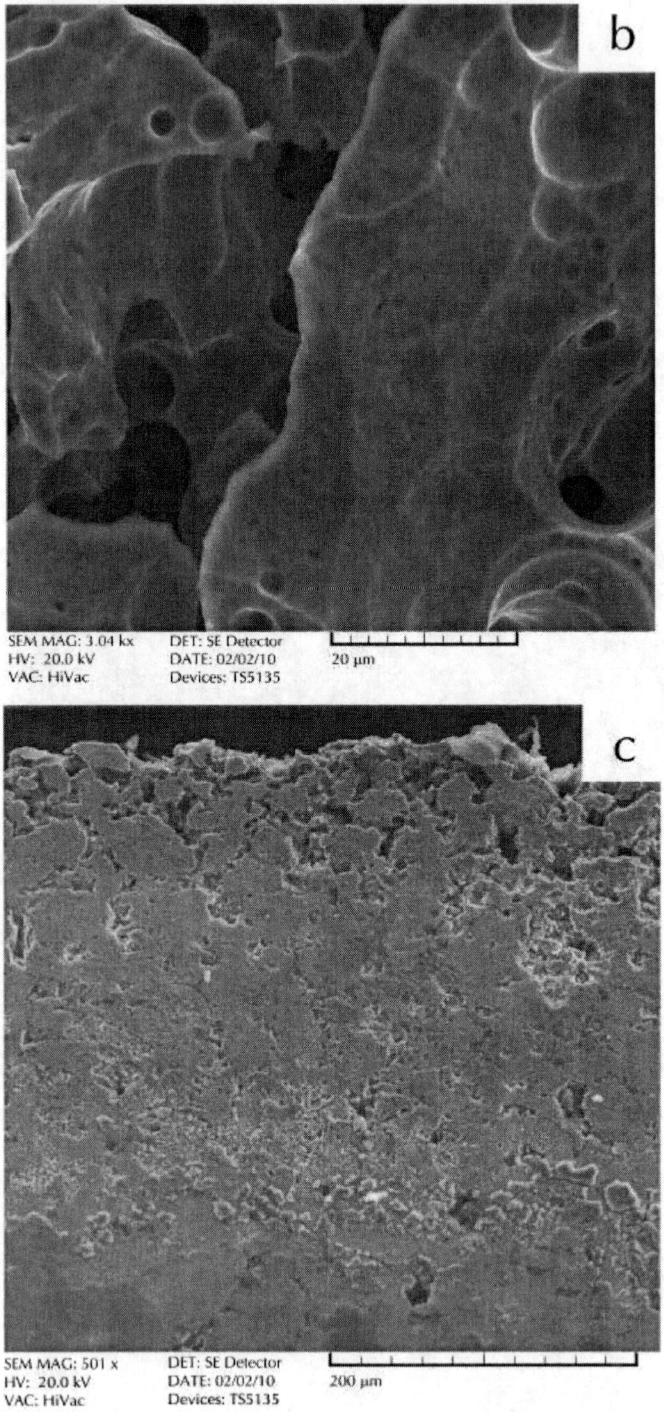

FIGURE 8 *(Continued)*

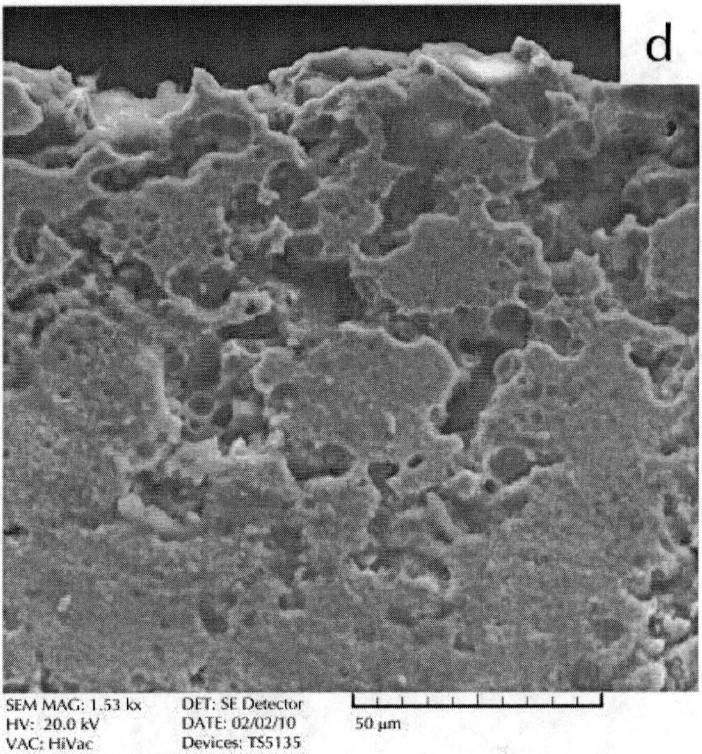

FIGURE 8 The SEM images of the Ti-6Al-4V sintered alloy after electrochemical etching in 1M H_3PO_4 + 2% HF electrolyte at 10 V for 300 min: (a) and (b) surface morphology (different magnifications), (c), and (d) cross section of the surface (different magnifications).

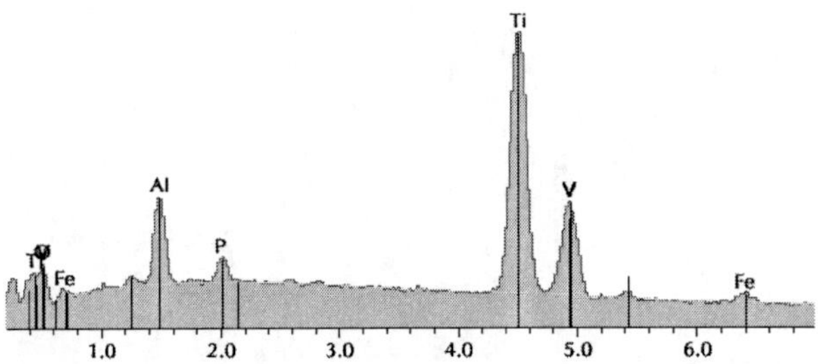

Element	Wt%	At%
Ti	74,46	57,7
Al	6,3	8,68
V	3,4	2,47
P	1,39	1,67
Cl	1,72	1,81
Mg	0,45	0,68
O	11,38	26,39
Fe	0,91	0,6
Total	100	100

FIGURE 9 The EDS spectrum of the Ti-6Al-4V sintered alloy after electrochemical etching in 1M H_3PO_4 + 2% HF electrolyte at 10 V for 30 min.

The porosity of the nanocrystalline Ti-6Al-4V and Ti-15Zr-4Nb alloys was investigated by image analysis software (Figure 10) and accelerated surface area and porosimetry analyzer (ASAP) (Figure 11 and Table 1) [23].

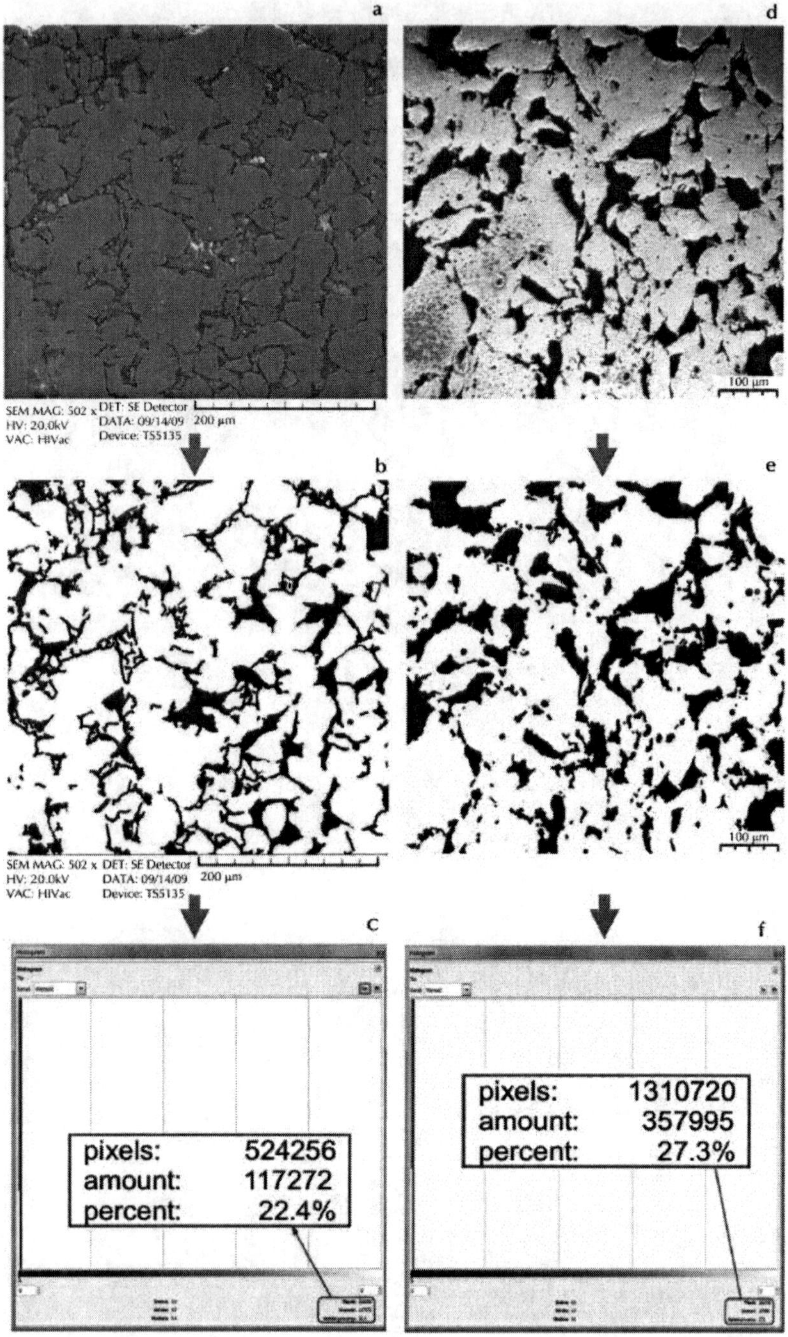

FIGURE 10 Overall porosity of the Ti-6Al-4V (a, b, and c) and Ti-15Zr-4Nb (d, e, and f) nanocrystalline sinters; microscope images (a and d), dual tone images (b and e), porosity content spectrum (c and f).

Because, large isolated pores are located inside the sinters, the good method for porosity analysis is a software method (Figure 10). For the Ti-6Al-4V (Figure 10 (a, b, and c)) and Ti-15Zr-4Nb (Figure 10 (d, e, and f)) after sintering the pore volume reaches 22.4 and 27.3%, respectively. The results are consistent with those from density measurements, mentioned earlier. It is very promising that the porosity of the Ti-15Zr-4Nb is higher than the Ti-6Al-4V. For example, Oh found that 30% porosity results in Young modulus close to those of the human cortical bone [26]. The implants with porosity should induce smaller stress and suppress damage at the implant/bone interface.

The porosity results after additional electrochemical etching are shown on Figure 11 and in Table 1. Because of the interconnected pore formation during the etching, the ASAP is a more accurate method of the porosity measurements. For the Ti-6Al-4V alloy increase of the etching time from 30 to 300 min results in increase of the average pore size (micro and mesopores with size<100 nm) and total volume of the pores, while the BET surface, micropore surface, and volume decreases. In the case of the Ti-15Zr-4Nb alloy two orders higher porosity parameters were achieved in comparison to the Ti-6Al-4V alloy, with additionally smaller average micropore size, too.

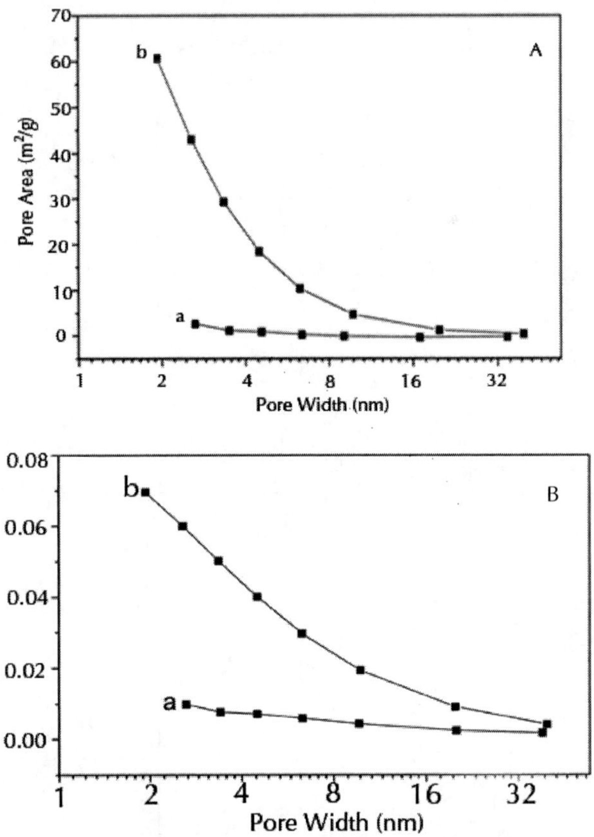

FIGURE 11 Pore surface area (A) and pore volume (B) of the electrochemically etched (1M H_3PO_4 + 2% HF; 10 V/30 min) Ti-6Al-4V (a) and Ti-15Zr-4Nb (b) alloys.

TABLE 1 The BET/BJH (BET: Brunauer-Emmett-Teller; BJH: Barrett-Joyner-Halenda) porosity parameters for sintered nanocrystalline alloys, etched at 10 V for 30 and 300 min in 1 MH_3PO_4 + 2% HF electrolyte.

Alloy	Etching time [min]	BET surface area [m^2/g]	Micropore surface area [m^2/g]	Micropore volume [cm^3/g]	Total volume of pores (micro- and mesopores <100 nm) [cm^3/g]	Average pore size [nm]
Ti-6Al-4V	30	0.8503	0.8347	0.000424	0.00116	5.46
Ti-6Al-4V	300	0.3087	0.2826	0.000142	0.00130	16.87
Ti-15Zr-4Nb	30	94.8676	87.3278	0.081295	0.08297	3.50

The SEM image sample surface analysis and porosity measurements lead to conclusion that smaller micro and mesopores are presented on the macropore walls (see Figure 7 and Figure 8). The ASAP analysis, made on the unetched samples did not reveal micro and mesopores with diameter <100 nm—so the surface is relatively continous and flat (in the ASAP analysis the large pores observed on Figure 6 were not taken into account).

5.3.2 Electrochemical Cathodic CaP Deposition

The next stage of the Ti alloys surface biofunctionalization, with respect to the implant applications, is a deposition of the calcium phosphate compounds. The calcium phosphate was cathodically deposited on the porous etched surface for time of 60 min at the potential –0.5 V (Figure 12), –1.5 V (Figure 13), –3.0 V (Figure 14), –5.0 V (Figure 15), and –10 V (Figure 16) *versus* OCP using a mixture of 0.042M $Ca(NO_3)_2$ + 0.025 $(NH_4)_2HPO_4$ + 0.1M HCl electrolyte.

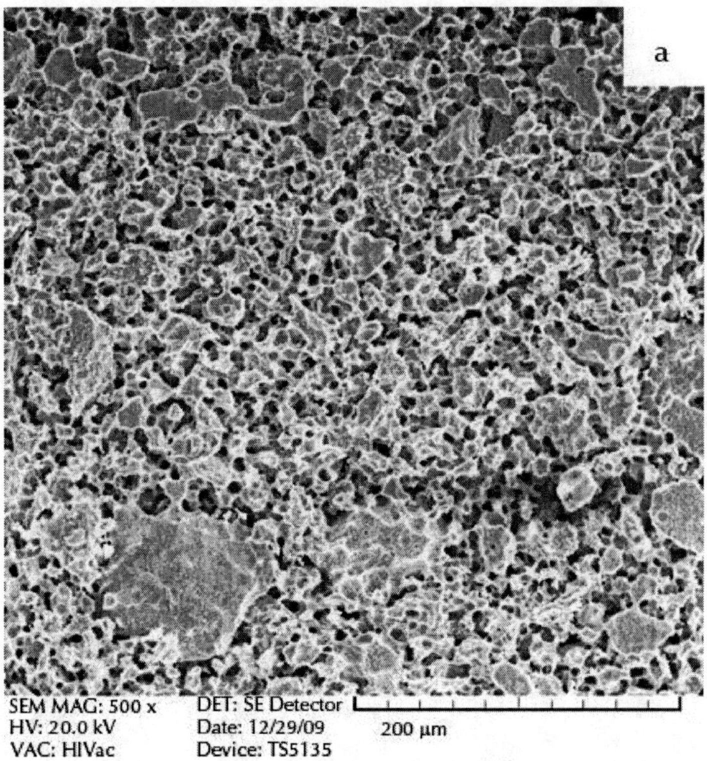

FIGURE 12 *(Continued)*

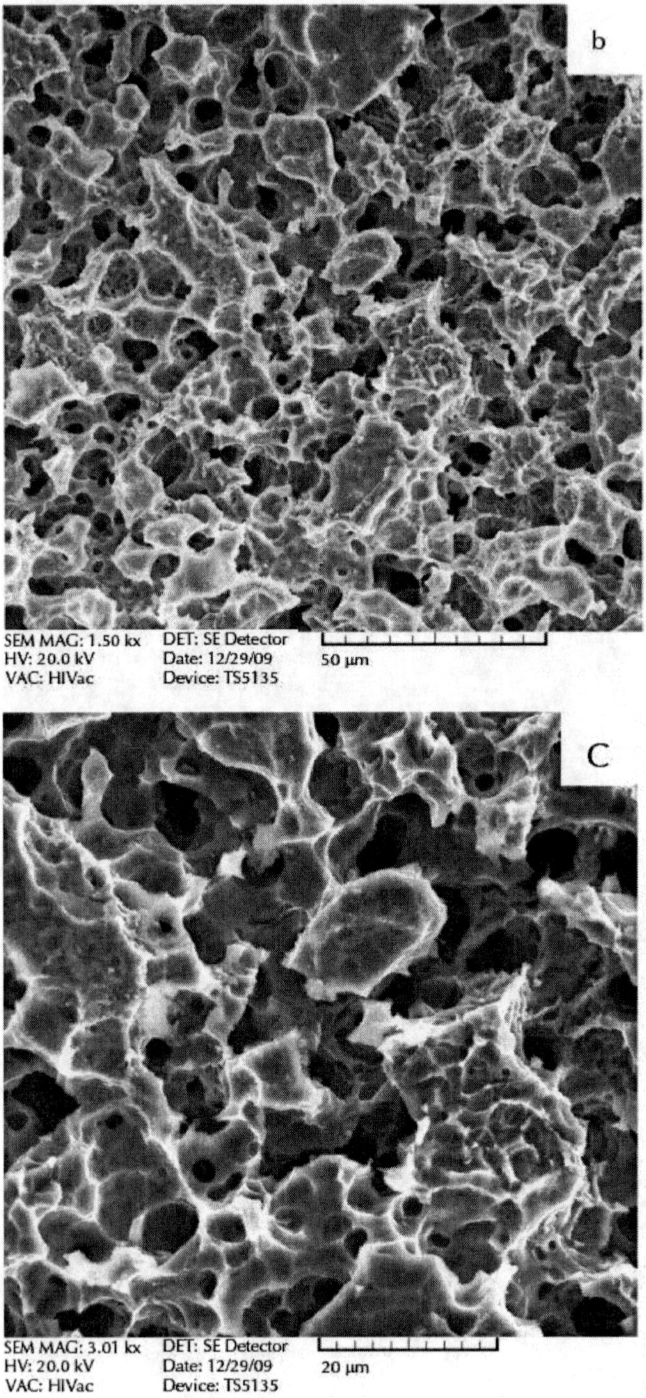

FIGURE 12 *(Continued)*

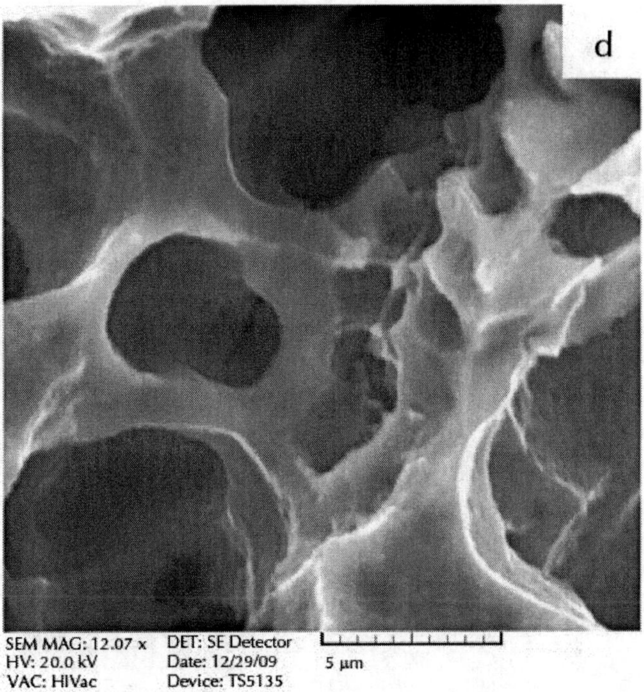

FIGURE 12 The SEM images of the CaP deposited (–0.5 V/60 min) on the nanocrystalline porous Ti-6Al-4V (a→d: sequence of images with increasing magnification).

The depositing of the CaP layer at –0.5 V (Figure 12) slightly changes the surface morphology. Because of the relatively low cathodic potential, low current density and hence, low stream of calcium and phosphate ions flowing into the surface. The formed calcium-phosphate layer has scaffold morphology, with pore diameter in the range from 0.1 to 15 µm. So, the pores in the etched alloy surface are only partly filled with CaP. Increase of the depositing potential to –1.5 V (Figure 13) lead to more continuous but slightly cracked layer, which fully covered the metallic background. Increase of the depositing potential to –3.0 V (Figure 14) leads to a significant crack in the deposited layer. The cracking is related to higher stress at the Ti/CaP interface. The depositing of the CaP at –5.0 V (Figure 15) results in transformation of the layer to highly cracked lamellar grains with high roughness. At –10 V the CaP layer is more continuous and homogenous (Figure 16). Unfortunately, the layer has less adhesion to the surface and part of the layer falls down during the sample removing from the electrolyte and drying in the stream of nitrogen. The weak adhesion of the CaP layer could be related to a significant hydrogen emission at the electrode/electrolyte interface and large thickness of the deposited material.

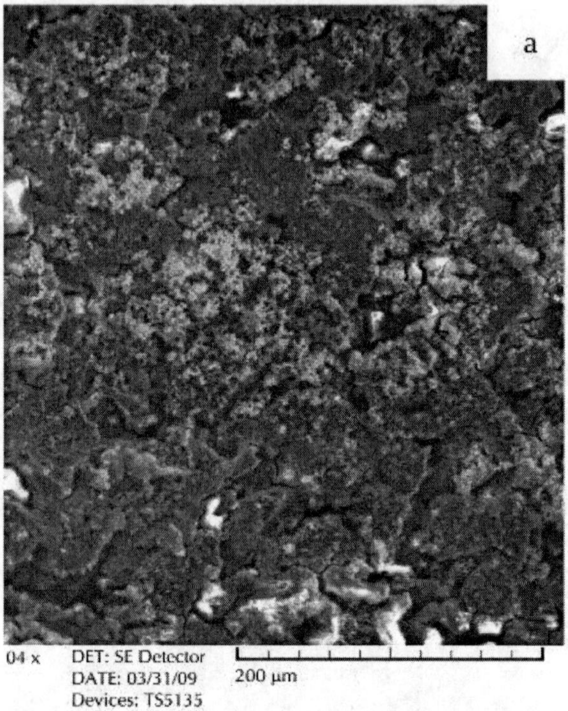

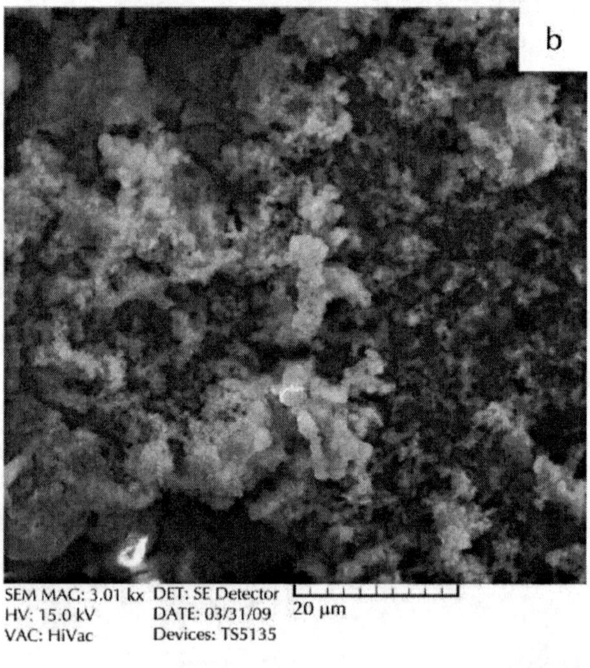

FIGURE 13 *(Continued)*

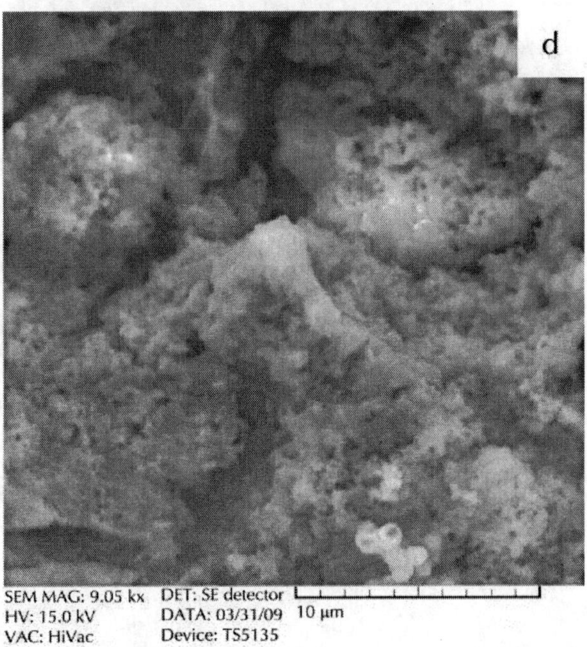

FIGURE 13 The SEM images of the CaP deposited (–1.5 V/60 min) on the nanocrystalline porous Ti-6Al-4V (a→d: sequence of images with increasing magnification).

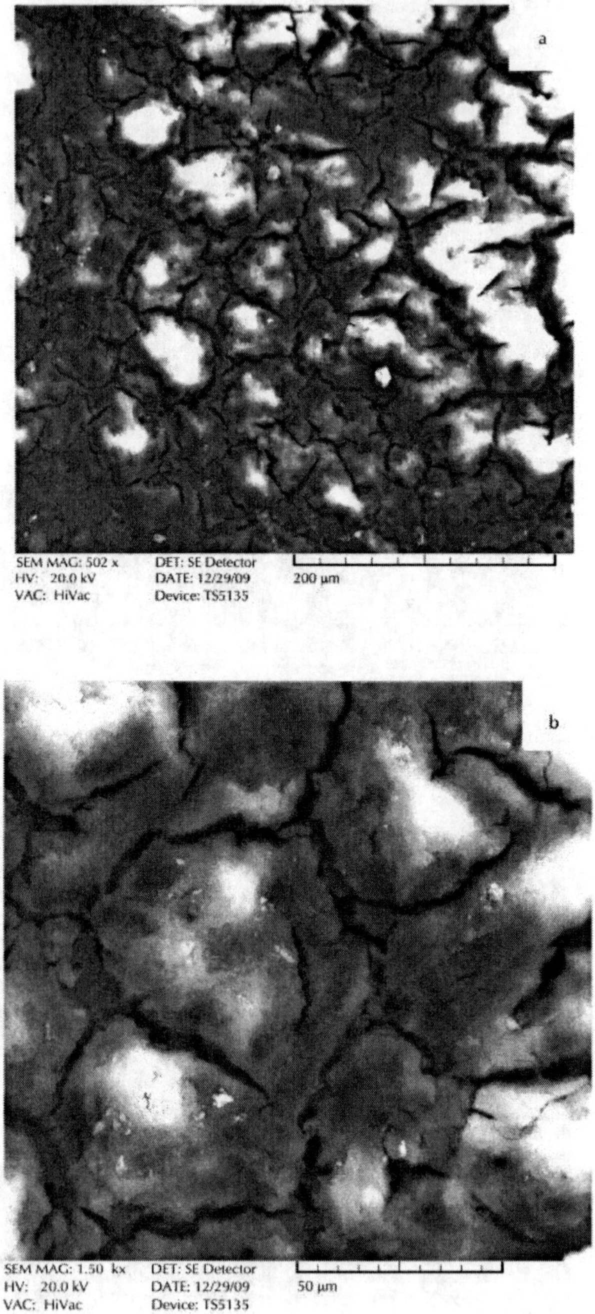

FIGURE 14 *(Continued)*

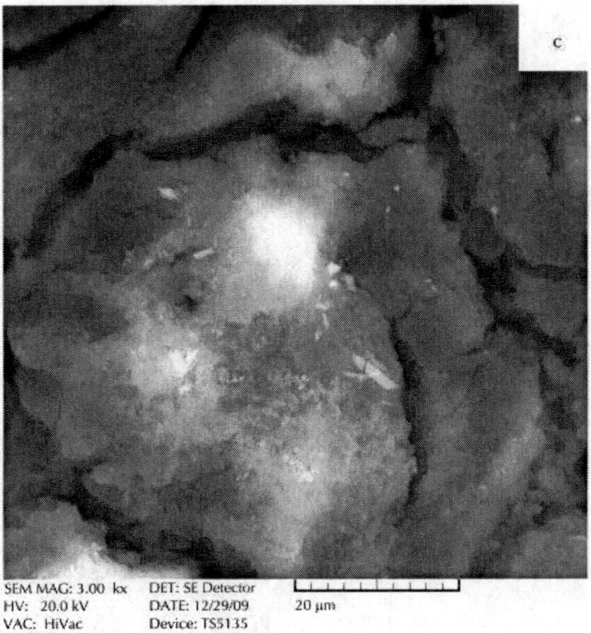

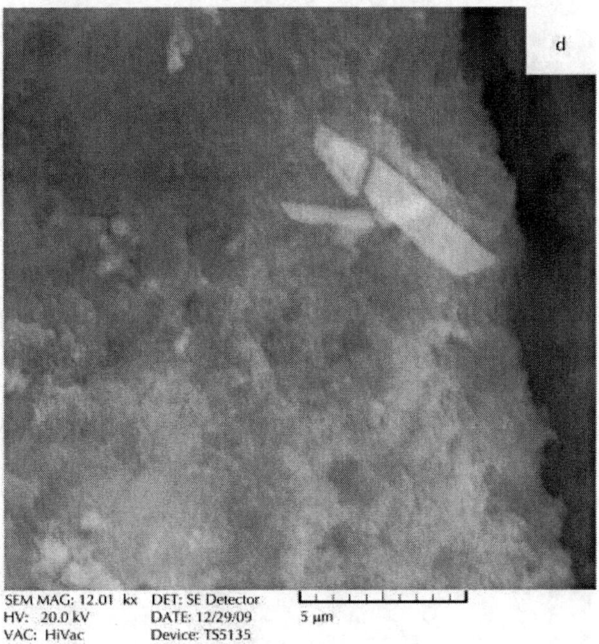

FIGURE 14 The SEM images of the CaP deposited (–3.0 V/60 min) on the nanocrystalline porous Ti-6Al-4V (a→d: sequence of images with increasing magnification).

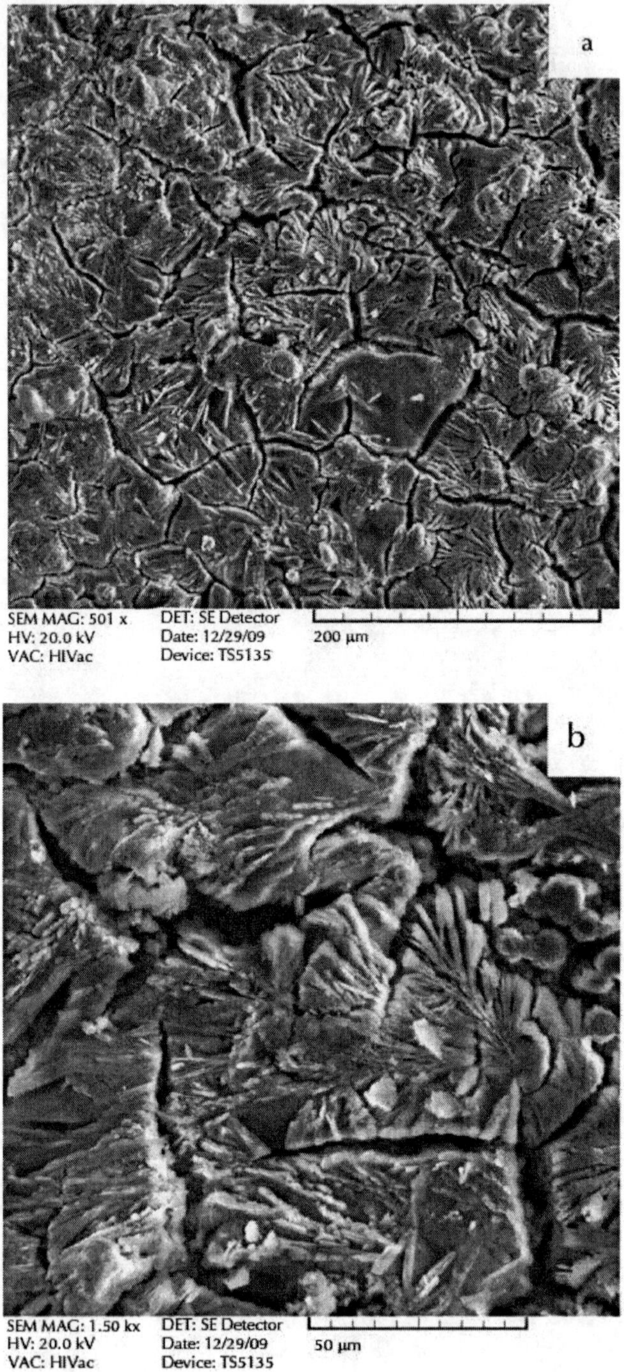

FIGURE 15 *(Continued)*

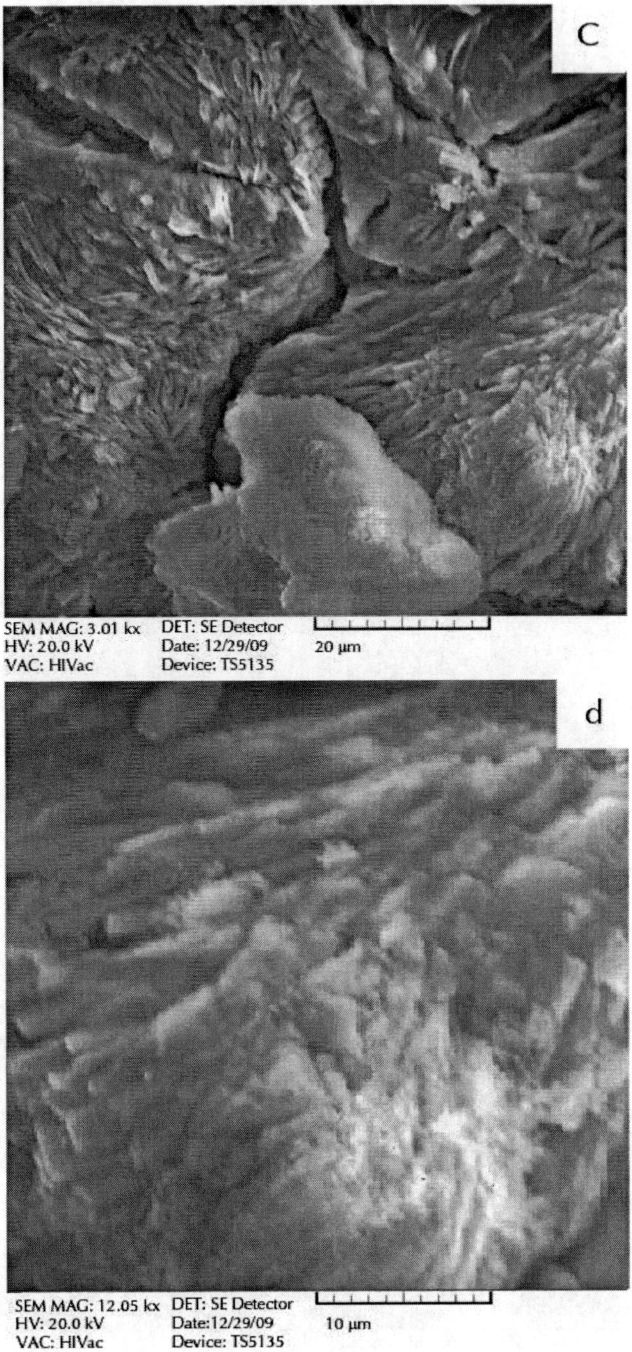

FIGURE 15 The SEM images of the CaP deposited (–5.0 V/60 min) on the nanocrystalline porous Ti-6Al-4V (a→d: sequence of images with increasing magnification).

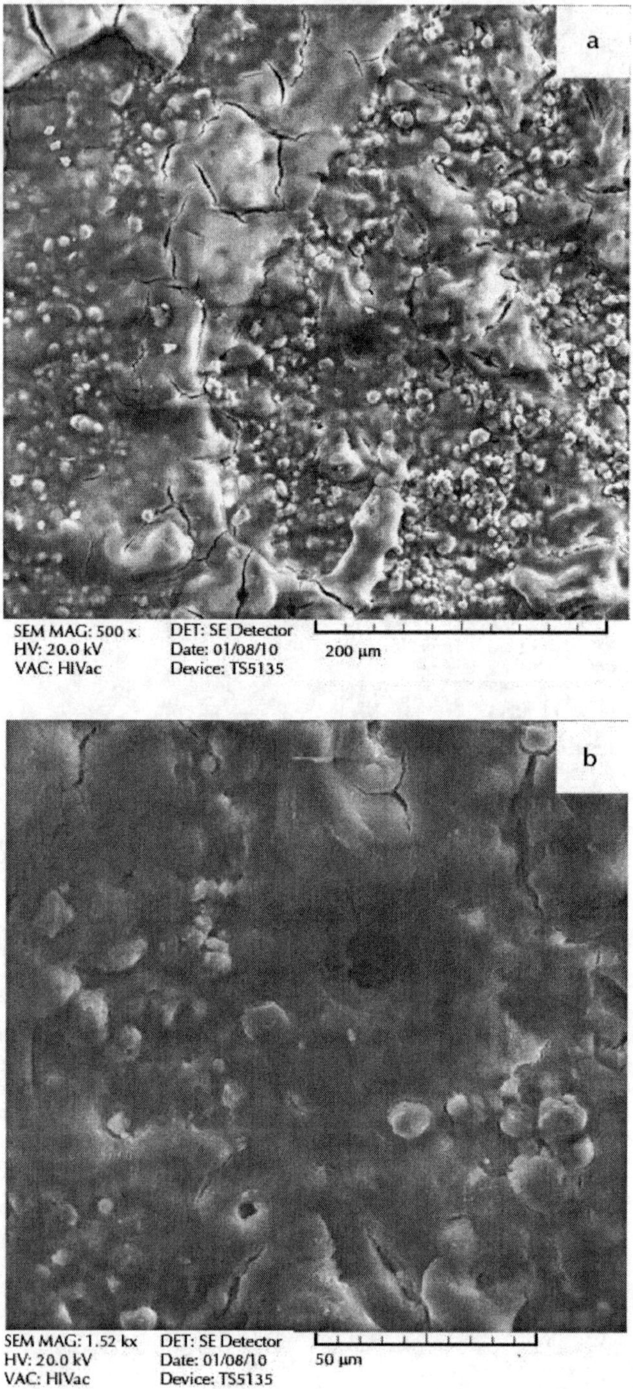

FIGURE 16 *(Continued)*

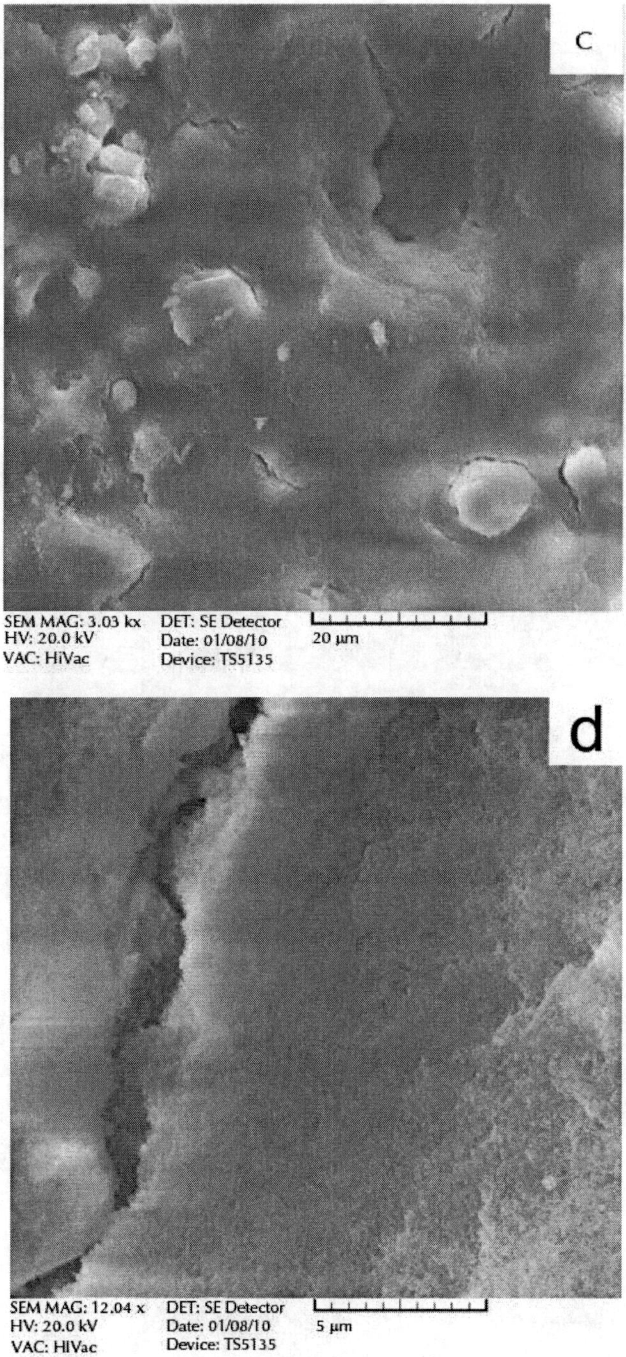

FIGURE 16 The SEM images of the CaP deposited (–10.0V/60 min) on the nanocrystalline porous Ti-6Al-4V (a→d: sequence of images with increasing magnification).

The chemical composition analysis of the CaP deposited at different potentials (according to Figures 12–16) on the Ti-6Al-4V porous surface is shown on Figure 17.

a)

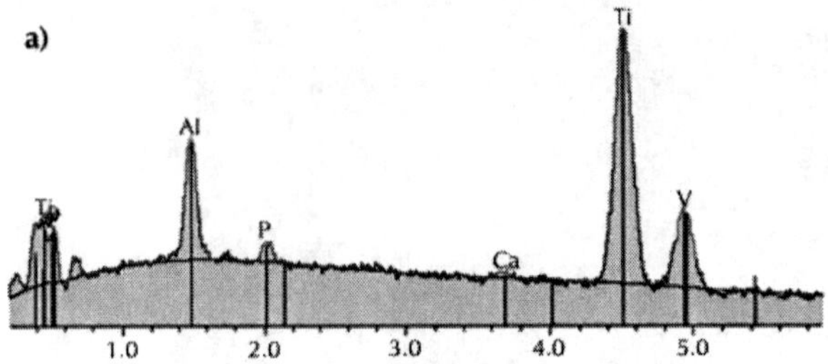

Element	Wt%	At%
Ti	78.63	61.06
Ca	0.29	0.27
P	0.92	1.1
O	11.33	26.34
Al	7.39	10.15
V	1.47	1.07
Total	**100**	**100**

b)

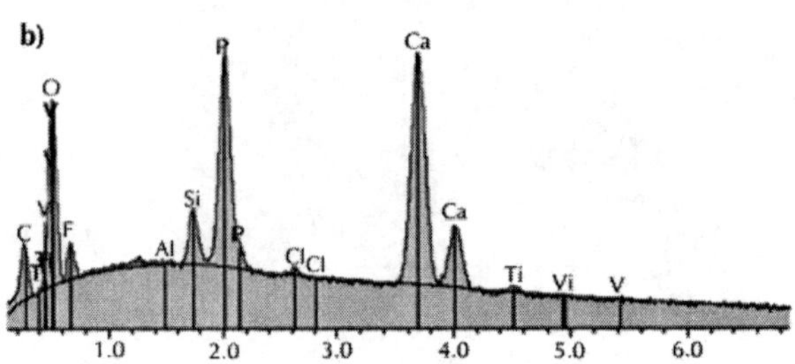

Element	Wt%	At%
Ti	0.23	0.10
Ca	26.30	12.98
P	15.83	10.11
O	37.01	45.76
Al	0.07	0.05
V	0.00	0.00
Si	1.54	1.09
F	1.98	2.06
Cl	0.19	0.11
C	16.85	27.76
Total	**100**	**100**

C)

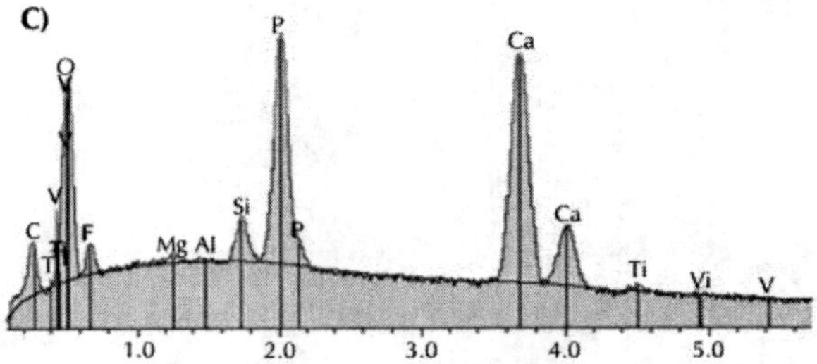

Element	Wt%	At%
Ti	0.24	0.10
Ca	24.89	12.24
P	17.17	10.92
O	38.24	47.11
Al	0.05	0.04
V	0.00	0.00
Si	1.19	0.83
F	1.68	1.74
Cl	16.37	26.86
Mg	0.19	0.15
Total	**100**	**100**

d)

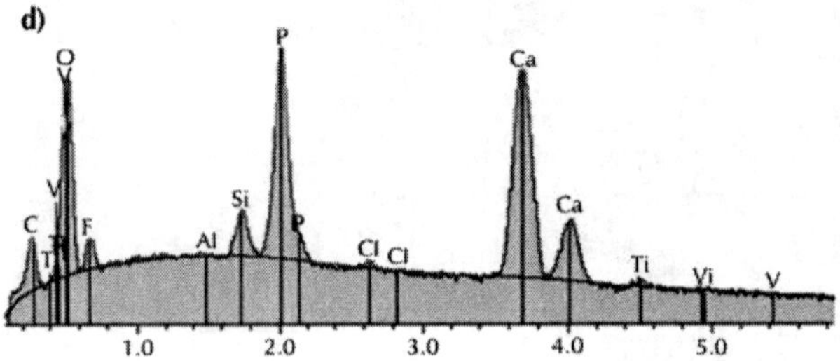

Element	Wt%	At%
O	39.46	44.43
Ti	0.22	0.1
V	0	0
Al	0.05	0.03
P	17.3	10
Ca	25.55	13.03
C	15.85	22.77
Si	1.44	9.58
Cl	0.13	0.06
Total	**100**	**100**

FIGURE 17 *(Continued)*

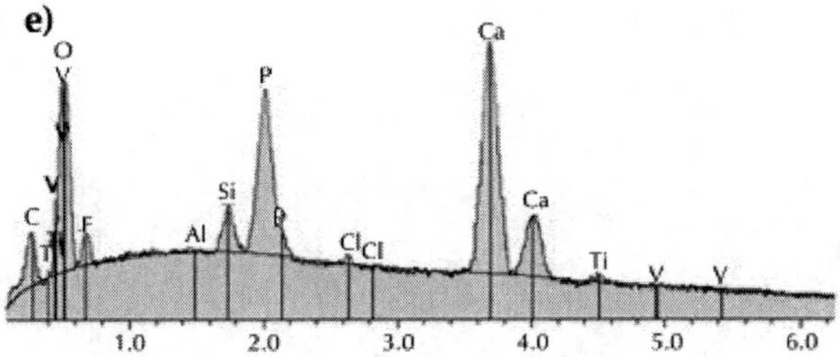

Element	Wt%	At%
O	39.51	43.98
Ti	0.00	0.00
V	0.03	0.01
Al	0.05	0.03
P	10.3	5.95
Ca	22.85	10.15
C	26.71	39.57
Si	0.44	0.28
Cl	0.11	0.05
Total	100	100

FIGURE 17 The EDS analysis of the CaP deposited on porous Ti-6Al-4V surface at: –0.5 V (a), –1.5 V (b), –3 V (c), –5 V (d), and –10 V (e); the spectra (a–e) corresponds to Figures 12–16, respectively.

The Ca/P ratio in the ceramic layer changes with the depositing potential. After deposition at –0.5 V, –1.5 V, –3 V, –5 V, and –10 V, the Ca/P atomic ratio is 0.24, 1.28, 1.12, 1.30, and 1.71, respectively, leading to HA layer composition (Ca/P = 1.66) deposited at the potential range between –5 V and –10 V. Only in the (a) spectrum (sample with CaP deposited at –0.5 V) the significant amount of Ti, Al, and V from the metallic background was recorded, which is in agreement with the data observed on Figure 12. At low depositing potential the CaP do not fully cover the Ti-6Al-4V alloy surface. The other element traces are the contaminations introduced from the reagents of the depositing solutions or from the sample holder's silicone seal.

Using the image analysis software the surface porosity was determined (Figure 18 columns). The initial surface porosity after electrochemical etching is estimated to be 49.3%. After additional CaP deposition, the surface porosity decreases with increasing depositing cathodic potential (increasing flowing charge density and amount of the CaP), reaching of about 4.3% porosity for –10 V depositing potential. Significant differences in porosity appears between –0.5 V and –3 V (as observed on Figures 12-14). The CaP layer thickness, measured on the samples cross sections, increases with increasing depositing potential (Figure 18—solid line). For –10 V the CaP layer thickness reaches 90 μm. The CaP growing rate is estimated to be 9 μm/V. Shift of the depositing potential to more negative value results in thicker CaP layer but simultaneously more hydrogen emission occurs (hydrogen bubbles formed at the electrode/electrolyte interface). It is obvious, that increase of potential to more negative value results in higher current and more flowing charge densities, leading finally to larger amount of the deposited CaP compounds. Because of the low adhesion and flat morphology of the CaP deposited at –10 V, the most promising layer is deposited below –10 V, for example at –5 V.

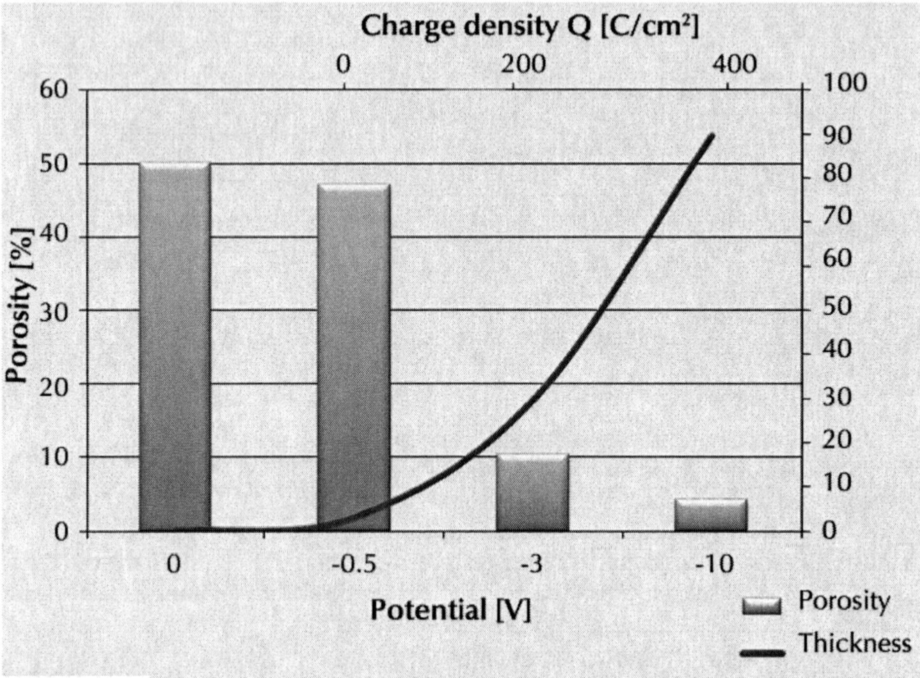

FIGURE 18 The CaP surface porosity and thickness as a function of the CaP depositing potential and charge density for the nanocrystalline Ti-6Al-4V alloy; porous sample, after electrochemical etching without CaP is indicated by column at 0 V.

The comparable results were obtained for the porous etched nanocrystalline Ti-15Zr-4Nb alloys. The CaP layer deposited at –1.5 V is built from lamellas, resulting in rough surface morphology (Figure 19). The depositing of the CaP layer on the flat polished Ti-15Zr-4Nb nanocrystalline alloy surface results in significantly different morphology in comparison to the shown results (Figure 20). On the polished surface mainly loose-lying particles (mainly spherical) are deposited. The CaP particles do not have possibility to anchor to the flat surface, so the adhesion of the CaP layer is rather poor.

When the pores and pits are present in the surface, the CaP effectively grows inside them (Figure 21). For example, after CaP deposition at –0.5 V, the CaP layer thickness (layer on the flat part of the alloy) is estimated to be about 2–5 μm, while the pores are almost filled (but still some free space is observed) and the local thickness of the CaP in the pores is up -60 μm (depth of the pores). It means that the pores are preferentially filled by CaP in the electrochemical deposition process. The CaP mass transport is enhanced at the surface pits and pores.

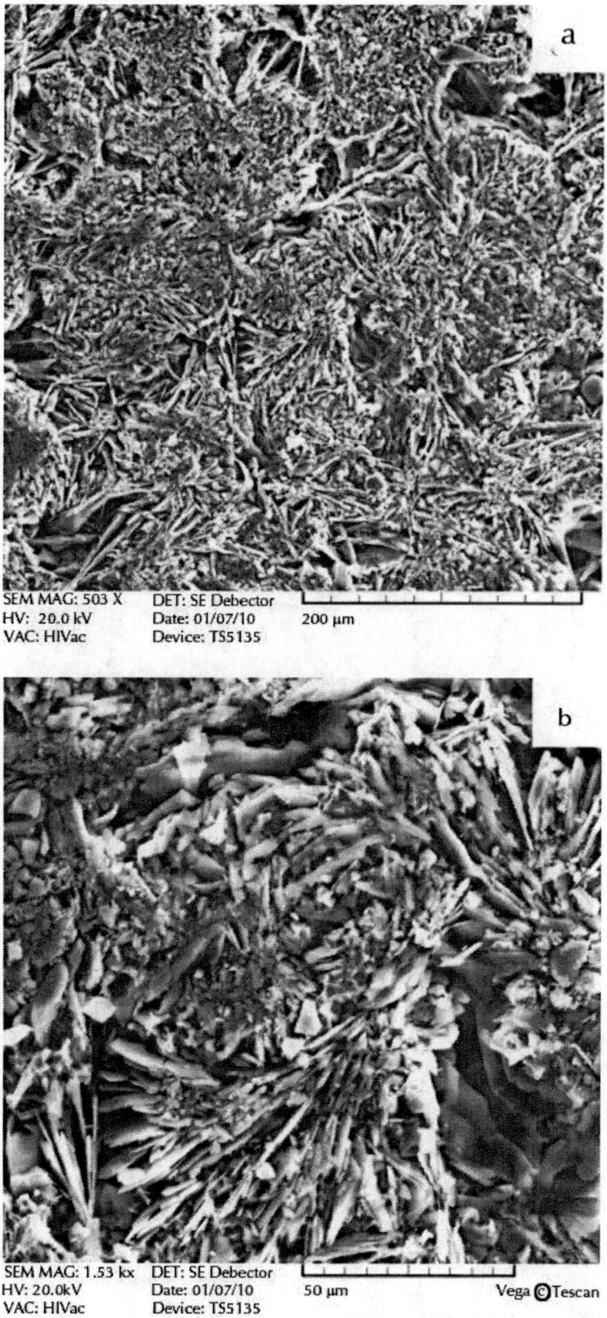

FIGURE 19 *(Continued)*

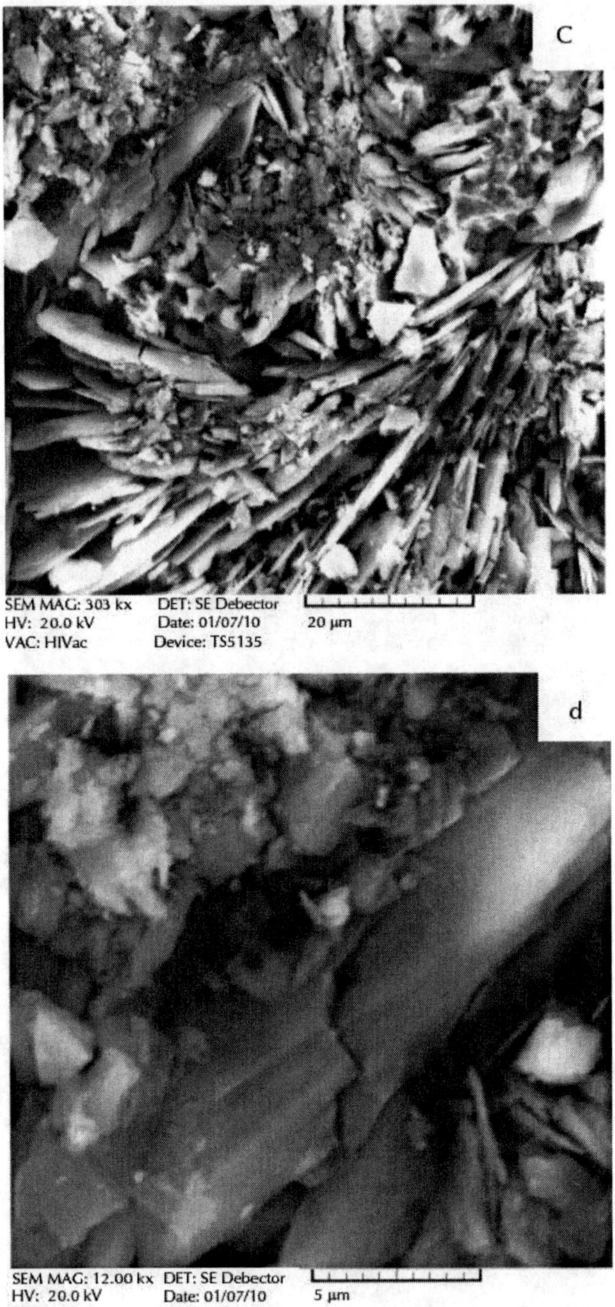

FIGURE 19 The SEM images of the CaP deposited (–1.5 V/60 min) on the nanocrystalline porous Ti-15Zr-4Nb (a→d: sequence of images with increasing magnification).

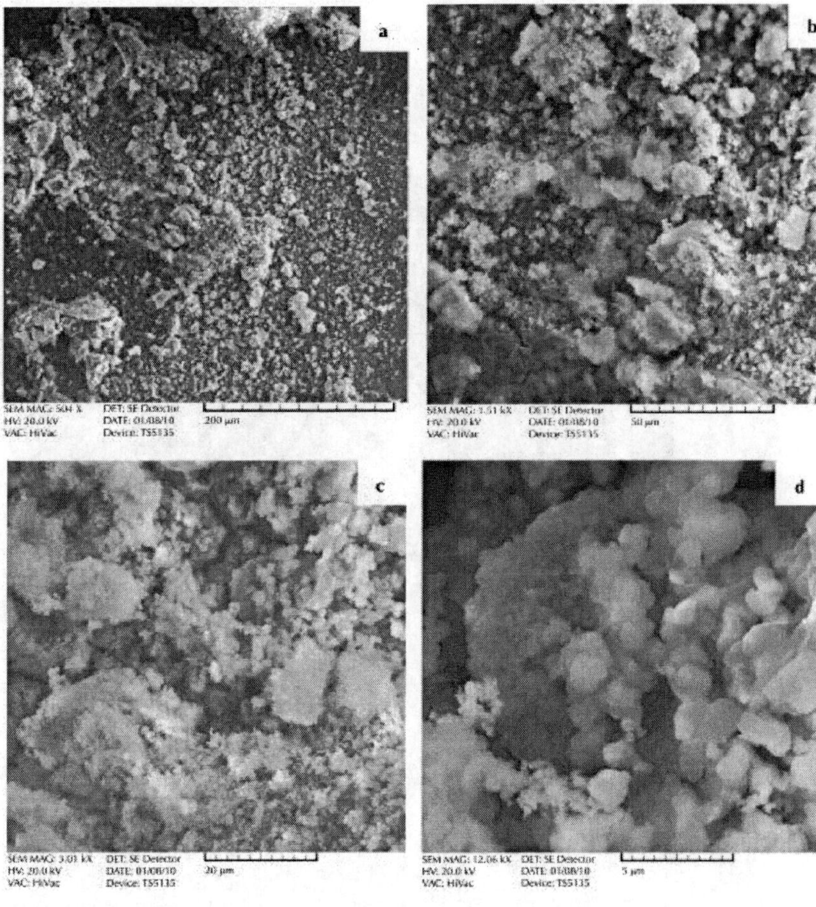

FIGURE 20 The SEM images of the CaP deposited (–1.5V/60 min) on the nanocrystalline polished Ti-15Zr-4Nb (a→d: sequence of images with increasing magnification).

We suggest, that a possible mechanism of the CaP (HA) growth is related to the electric field enhancement, occurring on the surface irregularities, such as, pores and surrounding hillocks (Figure 22). The electric field lines are concentrated on these specific surface features, enhancing ions flow to these places (Figure 22 (a)). Pore walls, which are perpendicular to the surface, are rather unaffected at the initial stage of the deposition process, especially if they are smooth, without irregularities. If the pore walls consist of protrusions, then the CaP grows on them. The increasing of the CaP (HA) cathodic deposition potential (as well as time) leads then to the pore filling and semi-continuous cracked layer formation (Figure 22 (b)). The more pronounced increase of the depositing potential leads to the fully filled pores and formation of homogenous CaP (HA) layer (Figure 22 (c)). The significant shift of the cathodic potential to the more negative value leads to electric field enhancement, resulting in enhancing of the Ca^{2+} and the HPO_4^{2-} (or PO_4^{3-}) ions flow to the specific surface features, where the chemical reactions leads to the respective CaP compound formation, for example HA.

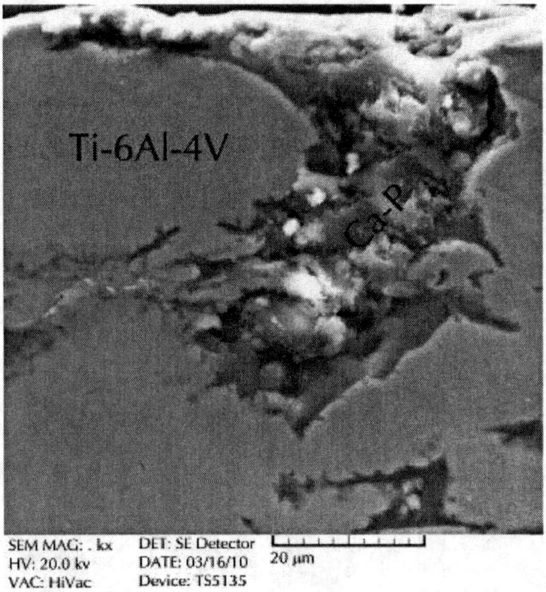

FIGURE 21 Cross section SEM image of the porous nanocrystalline Ti-6Al-4V alloy after CaP (HA) deposition for 60 min at –0.5 V.

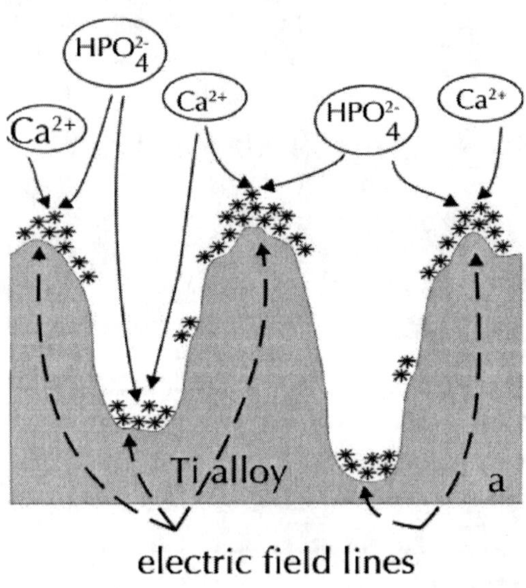

FIGURE 22 *(Continued)*

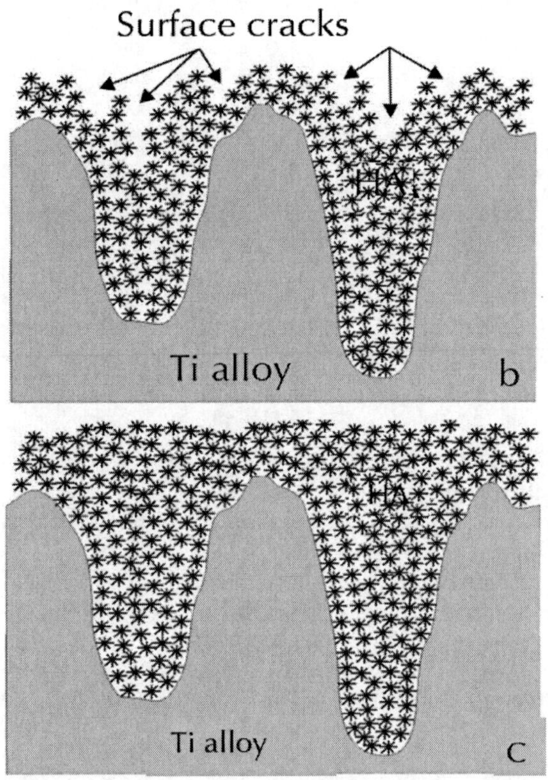

FIGURE 22 Scheme of the CaP (HA) layer growth mechanism on the porous surface during electrochemical cathodic deposition; a→c – stages of the HA growth.

5.4 CORROSION RESISTANCE OF THE TIALV AND TIZRNB ALLOYS

The corrosion resistance of the biomaterials is a very important factor, which can decide about application of the material in the aggressive human body environment. In tests applied potentiodynamic method, when the samples were immersed in Ringer's electrolyte consisting significant amount of chloride (simulated body fluid with composition: NaCl = 9 g/l, KCl = 0.42 g/l, $CaCl_2$ = 0.48 g/l, $NaHCO_3$ = 0.2 g/l). The sample was charged from –1 to 2.5 V *versus* OCP with scan rate of 0.5 mV/s. From the corrosion curves (Figure 23), using the Tafel extrapolations, the corrosion current density (I_{corr}) and corrosion potential (E_{corr}) were estimated (Table 2). The corrosion resistance of the untreated Ti-6Al-4V nanocrystalline alloy is lower than the Zr containing alloys (I_{corr} is higher for the Ti-6Al-4V). After electrochemical etching, the corrosion current decreases (except Ti-15Zr-4V alloy). The subsequent CaP deposition does not significantly change the corrosion current, even at different depositing potential (probably due to layer's properties: the CaP layer is cracked and belongs to isolating materials). Taking into account all our corrosion experiments, we can generally say that the corrosion resistance increases after anodic etching and does not significantly change after the CaP deposition.

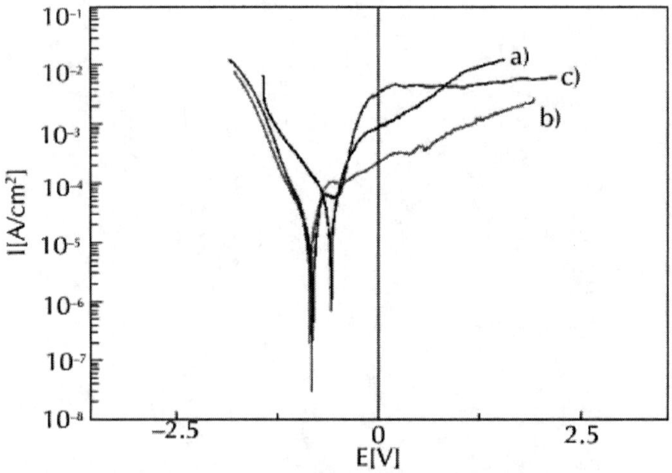

FIGURE 23 Corrosion curves for the nanocrystalline Ti-6Al-4V (a), nanocrystalline Ti-15Zr-4Nb (b), and nanocrystalline Ti-6Zr-4Nb (c); all samples without etching and CaP deposition.

TABLE 2 Corrosion current density I_{corr} and corrosion potential E_{corr} of the nanocrystalline Ti alloys before and after electrochemical etching (1M H_3PO_4 + 2% HF; 10 V/30 min) as well as after additional CaP deposition (60 min) at different potentials.

Material	I_{corr} [A/cm²]	E_{corr} [V]
Ti-6Al-4V before etching	9.19×10^{-6}	–0.58
Ti-6Al-4V after etching	4.28×10^{-6}	–0.36
Ti-6Al-4V after etching with CaP (–0.5 V)	9.35×10^{-6}	–0.94
Ti-6Al-4V after etching with CaP (–1.5 V)	1.88×10^{-5}	–0.85
Ti-6Al-4V after etching with CaP (–3.0 V)	7.57×10^{-6}	–0.94
Ti-6Al-4V after etching with CaP (–5.0 V)	7.14×10^{-6}	–0.85
Ti-6Al-4V after etching with CaP (–10.0 V)	1.88×10^{-5}	–0.85
Ti-15Zr-4V before etching	1.85×10^{-6}	–0.84
Ti-15Zr-4V after etching	3.12×10^{-5}	–0.85
Ti-15Zr-4V after etching with CaP (–1.5 V)	1.10×10^{-5}	–0.99
Ti-6Zr-4V before etching	2.27×10^{-6}	–0.82
Ti-6Zr-4Nb after etching	4.43×10^{-6}	–0.75
Ti-6Zr-4Nb after etching with CaP (–1.5 V)	5.74×10^{-6}	–0.89

5.5 CONCLUSION

In this study formation of porous nanocrystalline Ti-6Al-4V and (Al, V)free alloys with composition of the Ti-15Zr-4Nb and Ti-6Zr-4Nb prepared by MA, pressing, sintering, and subsequent electrochemical functionalization was described. The mechanically alloyed powders are composed from nanograins and sub-grains, formed by high plastic deformations, or even amorphous areas. The sintering process done at high temperature results in grain size growth but due to the composition and refinement effect of the Zr, the Ti-15Zr-4Nb and Ti-6Zr-4Nb alloys are more resistant to grain growth than the Ti-6Al-4V. The ultrafine structure affects the etching process, improving porosity, which are two orders of magnitude larger for the Ti-15Zr-4Nb in comparison to the Ti-6Al-4V. The electrolyte penetrates sinters through the grain boundaries, resulting in effective material removing and pores formation.

On the porous background bioactive CaP layer was deposited. Increase of the deposition potential leads to more calcium and phosphate ions flow to the surface, which results in significant different CaP layer morphologies. At low depositing potential –0.5 V the CaP is porous, thin and does not fully cover the metallic background. The increase of the depositing potential by a few volts (up to –10 V) results in thick continuous CaP layer. As suggested, the electric field enhancement on the surface pores and surrounding hillocks plays a key role in the CaP growth, leading to preferential CaP deposition.

Due to obtained properties, the prepared porous biofunctionalized nanocrystalline Titype alloys could be a possible candidate for the hard tissue implant applications.

KEYWORDS

- **Biomaterials**
- **Electrochemical etching**
- **Hydroxyapatite**
- **Lamellar morphology**
- **Mechanical alloying**
- **Ti alloys**

ACKNOWLEDGMENT

The financial support of the Polish Ministry of Education and Science under the contract No N N507 277536 is gratefully acknowledged.

REFERENCES

1. Noort, R. V. Titanium: The implant material of today. *J. Mater. Sci.*, **22**, 3801–3811 (1987).
2. Davidson, J. A., Mishra, A. K., Kovacs, P., and Poggie, R. A. New surface-hardened, low-modulus, corrosion resistant Ti-13Nb-13Zr alloy for total hip arthroplast. *Biomed. Mat. Eng.*, **4,** 231–243 (1994).
3. Lugowski, S. J., Smith, D. C., McHugh, A. D., and Loon, V. Release of metal ions from dental implant materials in vivo Determination of Al, Co, Cr, Mo, Ni, V, and Ti in organ tissue. *J. Biomed. Mat. Res.*, **25**, 1443–1458 (1991).

4. Okazaki, Y., Rao, S., Ito, Y., and Tateishi, T. Corrosion resistance, mechanical properties, corrosion fatigue strength and cytocompatibility of new Ti alloys without Al and V. *Biomaterials*, **19**, 1197–1215 (1998).
5. Santos, L. V., Trava-Airoldi, V. J., Corat, E. J., Nogueira, J., and Leite, N. F. DLC cold welding prevention films on a Ti6Al4Valloy for space applications. *Surf. Coat. Techn.*, **200**, 2587–2593 (2006).
6. Rack, H. J. and Quazi, J. I. Titanium alloys for biomedical applications. *Mat. Sci. Eng.*, **C26**, 1269–1277 (2006).
7. Saji, V. S. and Choe, H. Ch. Electrochemical corrosion behavior of nanotubular Ti–13Nb–13Zr alloy in Ringer's solution. *Corr. Sci.*, **51**, 1658–1663 (2009).
8. Niinomi, M. Cyto-toxicity and fatigue performance of low rigidity titanium alloy, Ti–29Nb–13Ta–4.6Zr, for biomedical applications. *Biomaterials*, **24**, 2673–2683 (2003).
9. Oh, I. H., Nomura, N., Masahashi, N., and Hanada, S. Mechanical properties of porous titanium compacts prepared by powder sintering. *Scripta. Mat.*, **49**, 1197–1202 (2003).
10. Jakubowicz, J. and Adamek, G. Preparation and properties of mechanically alloyed and electrochemically etched porous Ti-6Al-4V. *Electrochem. Commun.*, **11**, 1772–1775 (2009).
11. Niespodziana, K., Jurczyk, K., Jakubowicz, J., and Jurczyk, M. Fabrication and properties of titanium—hydroxyapatite nanocomposites. *Mat. Chem. Phys.*, **123**, 160–165 (2010).
12. Kim, H. M., Miyaji, F., Kokubo, T., and Nakamura, T. Preparation of bioactive Ti and its alloys via simple chemical surface treatment. *J. Biomed. Mat. Res.*, **32**, 409–417 (1996).
13. Adamek, G. and Jakubowicz, J. Mechanoelectrochemical synthesis and properties of porous nano-Ti-6Al-4V alloy with hydroxyapatite layer for biomedical applications. *Electrochem. Commun.*, **12**, 653–656 (2010).
14. Jakubowicz, J., Jurczyk, K., Niespodziana, K., and Jurczyk, M. Mechanoelectrochemical synthesis of porous Ti-based nanocomposite biomaterials. *Electrochem. Commun.*, **11**, 461–465 (2009).
15. Yang, B., Uchida, M., Kim, H. M., Zhang, X., and Kokubo, T. Preparation of bioactive titanium metal via anodic oxidation treatment. *Biomaterials*, **25**, 1003–1010 (2004).
16. Narayanan, R., Seshadri, S. K., Kwon, T. Y., and Kim, K. H. Electrochemical nano-grained calcium phosphate coatings on Ti-6Al-4V for biomaterial applications. *Scripta. Mat.*, **56**, 229–232 (2007).
17. Sgambato, A., Cittadini, A., Ardito, R., Dardeli, A., Facchini, A., Pria, P. D., and Colombo, A. Osteoblast behavior on nanostructured titanium alloys. *Mat. Sci. Eng.*, **C23**, 419–423 (2003).
18. Kuo, M. C. and Yen, S. K. The process of electrochemical deposited hydroxyapatite coatings on biomedical titanium at room temperature. *Mat. Sci. Eng.*, **C20**, 153–160 (2002).
19. Oh, S. and Jin, S. Titanium oxide nanotubes with controlled morphology for enhanced bone growth. *Mat. Sci. Eng.*, **C26**, 1301–1306 (2006).
20. Oh, I. H., Nomura, N., Chiba, A., Murayama, Y., Masahashi, N., Lee, B. T., and Hanada, S. Microstructures and bond strengths of plasma-sprayed hydroxyapatite coatings on porous titanium. *J. Mat. Sci. Mat in Med.*, **16**, 635–640 (2005).
21. Xiao, X. F., Liu, R. F., and Zheng, Y. Z. Hydoxyapatite/titanium composite coating prepared by hydrothermal–electrochemical technique. *Mat. Lett.*, **59**, 1660–1664 (2005).
22. Raja, K. S., Mishra, M., and Paramguru, K. Deposition of calcium phosphate coating on nanotubular anodized titanium. *Mater. Lett.*, **59**, 2137–2141 (2005).
23. Adamek, G. and Jakubowicz, J. Microstructure of the mechanically alloyed and electrochemically etched Ti-6Al-4V and Ti-15Zr-4Nb nanocrystalline alloys. submitted to *Mat. Chem. Phys.*, doi: 10.1016/j.matchemphys.2010.08.057.
24. Itälä, A. I., Ylänen, H. O., Ekholm, C., Karlsson, K. H., and Aro, H. T. Pore diameter of more than 100 μm is not requisite for bone in growth in rabbits. *J. Appl. Biomat.*, **58**, 679–683 (2001).
25. Webster, T. J. and Ejiofor, J. U. Increased osteoblast adhesion on nanophase metals Ti, Ti6Al4V, and CoCrMo. *Biomaterials*, **25**, 4731–4739 (2004).
26. Oh, I. H., Nomura, N., Masahashi, N., and Hanada, S. Mechanical properties of porous titanium compacts prepared by powder sintering. *Scripta Mat.*, **49**, 1197–1202 (2003).

6 Synthesis of Water-dispersible Carbon Nanotube-fullerodendron Hybrids

Toyoko Imae, Kumi Hamada, Yu Morimoto, and Yutaka Takaguchi

CONTENTS

6.1 Introduction 119
6.2 Experimental.... 121
6.2.1 Materials.... 121
6.2.2 Preparation of Anthryl Dendron G2.0-Oh.... 121
6.2.3 Preparation of Anthryl Dendron G2.0-Oac 121
6.2.4 Preparation of Fullerodendron C60(G2.0-Oac).... 121
6.2.5 Preparation of Hydroxyl-terminated Fullerodendron C60(G2.0-Oh) ... 122
6.2.6 Preparation of Hybrids of Mwcnt and Fullerodendron (Fd) 122
6.2.7 Measurements.... 122
6.3 Discussion and Results.... 122
6.3.1 Synthesis of Hydroxyl-terminated Fullerodendron.... 122
6.3.2 Synthesis of Mwcnt-fullerodendron Hybrids 123
6.3.3 Characterization of Mwcnt-fullerodendron Hybrids.... 124
6.3.4 Dispersibility of Mwcnt-fullerodendron Hybrids.... 129
6.4 Conclusion.... 129
Keywords.... 130
Acknowledgment.... 130
References.... 130

6.1 INTRODUCTION

Fullerodendrons C_{60}(Gn-OAc) (n = 1,2) were produced by Diels–Alder reaction of anthryl dendrons Gn-OAc (n = 1, 2) with fullerene C_{60}. After deprotection of the terminal acetates, fullerodendrons were chemically immobilized by ester bond on acidified multiwall carbon nanotubes (MWCNT) to produce MWCNT-fullerodendron hybrids. The synthesis process was confirmed by FTIR measurements, and the shell formation on MWCNT was visually determined from transmission electron microscopic (TEM) images, indicating the structural arrangement of fullerene moiety at the outermost layer. The coverage of fullerodendron on MWCNT was evaluated from thermogravimetric measurement to be 43 and 52 wt% for n = 1 and 2, respectively. The hybrids were preferably dispersible in both polar and nonpolar mediums.

Carbon-based compounds, fullerene and carbon nanotube (CNT), are ones of most focused materials because of their prominent characteristic properties, such as structural, electronic, electromagnetic, and mechanical properties, which are valuable for their applications [1-10]. However, both carbon-based compounds are poorly soluble/dispersible in any mediums and, therefore, their purification, separation and chemical reaction are not easy. Thus, many efforts are paid for this task: The subject is going to be solved by the hybridization of fullerene and CNT with organic molecules or polymers [11, 12].

One of organic compounds adopted for hybridizing with fullerene is a dendrimer or dendron, which is a branched polymer and possesses a strictly controlled structure with multiple functional groups on its periphery. This type of hybrid called fullerodendrimer or fullerodendron [13, 14] attained the solubility in mediums and was used for the fabrication of silver particle/fullerene nanocomposites and for the preparation of Langmuir Blodgett films with the two-dimensional (2D) ordering of fullerene to be incorporated in the field-effect transistor [15-17].

The hybridization of CNT with organic compounds was accomplished by using pyrene derivatives, anthracene dendron, [18] fullerodendron, [19] enzyme (glucose oxidase), [20] protein, [21] DNA, [22] surfactant, [23] and polymer-metal ion complex [24] as dispersants and adsorbates. Different from the non-covalent functionalization process or the spontaneous dissolution, CNT hybrids were also produced by the covalent-bonding with hyperbranched polymer [25, 26] or dendrimer [27, 28] and by the radical polymerization [29]. The CNT hybrids were ascertained to be dispersible in mediums [18-21, 25-30] Such hybrids were used for building of layer-by-layer assembly [20] or self-assembled layer [22, 23] as a sensor/device and one-dimensional (1D) fabrication of metal and metal oxide nanoparticles [24, 26, 27] and $CaCO_3$ crystals [31] on CNT. Moreover, the synergistic turning of CNT was targeted as a supporter of Pt nanoparticle-encapsulated dendrimer [30].

In the present work, the fabrication of fullerodendron on CNT was carried out by using the coupling reaction of amine terminal on a dendrimer with carboxylic acid functional group on a CNT. As a result of such fabrication, CNTs became dispersible in both polar and nonpolar mediums. What is most important is that the hybrid material of fullerene and CNT is accomplished. The hybridization of fullerenes on a CNT gives us a merit on a design and fabrication of novel nanostructure, since fullerenes are arranged on a CNT. Resultantly, this architecture will develop the new field of applications and be valuable as a component unit of highly organized systems, since the specific characters of component fullerene, CNT, and dendron can be utilized.

6.2 EXPERIMENTAL

6.2.1 Materials

The MWCNT was purchased from Wako Pure Chemical Industries, Ltd. N,N-dimethylformamide (DMF) (Tokyo Kasei Industries, Ltd.) was dehydrated upon refluxing on molecular sieves. Ultra pure water (Milli-Q Labo, Millipore) was used through whole measurements. Other commercial reagents were used as received.

6.2.2 Preparation of Anthryl Dendron G2.0-OH

Anthryl dendron (G1.5-COOMe) was prepared according to the procedure reported [32]. A solution of G1.5-COOMe (418 mg, 0.50 mmol) and 2-aminoethanol (7.3 cm^3, 0.12 mmol) in methanol (30 cm^3) was stirred at 45°C for one day under the nitrogen atmosphere. After the removal of the solvent under the reduced pressure at 45°C, the residue was purified by HPLC (LC 918, Japan Analytical Industry, Co. Ltd.) by the use of gel permeation columns (Jaigel GS-320) and methanol as an eluent to produce anthryl dendron (G2.0-OH) (269 mg, 0.28 mmol) as a light yellow oil in 57% yield: ^{1}H-NMR (300 MHz, CD_3OD) δ 2.23 (8H, t, *J* = 4.4 Hz), 2.39-2.43 (8H, m), 2.61 (8H, t, *J* = 4.4 Hz), 2.70–2.75 (2H, m), 2.82 (4H, t, *J* = 4.4 Hz), 3.13 (4H, t, *J* = 4.0 Hz), 3.21–3.30 (39H, m), 3.53 (11H, t, *J* = 4.0 Hz), 7.47–7.52 (2H, m), 7.83 (1H, d, *J* = 6.0 Hz), 8.01–8.09 (3H, m), 8.49 (1H, s), 8.57 (1H, s), 8.59 (1H, s); IR (neat) 3336 cm^{-1}: MALDI-TOF-MS for $C_{47}H_{72}N_{10}O_{11}$: *m/z* calcd., 953.54 [MH^+]; found: 953.43.

6.2.3 Preparation of Anthryl Dendron G2.0-Oac

A solution of dendron G2.0-OH (64 mg, 0.067 mmol) in *N,N*-dimethylformamide (DMF) (1.0 cm^3) was added acetic anhydride (Ac_2O) (0.21 cm^3, 2.7 mmol), 4-dimethylaminopyridine (DMAP) (33 mg, 0.27 mmol), and pyridine (0.25 cm^3, 3.2 mmol). After stirring for 2 hr at 45°C under the nitrogen atmosphere, the volatile molecules were evaporated. The residue was purified by HPLC (LC 918) by the use of gel permeation columns (Jaigel GS-320) and methanol as an eluent to produce anthryl dendron G2.0-OAc (75 mg, 0.067 mmol) as a light-yellow oil in 73% yield: ^{1}H-NMR (300 MHz, $CHCl_3$) δ 2.05 (14H, s), 2.16 (4H, t, *J* = 4.0 Hz), 2.23 (8H, t, *J* = 4.0 Hz), 2.36 (4H, t, *J* = 4.0 Hz), 2.50 (8H, t, *J* = 4.0 Hz), 2.65-2.82 (9H, m), 3.11–3.13 (4H, m), 3.41–3.47 (9H, m), 3.69-3.70 (2H, m), 4.12 (9H, t, *J* = 4.0 Hz), 7.13 (2H, t, *J* = 4.0 Hz), 7.27–7.29 (4H, m), 7.50–7.53 (2H, m), 7.92 (1H, d, *J* = 4.0 Hz), 7.98–8.02 (3H, m), 8.34 (1H, t, *J* = 3.0 Hz), 8.44 (1H, s), 8.52 (1H, s), 8.70 (1H, s); IR (neat) 1736 cm^{-1}; MALDI-TOF-MS Found: *m/z* 1121.77. Calcd for $C_{55}H_{80}N_{10}O_{15}$: [MH^+], 1122.28.

6.2.4 Preparation of Fullerodendron C_{60}(G2.0-Oac)

A mixture of C_{60} (72 mg, 0.10 mmol) and G2.0-OAc (75 mg, 0.067 mmol) in *o*-dichlorobenzene (7 cm^3) was stirred for three days at 40°C under the nitrogen atmosphere. After removal of the solvent, the residue was purified by a silica gel column chromatography (eluent, chloroform/methanol = 20/1) to produce fullerodendron C_{60}(G2.0-OAc) (23 mg, 0.013 mmol) as a brown oil in 19% yield: ^{1}H NMR (300 MHz, $CDCl_3$) δ 2.01 (18H, s), 2.10 (5H, br), 2.20–2.35 (17H, m), 2.51–2.60 (8H, m), 2.72–2,74 (2H, m), 3.05 (4H, d, *J* = 3.0 Hz), 3.36–3.63 (11H, m), 4.05–4.14 (9H, m), 5.77 (1H, s), 5.79 (1H, s), 6.87 (2H, br), 7.26 (4H, t, *J* = 3.0 Hz), 7.40–7.44 (2H, m), 7.66–7.72

(3H, m), 7.95 (1H, d, $J = 6.0$ Hz), 8.36 (1H, t, $J = 6.0$ Hz), 8.40 (1H, s); ^{13}C NMR (75 MHz, $CDCl_3$) δ 10.4, 10.9, 14.0, 20.9, 23.0, 23.7, 28.9, 29.7, 30.3, 34.2, 38.3, 38.7, 47.5, 49.7, 63.0, 65.0, 67.5, 68.1, 70.5, 76.6, 77.0, 77.2, 77.4, 83.7, 86.4, 87.5, 89.6, 91.0, 97.4, 101.4, 102.6, 107.3, 112.6, 128.8, 130.9, 132.4, 136.9, 138.5, 139.9, 141.9, 144.1, 145.3, 146.2, 146.4, 147.6, 153.9, 163.1, 167.8, 171.1, 172.1, 172.9, 175.6, 182.2, 186.6, 188.0, 189.0, 196.2; IR (neat) 1734 cm^{-1}; MALDI-TOF-MS for $C_{115}H_{80}N_{10}O_{15}$: *m/z* calcd., 1838.72 [MH$^+$]; found: 1839.56.

6.2.5 Preparation of Hydroxyl-terminated Fullerodendron C_{60}(G2.0-Oh)

To a solution of dendron C_{60}(G2.0-OAc) (23 mg, 0.013 mmol) in THF (3 cm^3), 1 N NaOH in methanol (0.5 cm^3) was added dropwise. After stirring for 6 hr, the solvent was evaporated. The residue was reprecipitated in THF, and further purified by a Amberlite IR120B column to produce fullerodendron C_{60}(G2.0-OH) (18 mg, 0.011 mmol) in 98% yield: IR (neat) 3363 cm^{-1}; MALDI-TOF-MS for $C_{107}H_{72}N_{10}O_{11}$: *m/z* calcd., 1670.50 [MH$^+$]; found: 1671.52.

6.2.6 Preparation of Hybrids of MWCNT and Fullerodendron (Fd)

The MWCNT (20 mg) was refluxed at 70°C in a concentrated HNO_3 (10 cm^3) for 12 hr, filtrated with a Millipore polycarbonate membrane (pore size 0.1 μm), washed with water, and dried at room temperature in atmosphere [27, 28]. For the preparation of MWNT-COOH, the purified MWCNT was refluxed with a mixture of $H_2NO_3 : H_2SO_4 = 1 : 1$ for 12 hr at 70°C, sonicated in an ultrasonic bath for 10 min, filtered with a Millipore polycarbonate membrane (pore size 0.1 μm), and then washed with water. The obtained MWCNT-COOH was dried at room temperature in atmosphere. MWNT-COCl was synthesized by refluxing MWNT-COOH with $SOCl_2$ at 65°C for 10 hr and evaporating $SOCl_2$ by N_2 gas under heating. For the preparation of MWNT-FD, the mixture of MWNT-COCl, FD(G1 or G2) and DMF were refluxed for 24 hr at 60°C under N_2 atmosphere. The obtained product was filtrated with a Millipore PTFE membrane (pore size 0.22 μm) and washed with methanol, and then dispersed in methanol.

6.2.7 Measurements

The FTIR absorption spectra were recorded on a BioRad FTS-575C instrument with a cryogenic mercury cadmium telluride (MCT) detector. KBr pellets including samples were prepared or methanol solutions of samples were dried on CaF_2 plates. The TEM images were taken on an H-7000 with a CCD camera. Methanol dispersion was dried on a carbon-coated cupper grid. Thermal gravimetric analytical (TGA) measurements were carried out from room temperature to 700°C at a heating speed of 1°C/min on a Rigaku 8078G2.

6.3 DISCUSSION AND RESULTS

6.3.1 Synthesis of Hydroxyl-terminated Fullerodendron

Fullerodendron C_{60}(G2.0-OH) was synthesized as shown in Figure 1. Anthryl dendron G1.5-COOMe possessing methyl ester groups at the terminals was prepared as described [32]. End group transformation of G1.5-COOMe was accomplished easily by the treatment of the terminal ester groups with 2-aminoethanol to produce anthryl dendron G2.0-OH at 57% yield. Subsequently, the terminal hydroxyl groups were protected as acetates by heating with acetic anhydride and 4-dimethylaminopyridine in DMF to obtain anthryl dendron G2.0-OAc. Diels-Alder reaction of the anthryl dendron G2.0-OAc with C_{60} produced fullerodendron C_{60}(G2.0-OAc) at 19% yield under the same reaction conditions reported [32]. Finally, deprotection of the terminal acetates by the treatment with NaOH gave rise to fullerodendron C_{60}(G2.0-OH) (98% yield), which has hydroxyl groups at the terminals. The chemical structure of C_{60}(G2.0-OAc) and C_{60}(G2.0-OH) was confirmed by ^{1}H and ^{13}C NMR spectroscopies, IR absorption spectroscopy, and MALDI-TOF-Mass spectrometry.

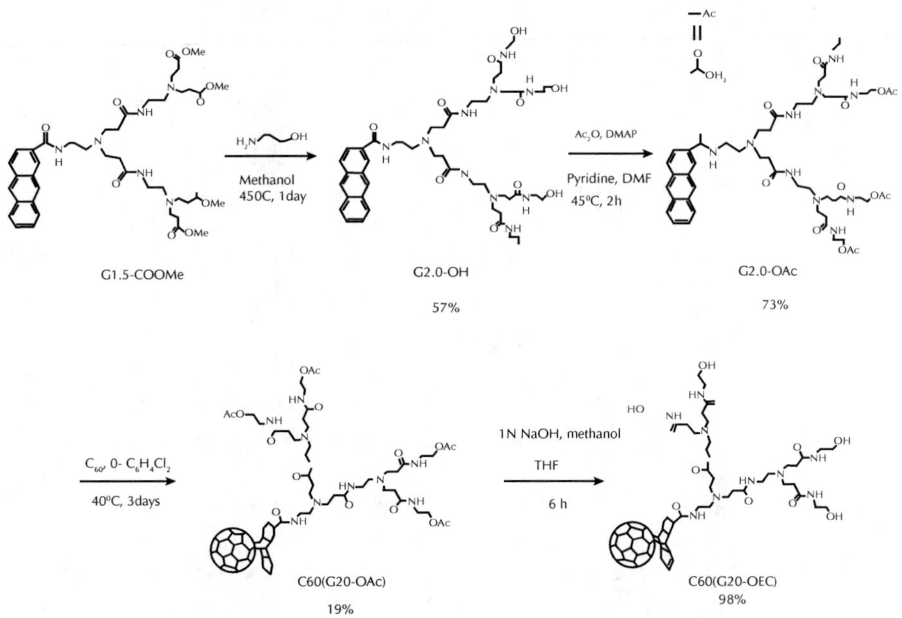

FIGURE 1 Schematic illustration of the synthesis of hydroxyl terminated fullerodendron.

6.3.2 Synthesis of MWCNT-fullerodendron Hybrids

The synthesis of MWCNT-fullerodendron hybrids followed the reported procedure as seen in Figure 2, where amine-terminated dendrimers were chemically bound on MWCNT, [27, 28] and the synthesis was similar to the binding of MWCNT with hydroxyl terminated polymer [33]. The optimum reaction conditions were determined as follows. The reflux time (12 hr) on the purification of MWNT was shorter than the literature (24 hr) [33] because of the breakdown of MWNT and the production of short bundles and amorphous materials. On the production of MWNT-COOH, number of carboxylic acid groups effectively increased with increasing the content of H_2SO_4, but the following filtration was difficult $HNO_3 : H_2SO_4 = 1 : 1$. The increase in the reaction time at least up to 12 hr stimulated the reaction, and the increase (to be double) in the concentration of reaction reagents was effective to inhibit the reduction of yield. The sonication after the production of MWNT-COOH is preferable for dispersing

MWCNT but must be short so as not to damage MWCNT. Since the MWCNT-COCl is unstable under the moisture, that is, the reverse reaction happens, N_2 gas must be bubbled in the reaction medium including fulleredendron. Moreover, the temperature above 60°C, where the decomposition (backward Diels-Alder reaction) of fulleroden-dron occurs, had to be avoided on the reaction.

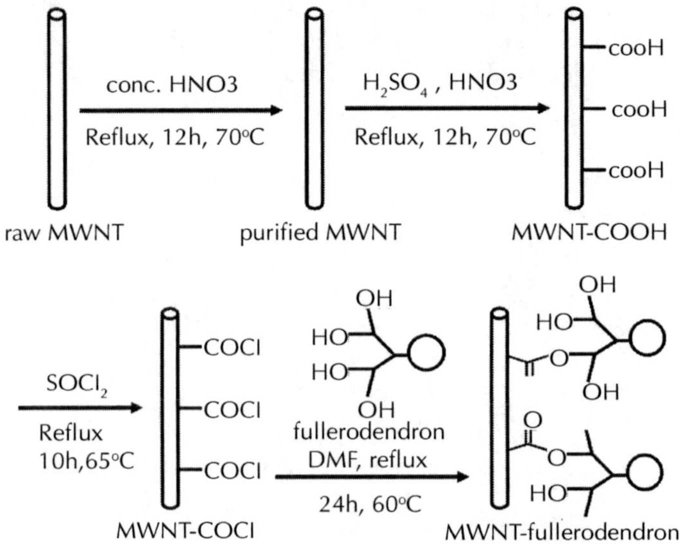

FIGURE 2 Schematic illustration of the synthesis of MWCNT-fullerodendron hybrids.

6.3.3 Characterization of MWCNT-fullerodendron Hybrids

Figure 3 shows IR absorption spectra of MWCNT-COOH, fullerodendrons (G1 and G2) and MWCNT-fullerodendron hybrids (G1 and G2). An IR absorption spectrum of MWNT-COOH is characterized by absorption bands of OH and C = O stretching and OH bending modes. After the induction of C_{60}(G1.0-OH) on MWCNT, characteristic bands of COOH disappeared. However, amide bands (A, B, I and II at 3262, 3083, 1661, 1538 cm^{-1}, respectively) and CH_2 stretching and bending bands appeared. These characteristic bands of dendron are also appeared in an absorption spectrum of free C_{60}(G1.0-OH). Additionally, an ester (C = O stretching) band was observed at 1735 cm^{-1} in a spectrum of MWCNT-FD(G1) hybrid. This confirms the chemical reaction of COOH group on MWCNT with terminal OH of fullerodendron. Similar IR absorption bands were observed even MWNT-C_{60}(G2.0-OH) hybrid and C_{60}(G2.0-OH), indicating the esterification reaction between MWNT-COOH and FD(G2) as well. Now it is noticed that the ester band is fairly weaker than amide bands, although the ratios of OH group against amide group are 3 : 2 and 4 : 7 for C_{60}(G1.0-OH) and C_{60}(G2.0-OH), respectively. This indicates that all OH groups in fullerodendron do not necessarily bind with MWCNT because of less amount of carboxylic acid content on MWCNT.

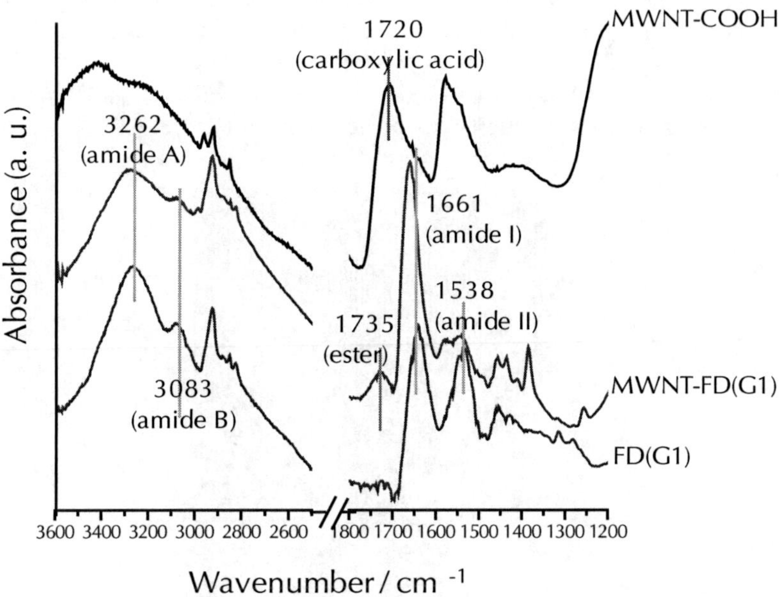

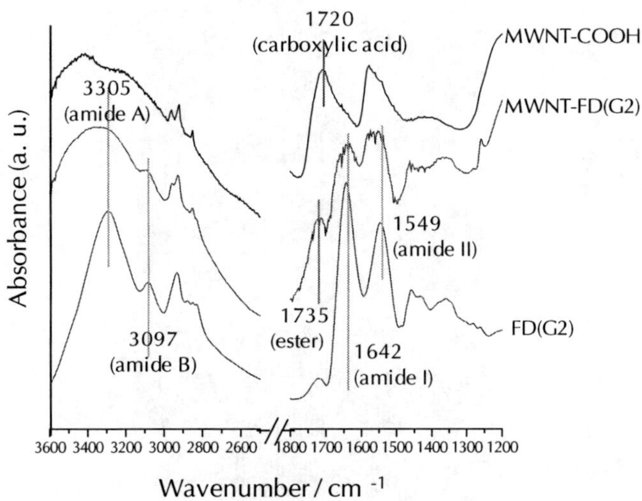

FIGURE 3 The IR absorption spectra of MWNT-COOH, MWNT-FD(G1, G2) and FD(G1, G2).

The TEM images of MWCNT, MWCNT-FD(G1), and MWNT-FD(G2) are shown in Figure 4. While a TEM image of MWCNT displayed the tubular texture, MWNT-FD(G1, G2) presented TEM textures with the thin shell on MWCNT. The wall thicknesses of ~3 and ~4 nm are comparable to sizes of FD (G1 and G2), respectively, indicating that the shell consists of fullerodendrons. Since some hydroxyl terminals of fullerodendron chemically bind on MWCNT, the probable structures of hybrids can be assumed as illustrated in Figure 5, where fullerene moiety must locate on the periphery of the shell on MWCNT. Different from the thin shell of dendrimer on MWCNT, [27, 28] the outermost ~1 nm layers of the fullerodendron shells on MWCNT in TEM images seems thick because of high electron density of fullerene.

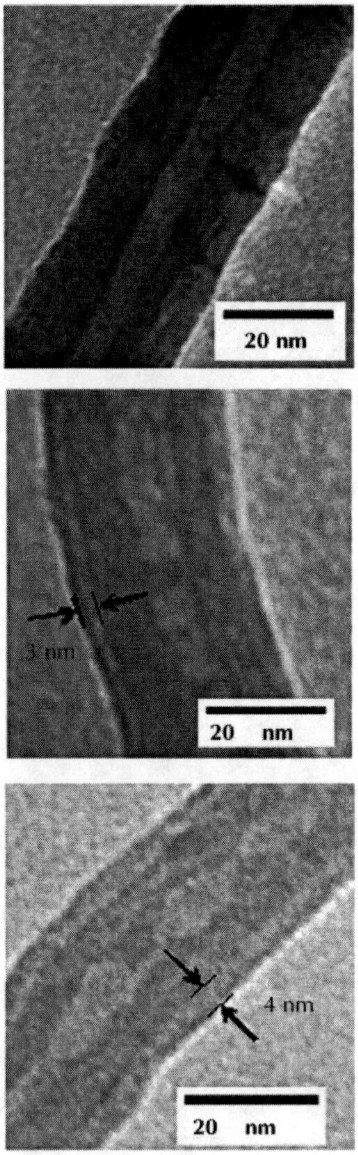

FIGURE 4 The TEM images of MWCNT (top), MWCNT-FD(G1) (middle), and MWNT-FD(G2) (bottom).

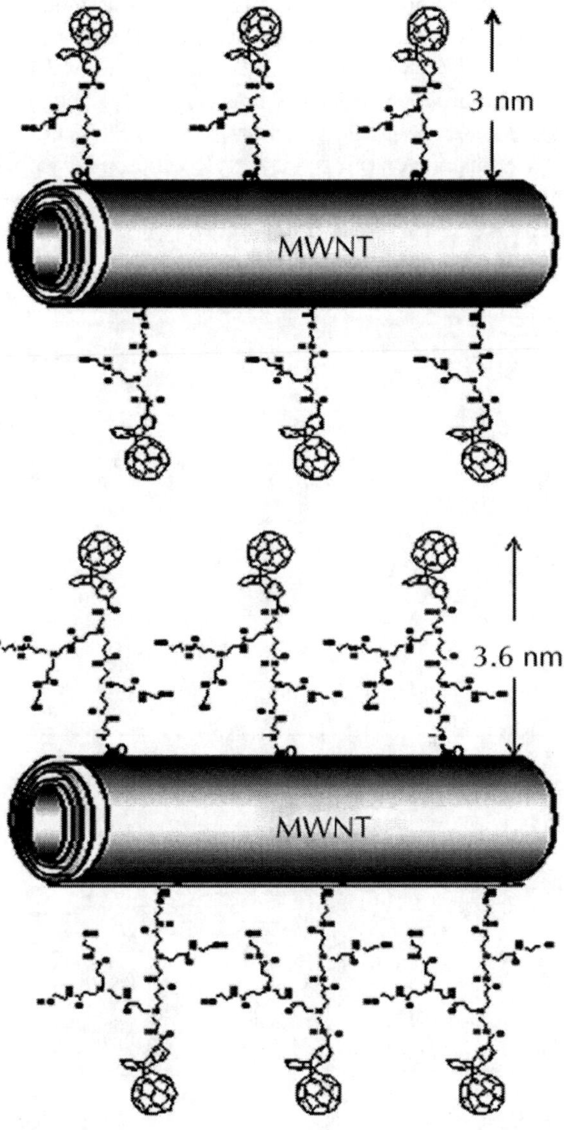

FIGURE 5 Probable structures of MWCNT-fullerodendron (C60(G2.0-OH), C60(G1.0-OH)) hybrids.

Figure 6 shows TGA curves of MWNT, MWNT-FD, and FD. The weight of MW-CNT decreased between 500 ~ 600°C, as already reported, [27] while the weight of FD came down to 90wt% at 150 ~ 300°C and FD was decomposed perfectly at ~ 450°C for G1 and ~ 500°C for G2. On the other hand, water and/or contaminants in MWNT-FD was burned below 300°C and the rest was completely decomposed at 600°C, indicating the coexistence of MWNT and FD for both G1 and G2. Then the wt% of fullerodendron bound on MWCNT can be calculated as weight losses at 450 and 500°C, where G1 and G2, respectively, are perfectly burned. The calculated weight of FD was 43 and 52 wt% in MWNT-FD(G1) and MWNT-FD(G2), respectively. It was reported that when MWCNT was oxidized, about 1% of carbon atoms of MWCNT possessed carboxylic acid residue [34]. If FD(G1) or FD(G2) (Mw 1214 or 1673) chemically binds one on one to carboxylic acid on MWCNT, contents of fullerodendrons is maximum 50 and 58 wt% for G1 and G2, respectively. These numbers are reasonable in comparison with observed values.

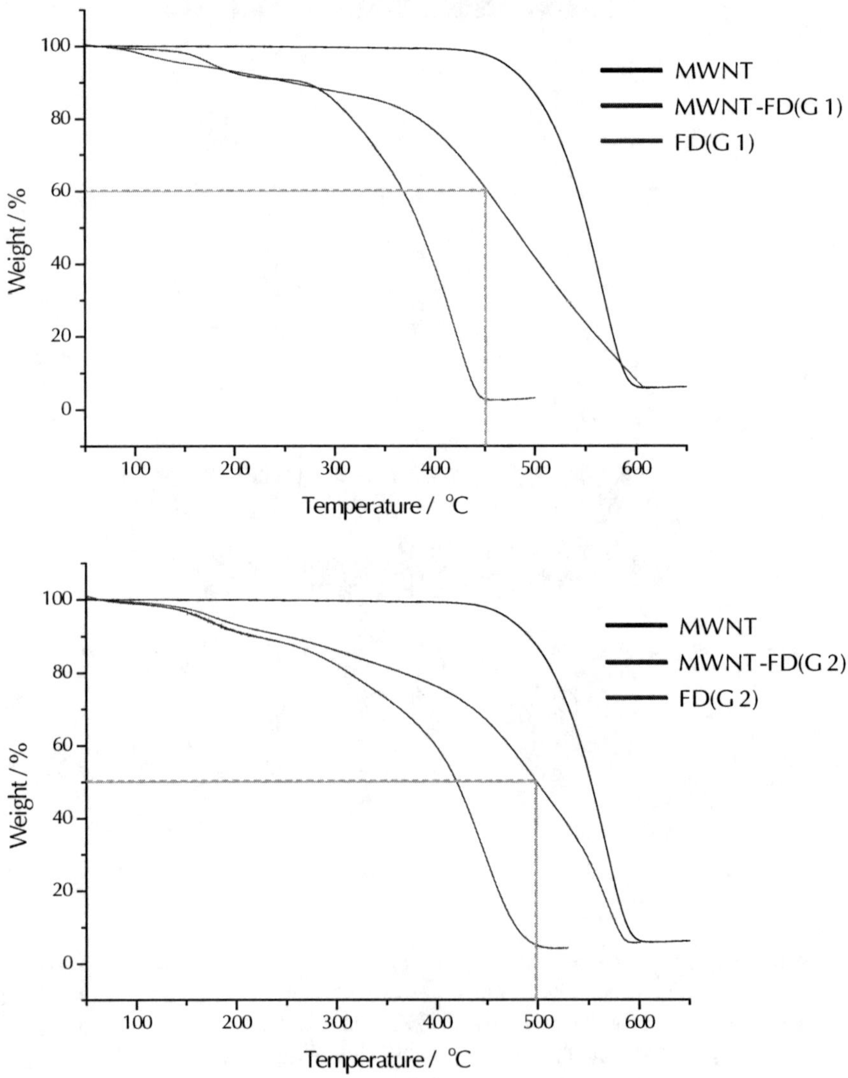

FIGURE 6 The TGA charts of MWNT, MWNT-FD(G1, G2) and FD(G1, G2).

6.3.4 Dispersibility of MWCNT-fullerodendron Hybrids

The dispersibility of MWCNT-fullerodendron hybrids and relating materials was determined by sonicating samples in solvents for 5 min in an ultrasonic bath and leaving one week. Figure 7 displays the dispersibility in methanol and water at one week after the sonication. MWCNT precipitated after 10 min in both medium. FD(G1) dispersed in methanol, while it precipitated partly after 30 min but dispersed mostly for a week in water. FD(G2) precipitated mostly in methanol and partly in water. The reason is that FD(G1) is more hydrophobic than FD(G2), because hydrophobic fullerene is major in FDG1 and hydrophilic dendron affects on the property of FDG2. Therefore, the former is familiar to methanol but the latter is not.

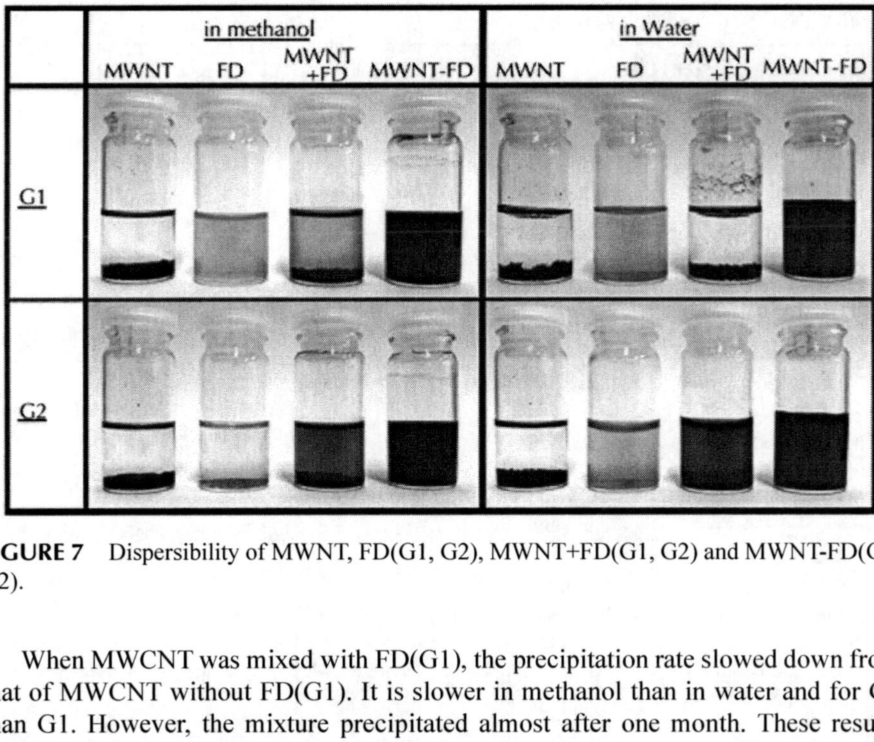

FIGURE 7 Dispersibility of MWNT, FD(G1, G2), MWNT+FD(G1, G2) and MWNT-FD(G1, G2).

When MWCNT was mixed with FD(G1), the precipitation rate slowed down from that of MWCNT without FD(G1). It is slower in methanol than in water and for G2 than G1. However, the mixture precipitated almost after one month. These results indicate that only mixing FD is not enough to disperse MWCNT. On the other hand, MWNT-FD(G1) and MWNT-FD(G2) dispersed stably in methanol and water for a week and no precipitates were detected. The dispersions were stable more than one month, although the mixtures were precipitated after one month. MWNT-FD(G1) and MWNT-FD(G2) were also dispersible in DMF, while they were partly precipitated in chloroform, THF, benzene and ethyl acetate.

6.4 CONCLUSION

In order to disperse MWCNT in mediums, especially, in polar mediums like water, the MWCNT was hybridized by covalent bonding with fullerodendron. The resultant hydrids consist of dendron inner shell and fullerene outermost layer on MWCNT, that is, dendron mediates the construction of fullerene layer on MWCNT. This situation can be compared with the case of noncovalent bonding of fullerodendron on single wall carbon nanotubes, whether fullelene is an adsorption site and directly adsorbs on CNT [18, 19].

The hybrids dispersed stably without any precipitates for long period in polar mediums like water and even organic mediums like methanol and DMF. Along with report, [27, 28] this report supports that PAMAM type dendrimers and dendrons validly play a role in increase of the affinity of MWCNT to mediums and the extension of the possible application of MWCNT in mediums. The applicable research can be extended by using water-dispersible MWCNT-fullereodendron hybrids.

KEYWORDS

- **Anthryl dendron**
- **Carbon nanotube**
- **Fourier-transform infrared**
- **Multiwall carbon nanotubes**
- **Thermal gravimetric analytical**

ACKNOWLEDGMENT

We appreciate Associate Professor H. Ikuta and Dr. G. Takami, Nagoya University, for their kind permission for using TGA instrument and useful discussion.

REFERENCES

1. Hebard, A. F., Rosseinsky, M. J., Haddon, R. C., Murphy, D. W., Glarum, S. H., Palstra, T. T. M., Ramirez, A. P., and Kortan, A. R. *Nature,* **350**, 600 (1991).
2. Tanigaki, K., Ebbesen, T. W., Saito, S., Mizuki, J., Tsai, J. S., Kubo, Y., and Kuroshima, S. *Nature*, **352**, 222 (1991).
3. Iijima, S. *Nature*, **354**, 56 (1991).
4. Mintmire, J. W., Dunlap, B. I., and White, C. T. *Phys. Rev. Lett.*, **68**, 631 (1992).
5. Hamada, N., Sawada, S., and Oshiyama, A. *Phys. Rev. Lett.*, **68**, 1579 (1992).
6. Saito, R., Fujita, M., Dresselhaus, G.., and Dresselhaus, M. S. *Phys. Rev. B*, **46**, 1804 (1992).
7. Iijima, S. and Ichihashi, T. *Nature*, **363**, 603 (1993).
8. Bethune, D. S., Kiang, C. H., DeVries, M. S., Gorman, G., Savoy, R., and Beyers, R. *Nature*, **363**, 605 (1993).
9. Tans, S. J., Devoret, M. H., Dai, H., Thess, A., Smalley, R. E., Geerligs, L. J., and Dekker, C. *Nature*, **386**, 474 (1997).
10. Ren, Z. F., Huang, Z. P., Xu, J. W., Wang, J. H., Bush, P., Siegal, M. P., and Provencio, P. N. *Science*, **282**, 1105 (1998).
11. Fréchet, J. M. J. and Tomalia, D. A. *Dendrimers and other dendritic polymers*. John Wiley & Sons, Ltd., Chichester (2001).
12. Newkome, G. R., Moorfield, C. N., and Vögtle, F. *Dendrimers and Dendrons*. Wiley-VCH, Weinheim (2001).
13. Wooley, K. L., Hawker, C. J., Fréchet, J. M. J., Wudl, F., Srdanov, G., Shi, S., Li, C., and Kao, M. J. *Am. Chem. Soc.*, **115**, 9836 (1993).
14. Nierengarten, J. F. *Chem. Eur. J.*, **6**, 3667 (2000).

15. Hirano, C., Imae, T., Tamura, M., and Takaguchi, Y. *Chem. Lett.*, **34**, 862 (2005).
16. Hirano, C., Imae, T., Fujima, S., Yanagimoto, Y., and Takaguchi, Y. *Langmuir*, **21**, 272 (2005).
17. Kawasaki, N., Nagano, T., Kubozono, Y., Sako, Y., Morimoto, Y., Takaguchi, Y., Fujiwara, A., Chu, C. C., and Imae, T. *Applied Physics Letters*, **91**, 243515 (2007).
18. Sandranayaka, A. S. D., Takaguchi, Y., Uchida, T., Sako, Y., Morimoto, Y., Araki, Y., and Ito, O. *Chem. Lett.*, **35**, 1188 (2006).
19. Takaguchi, Y., Tamura, M., Sako, Y., Yamamoto, Y., Tsuboi, S., Uchida, T., Shimamura, K., Kimura, S., Wakahara, T., Maeda, Y., and Akasaka, T. *Chem. Lett.*, **34**, 1608 (2005).
20. Wang, Y., Joshi, P. P., Hobbs, K. L., Johnson, M. B., and Schmidtke, D. W. *Langmuir*, **22**, 9776 (2006).
21. Karajanagi, S. S., Yang, H., Asuri, P., Sellitto, E., Dordick, J. S., and Kane, R. S. Langmuir, **22**, 1392 (2006).
22. Kocharova, N., Ääritalo, T., Leiro, J., Kankare, J., and Lukkari, J. *Langmuir*, **23**, 3363 (2007).
23. Zangmeister, R. A., Maslar, J. E., Opdahl, A., and Tarlov, M. J. *Langmuir*, **23**, 6252 (2007).
24. Hu, X., Wang, T., Wang, L., Guo, S., and Dong, S. *Langmuir*, **23**, 6352 (2007).
25. Hong, C. Y., You, Y. Z., Wu, D., Liu, Y., and Pan, C. Y. *Macromolecules*, **38**, 2606 (2005).
26. Hong, C. Y., You, Y. Z., and Pan, C. Y. *Chem. Mater.*,**17**, 2247 (2005).
27. Lu, X. and Imae, T. J. *Phys. Chem. C*, **111**, 2416 (2007).
28. Lu, X. and Imae, T. J. *Phys. Chem. C*, **111**, 8459 (2007).
29. Kong, H., Gao, C., and Yan, D. *J. Am. Chem. Soc.*, **126**, 412 (2004).
30. Vijayaraghavan, G. and Stevenson, K. J. *Langmuir*, **23**, 5279 (2007).
31. Li, W. and Gao, C. *Langmuir*, **23**, 4575 (2007).
32. Takaguchi, Y., Sako, Y., Yanagimoto, Y., Tsuboi, S., Motoyoshiya, J., Aoyama, H., Wakahara, T., and Akasaka, T. *Tetrahedron Lett.*, **44**, 5777 (2003).
33. Kong, H., Gao, C., and Yan, D. *J. Am. Chem. Soc.*, **126**, 412 (2004).
34. Mawhinney, D. B., Naumenko, V., Kuznetsova, A., Yates Jr., J. T., Liu, J., and Smalley, R. E. *Chem. Phys. Lett.* **324**, 213 (2000).

7 A Facile Sol-gel Method for the Preparation of N-doped TiO_2 Photocatalyst for Pollutant Degradation

P. P. Silija, Z. Yaakob, N. N. Binitha, P. V. Suraja, S. Sugunan, and M. R. Resmi

CONTENTS

7.1 Introduction 133
7.2 Experimental 134
7.2.1 Preparation of Photocatalysts 134
7.2.2 Catalyst Characterization 134
7.2.3 Photocatalytic Degradation 134
7.3 Discussion and Results 135
7.4 Conclusion 136
Keywords 137
Acknowledgment 137
References 137

7.1 INTRODUCTION

Photodegradation of organic and inorganic pollutants on semiconductors is a current topic of research. The TiO_2 has the advantages of its high chemical stability, nontoxicity, and relatively low-price, but a serious disadvantage would be that only UV light can be used for photocatalytic reactions. Therefore, it is of great interest to find ways to extend the absorption wavelength range of TiO_2 to visible region without the decrease of photocatalytic activity [1]. Asahi et al. reported that N-doped TiO_2 shows a significant shift of the absorption edge to the visible light region [2]. There are many reports for the successful preparation of nitrogen doped titania, [3, 4] The sol-gel method or a variation thereof, is the most commonly used chemical method in the literature [5].

Most N-TiO_2 samples were prepared by treating TiO_2 under NH_3 atmosphere at very high temperature which leads to the energy waste [6]. Hence, there is a growing interest of new facile approaches to the synthesis of N-TiO_2. The traditional surfactant templating method needs the removal of organic templates by thermal treatment, resulting in high energy consumption, environmental pollution and the agglomeration and collapse of pore structure in many cases [7]. In the present research, we use a simple surfactant free approach to the synthesis of N-TiO_2 photocatalysts with visible light response *via* sol-gel route. It is evident that urea is more environmentally friendly than NH_3. There are some reports using urea as the precursor for nitrogen doping [8]. The purpose of our research study is to use urea as N source to synthesize N-doped TiO_2 powder; in the presence of ethanol to form highly active anatase TiO_2.

Nitrogen doping is done on TiO_2 *via* sol-gel method. Urea acts as a template and also acts as the precursor for N-doping. This creates improved properties for N-doped systems when compared to the post preparation N-doping processes. Presence of ethanol as a co-solvent improves the surface area. The visible light absorbance is investigated using UV-vis DRS analysis. The visible light activity of various N-doped systems is used for photodegradation of methyleneblue (MB) under optimized conditions. Excellent pollutant degradability is shown by N-doped system prepared with N/Ti molar ratio of 20.

7.2 EXPERIMENTAL

7.2.1 Preparation of Photocatalysts

Room temperature sol-gel method is used for preparation. The 18.6 ml Titanium (IV) isopropoxide and 50 ml ethyl alcohol (Merck) were mixed (solution a), to this solution b, which contains 50 ml ethyl alcohol, 12.5 ml glacial acetic acid, and 6.25 ml distilled water was added. The resultant transparent solution was stirred for 3 hr. Urea–water–alcohol mixture with molar ratio 1:5:0.5 respectively was then slowly added to the above solution until the chosen N/Ti molar content was satisfied. It was again stirred for 3 hr, aged for 2 days and dried in an air oven at 80°C. Ground into fine powder and calcined at 400°C for 5 hr to obtain yellow colored N-doped TiO_2. The systems are designated as 0N/Ti, 6N/Ti, 10N/Ti, 16N/Ti, 20N/Ti, 24N/Ti, and 28N/Ti, whereas the numbers indicates the N/Ti molar ratio.

7.2.2 Catalyst Characterization

The XRD patterns of the samples were recorded on a Philips diffractometer employing a scanning rate of 0.02 s^{-1} with Cu Kα radiation. UV-Vis DRS of powder catalyst samples were carried out at room temperature using a Perkin Elmer Lamda-35 spectrophotometer. Surface area measurements were performed at liquid nitrogen temperature with a Micromeritics Tristar 3,000 surface area analyzer.

7.2.3 Photocatalytic Degradation

Photodegradation of MB is investigated. Degradation is done using Rayonet type Photoreactor with visible light having 16 tubes of 8 W (Associate Technica, India). The 50 ml of 10 mg/l MB was placed in the quartz tube, containing 4 g/l catalyst and is irradiated with visible light under continuous stirring for 6 hr. The MB concentration was analyzed using ESICO Microprocessor photocolorimeter model No. 1312.

7.3 DISCUSSION AND RESULTS

We had prepared a range of N-doped TiO_2 with various N/Ti molar ratio. The systems are characterized and the photodegradability is investigated using MB degradation. The XRD patterns of the systems shown in Figure 1(a) indicates that all systems contain anatase phase as the only phase which is reported as having the most photocatalytic activity [9]. The surface areas of different systems are given in Table 1. It can be seen that an increase in the amount of urea used for the N-doped TiO_2 preparation results in an increase in surface area. The TiO_2 without using urea template resulted in decreased surface area of 15.32 m^2/g and is expected to show least activity. Cheng et al [8] reported N-doped titania using urea as precursor *via* sol-gel method. They investigated the molar ratio of urea 0, 5, 10, 20, and 30% and the S_{BET} of urea/TiO_2 was 25.6, 41.1, 64.4, 46.4, and 32.8 m^2/g, respectively. In their study it can be seen that S_{BET} increased with the amount of urea to reach a maximum at 10% (Ti-N10) and then decreased with further increase of the amount of urea. In the present work the observed surface area was very high when compared to this report. We have selected the N/Ti molar ratios of 6, 10, 16, 20, 24, and 28. The presence of additional use of ethanol as a co-solvent for urea may be the reason for the improved surface area of the present study. Here the increase in N content increases surface area continuously.

TABLE 1 Surface area, absorption edges, band gap energy and photoactivity of various systems.

System	BET SA (m^2/g)	1st absorption edge (nm)	Band gap (eV)	2nd absorption edge (nm)	Band gap (eV)	MB Degradation (%)
0N/Ti	15.32	382.61	3.24	-	-	15.6
6N/Ti	49.50	410.91	3.02	-	-	28.9
10N/Ti	55.34	382.61	3.24	548.10	2.26	31.1
16N/Ti	68.50	385.10	3.22	545.40	2.27	40.0
20N/Ti	94.24	390.22	3.18	555.66	2.23	85.0
24N/Ti	105.69	395.51	3.14	529.85	2.34	66.7
28N/Ti	112.79	390.22	3.18	522.08	2.38	48.9

In UV-vis DRS of all the N containing samples, characteristic band for tetrahedral coordinated titanium appears at wavelength ~400 nm. The synthesized samples, except 0N/Ti and 6N/Ti showed excellent visible light absorption and two step absorption edges as evident from its spectrum in Figure 1(b). The second absorption edge around 520–555 nm is related to newly formed N 2p band which locates above O 2p valence band in $TiO_{2-x}N_y$ is in agreement with other recent reports [10]. The band gap of the samples was also determined. Table 1 summarizes the absorption edges and corresponding band gap values of the samples. From the results it is seen that the least band gap is shown by sample with 20N/Ti and is thus expected to have maximum visible light activity.

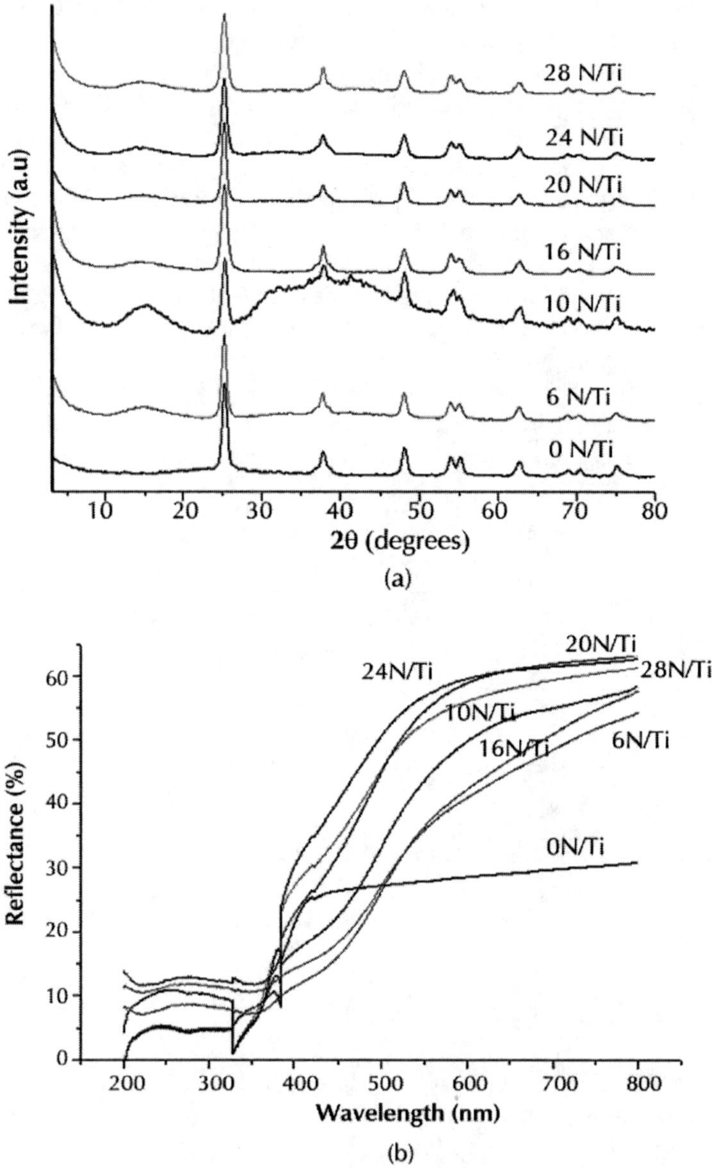

FIGURE 1 (a) XRD and (b) UV-vis DRS of N-doped systems with different N/Ti molar ratios.

The photodegradability of the present N-doped systems is investigated using MB degradation. To study the effect of N/Ti molar ratio on the degradability, systems prepared under similar conditions with different N/Ti molar ratio were used. Among the different catalysts, 20N/Ti system with N/Ti molar ratio 20 shows maximum activity. The result shown in Table 1 is in agreement with the lowest band gap energy of 20N/Ti system. Thus for the present N doped systems, more than surface area, low band gap energy of 20N/Ti contributes to their highest visible light activity.

7.4 CONCLUSION

In summary, we have demonstrated a facile and surfactant free method for the preparation of N-doped TiO_2. Effect of visible absorption and N/Ti molar ratio on the photocatalytic activity is investigated. Surface area of systems increases with nitrogen content. The effect of co solvent ethanol provided increased surface area when compared to similar reports. The photocatalysts show high visible light photocatalytic activity on the degradation of MB owing to their low band gap energy. The doped nitrogen species play a key role in expanding the photoactivity to visible light region and two step absorption edges are shown by systems with N/Ti molar ratio 10 and higher which are responsible for visible activity. The degradation of MB over the system with N/Ti ratio of 20 is found to be 85%.

KEYWORDS

- **N-doping**
- **Photocatalytic**
- **Photodegradation**
- **Semiconductors**
- **Sol-gel preparation**

ACKNOWLEDGMENT

The authors acknowledge the UKM, the grant number UKM-OUP- NBT-27-118/2009 for the financial support. The STIC, CUSAT, Cochin, India is acknowledged for XRD analysis.

REFERENCES

1. Sato, S. *Chem. Phys. Lett.*, **123**, 126–128 (1986).
2. Asahi, R., Morikawa, T., Ohwaki, T., Aoki, K., and Taga, Y. *Science.*, **293**, 269–271 (2001).
3. Yin, S., Aita, Y., Komatsu, M., Wang, J., Tang, Q., and Sato, T. *J. Mater Chem.*, **15**, 64(2005).
4. Yin, S., Ihara, K., Aita, Y., Komatsu, M., and Sato, T. *J. Photochem. Photobiol A.*, **179,** 105–114 (2006).
5. Di Valentin, C., Pacchioni, G., Selloni, A., Livraghi, S., and Giamello, E. *J. Phy. Lett. B.*, **109**, 11414–11419 (2005).
6. Irie, H., Watanabe, Y., and Hashimoto, K. *J. Phys. Chem. B.*, **107,** 5483–5486 (2003).
7. Wang, C. C. and Ying, J. Y. *Chem. Mater.*, **11**, 3113–3120 (1999).
8. Cheng, P., Deng, C., Gu, M., and Dai, X. *Mat Chem Phys.*, **107**, 77–81 (2008).
9. Watson, S., Beydoun, D., Scott, J., and Amal, R. *J. Nanoparticle Res.*, **6,** 193–207 (2004).
10. Asahi, R., Morikawa, T., Ohwaki, T., Aoki, K., and Taga, Y. *Science.*, **293**, 269–271 (2001).

8 TiO_2 Nanoparticle Thin Films Cast by Electrophoretic Deposition

Sameer V. Mahajan, Maria T. Colomer, Mario Borlaf, Rodrigo Moreno, and James H. Dickerson

CONTENTS

8.1 Introduction 139
8.2 Apparatus 140
8.3 Materials Synthesis 141
8.4 Discussion 141
8.5 Conclusion 145
Keywords 145
Acknowledgment 145
References 145

8.1 INTRODUCTION

The production of smooth, densely-packed, and continuous casts of colloidal nanoceramics into architectures for applications ranging from anti-corrosion coatings to transparent optical components and devices is a rapidly developing field. Post-processing procedures, such as, sintering that can form single crystalline or polycrystalline masses from the constituent materials and improves densification, robustness, and mechanical integrity, markedly can enhance the feasibility and durability of the dense-packing of the nanoparticles. To achieve high density materials post-sintering, we must explore various techniques that facilitate tight packing. Available methods to cast nanoparticle films directly onto targeted substrates, such as, Langmuir-Blodgett, layer-by-layer deposition, thermal evaporation, and spin coating, have recognized limitations, including the inability to achieve both large-scale nanoparticle ordering, poor chemical and structural robustness, and notably inferior film assembly rates [1-5]. Indeed, for nanoparticle composite films to be competitive with single crystalline materials, the facile and rapid production of homogeneous, densely packed, and topographically smooth nanoparticle films must be realized. Of the available schemes to fabricate such nanoparticle films, electrophoretic deposition (EPD) may be the only technique that includes superior deposition rate, size scalability, and densely-packed, smooth films. The superior packing density realized by EPD should engender high fracture strength in the films because of the size, shape or topology of the underlying substrate or surface. This chapter discusses the investigation of the EPD of TiO_2 ceramic nanoparticles (NPs).

We report the production of densely packed films of titanium dioxide (TiO_2) nanoparticles *via* EPD. The nanoparticles were characterized by transmission electron microscopy (TEM), whereas, aqueous suspensions of the materials were characterized by dynamic light scattering. Although we observed macroscopic cracks within the film upon drying in air, the ~100 μm sized flakes were comprised of densely-packed nanoparticles that suggest larger macroscopic films may be realized with further optimization of the nanoparticle suspensions and the EPD process.

8.2 APPARATUS

The EPD was conducted by insertion of 316 stainless steel (McMaster-Carr) electrodes (25.4 mm × 12.7 mm) into a suspension of TiO_2 NPs with an applied voltage. A schematic of the EPD setup is provided in Figure 1.

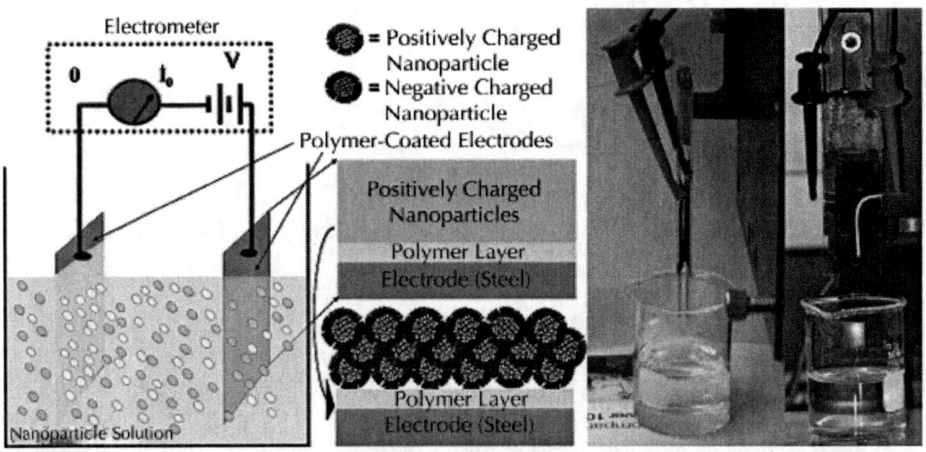

FIGURE 1 (Left) Schematic of the EPD setup to make freestanding ceramic NP films, with an optional sacrificial polymer layer indicated. (Middle and Right) Images of the electrodes and NP solution from my laboratory.

To cast the films, we used 316 stainless steel electrodes and conducted 5 min EPD experiments, applying 3 V across a 10 mm gap between the electrodes. The voltage was provided by a BK Precision Model 1787B Programmable DC Power Supply. The TEM characterization of nanocrystal samples was performed using a Philips CM20T microscope operating at 200 kV. Scanning electron microscopy (SEM) images of the films were acquired using a Hitachi S 4200 microscope. Dynamic light scattering measurements, to ascertain the hydrodynamic diameter of the NPs in solution, were carried out with a Nano ZS Model ZEN 3600 Malvern Zetasizer with a 633 nm laser.

8.3 MATERIALS SYNTHESIS

The TiO_2 NPs were made by mixing Ti(IV)-isopropoxide in deionized water at ratios of 1:100, 1:75, and 1:50 (Ti(IV)-isopropoxide:water), following a recently reported technique [6-7]. Figure 2 shows a typical TEM image of an individual TiO_2 nanocrystal, synthesized from the 1:75 molar ratios. We used the as prepared water-based sols for the EPD experiments.

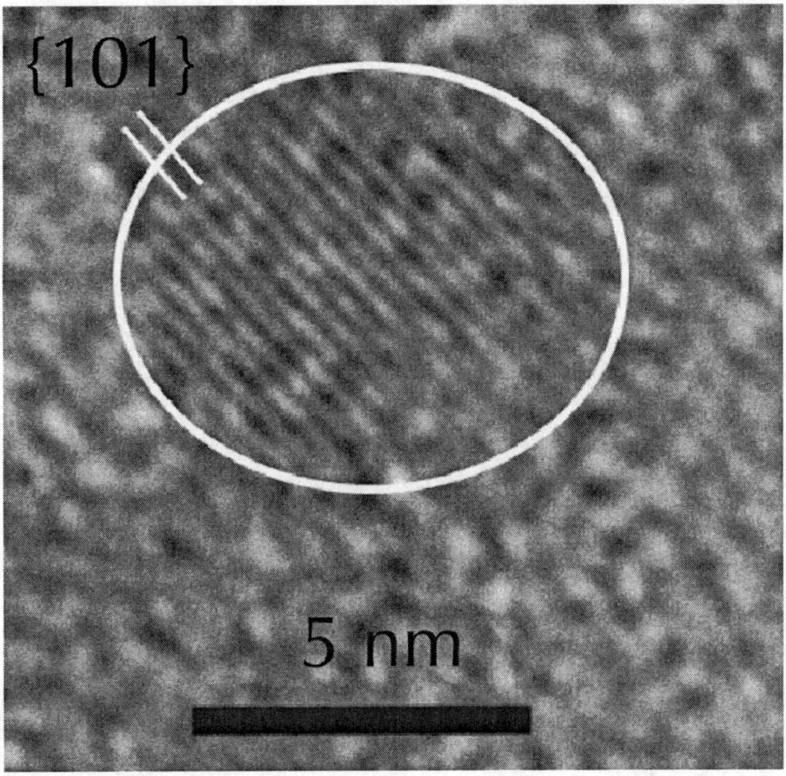

FIGURE 2 The TEM of a 1:75 molar ratio TiO_2 xerogel nanocrystal. Average diameter is 5.0 nm.

8.4 DISCUSSION

To ascertain what the dimensions of the NP aggregate constituents are within our suspensions prior to deposition, we performed dynamic light scattering using a Zetasizer Nano ZS (Malvern) fitted with an aqueous dip cell. The corresponding measurement of the diameter of the NPs, synthesized from the 1:75 molar ratios, as found in the colloidal suspension, is provided in Figure 3. To prepare the suspensions for this measurement, they were stabilized in the dark for over 24 hrs after preparation.

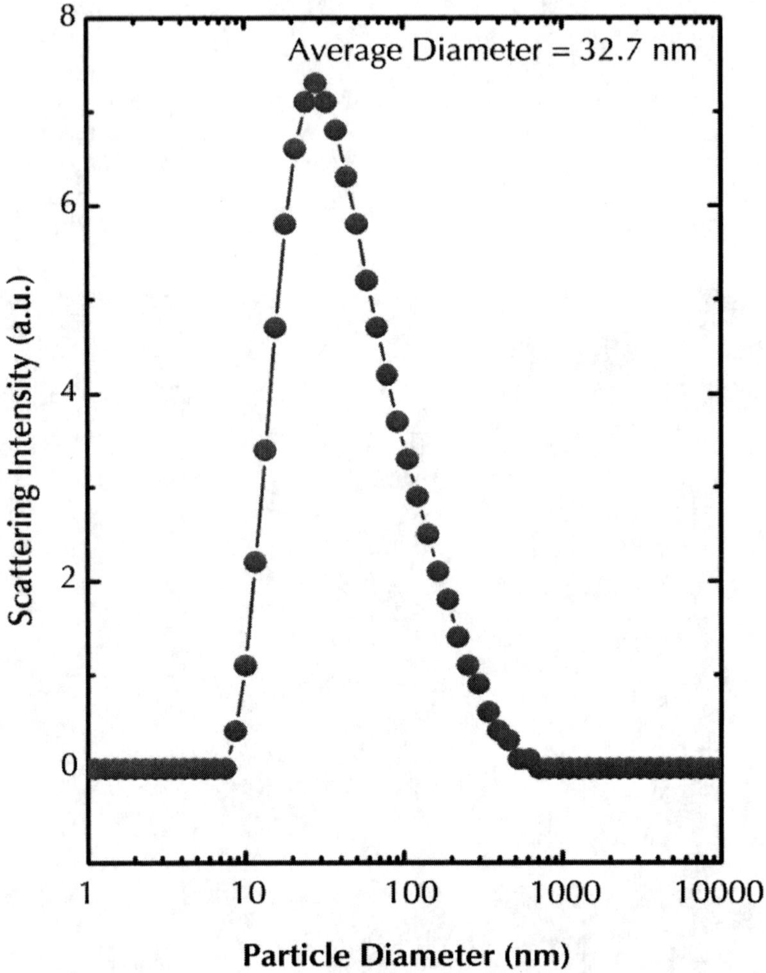

FIGURE 3 Dynamic light scattering to determine particle size of 1:75 molar suspensions. Average diameter of the suspended particles is ~33 nm.

The average diameter of the nanomaterials in suspension was ~32.7 nm, which comprises agglomerate whose diameter is approximately six NPs across. This affirms that these new TiO_2 nanomaterial are smaller both in the individual synthesized NP and in the individual constituent in suspension for EPD. We anticipated that these smaller constituents would yield films with greater uniformity and lower porosity. This produced an opaque white cast on the cathode, which appeared to be continuous to the naked eye. No deposition was visible on the anode. To confirm the composition of the material deposited on the electrode, we performed energy dispersive X-ray spectroscopy (EDS) on the electrodes, as seen in Figure 4. We attribute the titanium and oxygen signatures to the NPs and the iron, chromium, and nickel signals to the underlying substrate.

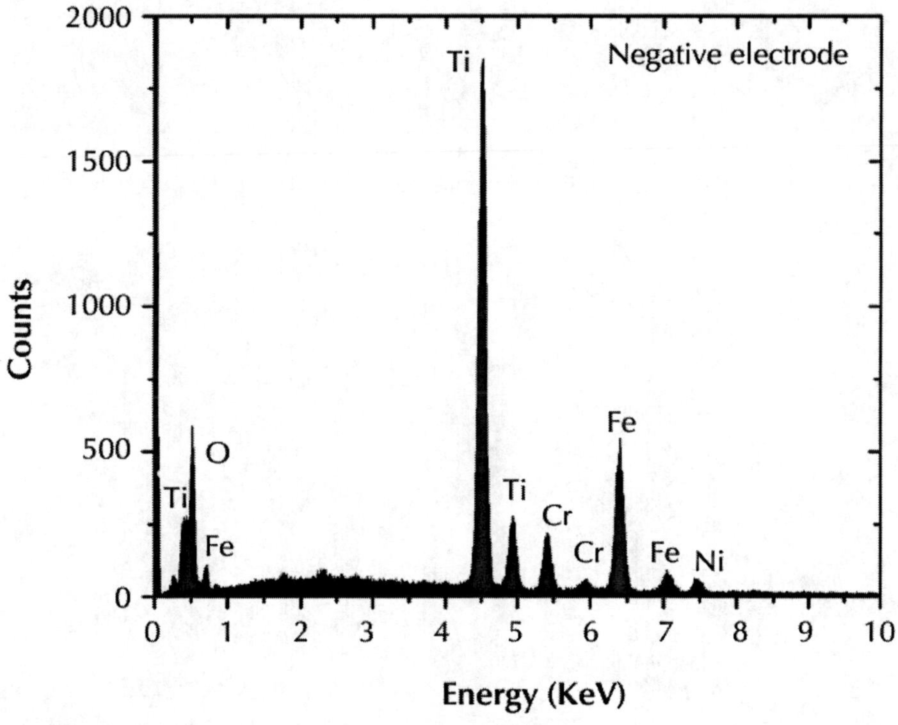

FIGURE 4 The EDS plot of the film on negative electrode. Ti and O are attributed to the NPs, and Fe, Cr, and Ni are attributed to the steel substrate.

To examine the integrity of the films, we took several SEM images at different magnifications. Figures 5 and 6 revealed microscopic cracks that formed after the substrate was removed from the NP suspension and after the film subsequently dried. Interestingly, the film was made up from an arrangement of 10–200 μm wide flakes. Inside of each of the flakes, the NPs appear to be relatively tightly-packed together (Figure 6). Within each flake, extremely small, densely packed agglomerates (approximately 30 nm in diameter) of the NPs can be observed. This yielded a formation of densely packed, continuous films of NPs; although they exist in small "flake-like films".

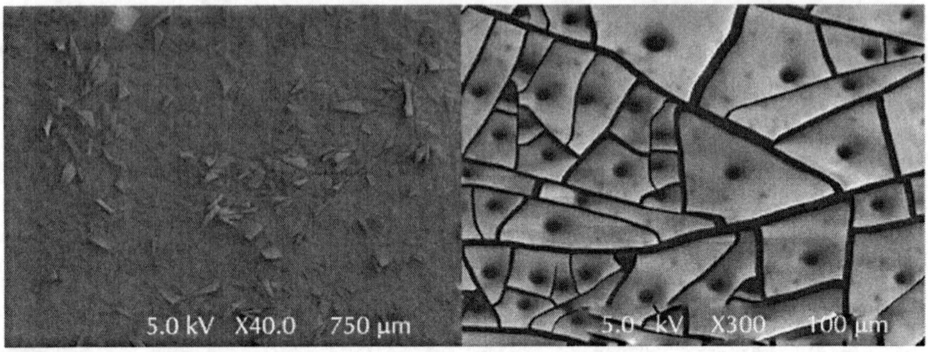

FIGURE 5 The (Left) SEM image of a negative electrode showing a film of TiO_2 NPs. (Right) SEM image of the flake-like TiO_2 film exhibiting cracks upon drying.

FIGURE 6 A high magnification SEM image of the film shows densely-packed NPs and small aggregates that were ~30 nm in diameter.

To suppress the cracking of the thick film, we explored using a soft polymeric surface (polystyrene) to mediate strain in the film induced by the evaporation of water during the drying steps. Although this has not yet resolved the film cracking issue, we anticipate that with the correct selection of polymer [perhaps PMMA (poly methyl methacrylate) or PLGA (*poly lactic-co-glycolic acid)] and* the solution-phase substitution a low surface tension polar solvent for water (which will help suppress the cracking), that fully continuous uncracked films will be achieved. We also are exploring whether the introduction of organic surfactants on the NPs surfaces and into the suspension could help eliminate strain-induced cracking of the films. Such waxy surfactant, which can be removed during a high temperature sintering step, can significantly facilitate the tight packing of NPs into films and can mediate the formation of freestanding NP films with high inter NP adhesion [8-9].

8.5 CONCLUSION

We successfully produced films of tightly-packed TiO_2 NPs, assembled by EPD. Although we observed macroscopic cracks within the film upon drying in air, the ~100 μm sized flakes were comprised of densely-packed NPs that suggest larger macroscopic films may be realized with further optimization of the EPD process. We anticipate that further optimization of the EPD suspension parameters with organic surfactants (introduced onto the NPs and into the suspensions) and the introduction of alternating current dielectrophoresis in partnership with our dc EPD system will yield the desired macroscopic, smooth, continuous ceramic NP films.

KEYWORDS

- **Ceramics**
- **Electrophoresis**
- **Electrophoretic deposition**
- **Nanocrystals**
- **Titania**

ACKNOWLEDGMENT

The authors thank to the United States Office of Naval Research, under Award N000140910523.

REFERENCES

1. Lee, D., Rubner, M. F., and Cohen, R. E., All Nanoparticle Thin Film Coatings. *Nano Letters*, **6**(10), 2305–2312 (2006).
2. Shchukin, D. G., Zheludkevich, M., Yasakau, K., Lamaka, S., Ferreira, M. G. S., and Mohwald, H., Layer-by-layer assembled nanocontainers for self-healing corrosion protection. *Advanced Materials*, **18**(13), 1672–1678 (2006).
3. Maenosono, S., Okubo, T., and Yamaguchi, Y., Overview of nanoparticle array formation by wet coating. *Journal of Nanoparticle Research*, **5**(1–2), 5–15 (2003).
4. Tian, Y. C. and Fendler, J. H., Langmuir-Blodgett film formation from fluorescence-activated, surfactant-capped, size-selected CdS nanoparticles spread on water surfaces. *Chemistry of Materials*, **8**(4), 969–974 (1996).
5. Islam, M. A. and Herman, I. P., Electrodeposition of patterned CdSe nanocrystal films using thermally charged nanocrystals. *Applied Physics Letters*, **80**(20), 3823–3825 (2002).

6. Colomer, M. T., Guzman, J., and Moreno, R., Determination of peptization time of particulate sols using optical techniques: Titania as a case study. *Chemistry of Materials*, **20**(12), 4161–4165 (2008).
7. Fazio, S., Guzman, J., Colomer, M., Salomoni, A., and Moreno, R., Colloidal stability of nano-sized titania aqueous suspensions. *Journal of the European Ceramic Society*, **28**(11), 2171–2176 (2008).
8. Hasan, S. A., Kavich, D. W., and Dickerson, J. H., Sacrificial layer electrophoretic deposition of free-standing multilayered nanoparticle films. *Chem. Commun.,* **25**, 3723–3725 (2009).
9. Lee, D., Jia, S., Banerjee, S., Bevk, J., Herman, I. P., and Kysar, J. W., Viscoplastic and Granular Behavior in Films of Colloidal Nanocrystals. *Phys. Rev. Lett.*, **98**(2), 026103 (2007).

9 Transforming Magnetic Photocatalyst to Magnetic Dye-adsorbent Catalyst

Satyajit Shukla

CONTENTS

9.1 Introduction 147
9.2 Experimental.......... 149
 9.2.1 Processing Magnetic Ceramic Particles 149
 9.2.2 Processing Magnetic Photocatalyst.......... 150
 9.2.3 Hydrothermal Treatment of Magnetic Photocatalyst 150
 9.2.4 Characterization.......... 151
 9.2.5 Dye-adsorption and Photocatalytic Activity Measurements 151
9.3 Discussion and Results 152
9.4 Conclusion.......... 164
Keywords 164
Acknowledgment.......... 165
References.......... 165

9.1 INTRODUCTION

Organic synthetic dyes find applications in various fields including textile, leather tanning, paper production, food technology, agricultural research, light-harvesting arrays, photo-electrochemical cells, and hair coloring. Due to the large-scale production, extensive use, and subsequent discharge of colored waste waters containing the toxic and non-biodegradable pollutants such as organic synthetic dyes, the latter are considered to be environmentally unfriendly and health hazardous. Moreover, they affect the sunlight penetration and the oxygen solubility in the water bodies, which in turn affect the underwater photosynthetic activity and life sustainability. In addition to this, due to their strong color even at lower concentrations, the organic synthetic dyes generate serious aesthetic issues in the waste water disposal [1-5].

As a consequence, powerful oxidation/reduction methods are required to ensure the complete decolorization and degradation of the organic synthetic dyes and their metabolites present in the waste water effluents. Over the last two decades, photocatalysis has been the area of rapidly growing interest for the removal organic synthetic dyes from the industrial effluents, which involves the use of semiconductor particles as photocatalyst for the initiation of the redox chemical reactions on their surfaces [6-9]. When the semiconductor oxide particle is illuminated with the radiation having energy comparable to its band gap energy, it generates highly active oxidizing/reducing sites, which can potentially oxidize/reduce large number of organic wastes. Metal-oxide and metal-sulfide semiconductors, such as titania (TiO_2) [6-9], zinc oxide (ZnO) [10], tin oxide (SnO_2) [11], zinc sulfide (ZnS) [12], and cadmium sulfide (CdS) [13] have been successfully applied as photocatalyst for the removal of highly toxic and non-biodegradable pollutants commonly present in air and waste water. Among them, TiO_2 is believed to be the most promising one since it is cheaper, environmentally friendly, non-toxic, highly photocatalytically active and stable to chemical and photo-corrosion. However, its effective application as a photocatalyst is hindered due to some of its major limitations. First, TiO_2 nanocrystallites trend to aggregate (or agglomerate) into large-sized nanoparticles, which affect its performance as a photocatalyst due to the decreased specific surface area. Secondly, it has lower absorption in the visible region, which makes it less effective in using the readily available solar energy. Third, the separation of photocatalyst from the treated effluent, *via* traditional sedimentation and coagulation approaches, has been difficult and time consuming.

The technologies such as adsorption on inorganic or organic matrices and microbiological or enzymatic decomposition have also been developed for the removal of organic synthetic dyes from the waste water to decrease their impact on the environment [5, 14-16]. However, the treatment of waste water containing organic synthetic dyes using these techniques is very costly and has lower efficiency in the color removal and mineralization. The adsorption method results in the generation of large amount of sludge, which causes further difficulties in the recovery of the photocatalyst for recycling the product as a catalyst. Therefore, further development of these techniques for the effective waste water treatment is essential.

In the literature, to overcome the major drawbacks associated with the photocatalytic degradation and adsorption mechanisms, the magnetic photocatalyst has been developed [17-26], which consists of a "core-shell" composite particle with a ceramic magnetic particle as a core and TiO_2-based photocatalyst particles as shell. Such magnetic nanocomposite, which possesses both the photocatalytic and magnetic properties, can be effectively separated from the treated solution using an external magnetic field. However, even the magnetic photocatalyst has been associated with several drawbacks. First, they show limited photocatalytic activity due to the presence of a core magnetic ceramic particle, which reduces the volume fraction of the photocatalyst available for the photo degradation. Second, the total time of dye-decomposition using the magnetic photocatalyst is substantially higher (few hours). Third, the dye-removal using the magnetic photocatalyst is based predominantly on the photocatalytic degradation mechanism. Forth, being an energy-dependent mechanism (that is, requiring an exposure to the ultraviolet (UV), visible, or solar radiation), the photocatalytic degradation is relatively an expensive process. Fifth, the dye-removal via other mechanism(s) such as the surface-adsorption, which is not an energy dependent process (that is, it can be carried out in the dark) has never been utilized for the magnetic photocatalyst. This has been mainly due to the non-suitability of the magnetic photocatalyst for the surface adsorption mechanism as a result of its lower specific surface area. Sixth, the techniques to enhance the specific surface area of the magnetic photocatalyst are not yet reported.

From these points of view, we demonstrate here the conversion of a magnetic photocatalyst having lower specific surface area to a "magnetic dye-adsorbent catalyst", having higher specific surface area, consisting a composite structure with the core of a magnetic ceramic particle and the shell of nanotubes of a dye-adsorbent material [27-29]. It is demonstrated here that, such conversion is accompanied by a concurrent change in the organic dye-removal mechanism from the photocatalytic degradation under the radiation exposure to the surface adsorption under the dark condition, which offers several advantages over the former.

The magnetic dye-adsorbent catalyst has been synthesized *via* hydrothermal processing of the magnetic photocatalyst followed by typical washing and thermal treatments. The magnetic dye-adsorbent catalyst consists of a core-shell nanocomposite with the core of a magnetic ceramic particle (such as mixed cobalt ferrite and hematite, and pure cobalt ferrite) and the shell of nanotubes of dye-adsorbing material (such as hydrogen titanate). The samples have been characterized for determining the phase structure, morphology, size, and magnetic properties using the X-ray and selected-area electron diffraction, transmission electron microscope, and vibrating sample magnetometer (VSM). The photocatalytic activity under the ultraviolet radiation exposure and the dye-adsorption under the dark condition have been measured using the methylene blue (MB) as a model catalytic dye-agent. It has been demonstrated that, the transformation of the magnetic photocatalyst to the magnetic dye-adsorbent catalyst is accompanied by a change in the mechanism of dye-removal from an aqueous solution from the photocatalytic degradation to the surface adsorption under the dark condition. It has been also shown that, due to its magnetic nature, the magnetic dye-adsorbent catalyst can be separated from the treated solution using an external magnetic field and the previously adsorbed dye can be removed from the surface of nanotubes *via* typical surface cleaning treatment, which make the recycling of the magnetic dye-adsorbent catalyst possible.

9.2 EXPERIMENTAL

9.2.1 Processing Magnetic Ceramic Particles

A mixed cobalt ferrite ($CoFe_2O_4$) and hematite (Fe_2O_3) (CFH) and pure-$CoFe_2O_4$ magnetic ceramic powders are first processed via polymerized complex technique [28, 30]. The 36.94 g of citric acid was dissolved in 40 ml of ethylene glycol as complexing agents and stirred to get a clear solution. 17 g of cobalt(II) nitrate ($Co(NO_3)_2.6H_2O$, 98+ %) and 47.35 g of iron(III) nitrate ($Fe(NO_3)_3.9H_2O$, 99.99+ %) (Sigma-Aldrich, India) were added and the solution was stirred for 1 hr followed by heating at 80°C for 4 hr. The yellowish gel obtained was charred at 300°C for 1 hr in a vacuum furnace. A black colored solid precursor was obtained which, after grinding, was heated at 600°C for 6 hr to obtain a mixed $CoFe_2O_4$-Fe_2O_3 magnetic powder. Further calcinations at 900°C for 4 hr resulted in the formation of pure $CoFe_2O_4$ magnetic powder. The selection of mixed $CoFe_2O_4$-Fe_2O_3 or pure-$CoFe_2O_4$ powder as a core magnetic material for the photocatalytic and dye-adsorption measurements was as per convenience.

9.2.2 Processing Magnetic Photocatalyst

An insulating layer of silica (SiO_2) was deposited on the surface of magnetic ceramic particles using the Stober process [28, 31]. To 2 g suspension of magnetic powder dispersed in 250 ml of 2-propanol (S.D. Fine-Chem Ltd., India), 1 ml of ammonium hydroxide (NH_4OH, 25 wt%, Qualigens Fine Chemicals, India) solution was slowly added. This was followed by the drop wise addition of 7.3 ml of tetraethylorthosilicate (TEOS, 98%, Sigma-Aldrich, India) and the resulting suspension was allowed to settle after stirring for 3 hr. The clear top solution was decanted and the powder was washed with 100 ml of 2-propanol and distilled water followed by drying in an oven at 60°C overnight.

In order to deposit nanocrystalline TiO_2, 2 g of SiO_2-coated $CoFe_2O_4$-Fe_2O_3 magnetic powder was suspended in a clear solution of prehydrolized titanium(IV) isopropoxide ($Ti(OC_3H_7)_4$, 98%, Sigma-Aldrich, India) precursor (4.73 g) dissolved in 125 ml of 2-propanol (Note: The prehydrolized precursor was obtained due to the reaction of pure-$Ti(OC_3H_7)_4$ with the atmospheric moisture over a prolonged period of time). To this suspension, a clear solution consisting 1.5 ml of distilled water ($R = 5$, defined as the ratio of molar concentration of water to that of the precursor), dissolved in 125 ml of 2-propanol, was added drop wise. The suspension was stirred for 10 hr, and after settling, the top solution was decanted. The powder was washed with 100 ml of 2-propanol and then dried in an oven at 80°C overnight followed by calcination at 600°C for 2 hr. The above procedure was utilized for $R = 10$ and 20 using the pure-alkoxide precursor, with a reduced concentration (0.5 g), and pure-$CoFe_2O_4$ as a core magnetic ceramic particle. In this case, the coating process was repeated twice to obtain relatively thicker TiO_2-coating.

9.2.3 Hydrothermal Treatment of Magnetic Photocatalyst

The magnetic photocatalyst, as processed above, was subjected to the hydrothermal treatment under highly alkaline condition followed by typical washing and thermal treatments [27, 28] 0.5 g of conventional magnetic photocatalyst was suspended in a highly alkaline aqueous solution containing 10 M sodium hydroxide (NaOH, 97%, S.D. Fine-Chem Ltd., India) filled up to 84 vol % of Teflon-beaker placed in a 200 ml stainless-steel (SS 316) vessel. The process was carried out in an autoclave (Amar Equipment Pvt. Ltd., Mumbai, India) at 120°C for 30 hr under an autogenous pressure. The hydrothermal product was washed once using 100 ml of 1 M hydrochloric acid (HCl, Qualigens Fine Chemicals, India) solution for 1 hr followed by washing multiple times with distilled water till the final pH of the filtrate was in between ~5–7. The washed powder was dried in an oven at 110°C overnight (dried-sample) and then calcined at 400°C for 1 hr (calcined-sample).

9.2.4 Characterization

The morphology of different samples at the nanoscale was examined using the transmission electron microscope (TEM, Tecnai G2, FEI, The Netherlands) operated at 300 kV. The selected area electron diffraction (SAED) patterns were obtained to confirm the crystallinity and the structure of different samples. The crystalline phases present were determined using the X-ray diffraction (XRD, PW1710, Phillips, and the Netherlands). The broad scan analysis was typically conducted within the 2-θ range of 10–80° using the Cu $K\alpha$ (λ = 1.542 Å) X-radiation. The magnetic properties of different samples were measured using a VSM attached to a Physical Property Measurement System (PPMS). The pristine samples were subjected to different magnetic field strengths (H) and the induced magnetization (M) was measured at 270K. The external magnetic field was reversed on saturation and the hysteresis loop was traced [28].

9.2.5 Dye-Adsorption and Photocatalytic Activity Measurements

The dye-adsorption measurements in the dark were conducted using the MB (>96%, S.D. Fine-Chem Ltd., India) as a model catalytic dye-agent. A 75 ml of aqueous suspension was prepared by dissolving 7.5 μMl^{-1} of MB dye and then dispersing 1.0 gl^{-1} of the catalyst powder in pure distilled water. The suspension was stirred in the dark and 3 ml sample suspension was separated after each 30 min time interval for total 180 min. The catalyst powder was separated using a centrifuge (R23, Remi Instruments India Ltd.) and the solution was used to obtain the absorption spectra using the UV-visible absorption spectrophotometer (UV-2401 PC, Shimadzu, Japan). The normalized concentration of surface-adsorbed MB dye was calculated using the equation of the form,

$$\%MB_{adsorbed} = \left(\frac{C_0 - C_t}{C_0} \right)_{MB} \times 100 \quad (1)$$

which is equivalent of the form,

$$\%MB_{adsorbed} = \left(\frac{A_0 - A_t}{A_0} \right)_{MB} \times 100 \quad (2)$$

where, C_0 and C_t correspond to the MB dye concentration at the start and after string time 't', under the dark-condition, with the corresponding absorbance of A_0 and A_t. In few experiments, the powder was used for the successive cycles of dye-adsorption, under the dark condition, to demonstrate its reusability as a catalyst. The dye-adsorption experiments were typically conducted under two different solution-pH (6.4 (neutral) and 10).

During the measurement of photocatalytic activity, the aqueous suspension of MB dye and the catalyst powder was stirred in the dark for 1 hr to stabilize the surface-adsorption of the former on the surface of latter. The suspension is then exposed to the UV-radiation, having the wavelength within the range of 200–400 nm peaking at 360 nm, in a Rayonet photoreactor (The Netherlands) and 3 ml sample suspension was separated after each 10 min time interval for total 60 min. The powder was separated using a centrifuge and the solution was used to obtain the absorption spectra. The normalized residual MB dye concentration was calculated using the equation of the form,

$$\%MB_{adsorbed} = \left(\frac{C_t}{C_0}\right)_{MB} \times 100 \tag{3}$$

which is equivalent of the form,

$$\%MB_{adsorbed} = \left(\frac{A_t}{A_0}\right)_{MB} \times 100 \tag{4}$$

where, C_0 and C_t correspond to the MB dye concentration just at the beginning of the UV-radiation exposure (that is, after stirring the suspension under the dark-condition for 1 hr) and after the UV-radiation exposure time of '*t*' with the corresponding absorbance of A_0 and A_t.

9.3 DISCUSSION AND RESULTS

The XRD broad-scan spectra, as obtained using the mixed $CoFe_2O_4$-Fe_2O_3 and pure-$CoFe_2O_4$ magnetic ceramic powders, are presented in Figure 1(a) and (b). The major peaks corresponding to $CoFe_2O_4$ and Fe_2O_3 are identified after comparison with the JCPDS card numbers 22–1086 and 33–663. The formation of magnetic ceramic powders consisting mixed $CoFe_2O_4$-Fe_2O_3 and pure-$CoFe_2O_4$ is, thus, confirmed *via* broad-scan XRD analyses.

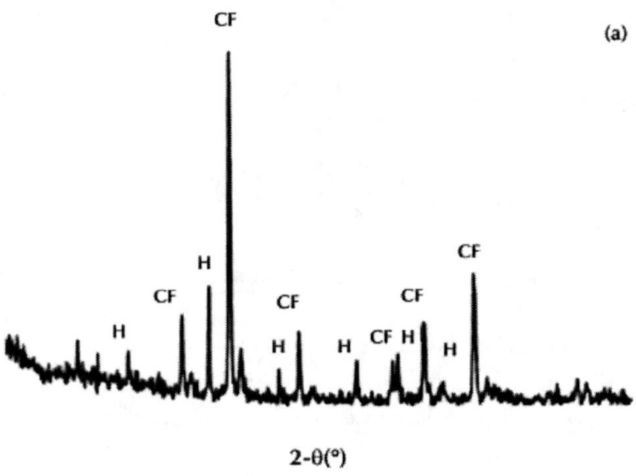

FIGURE 1 *(Continued)*

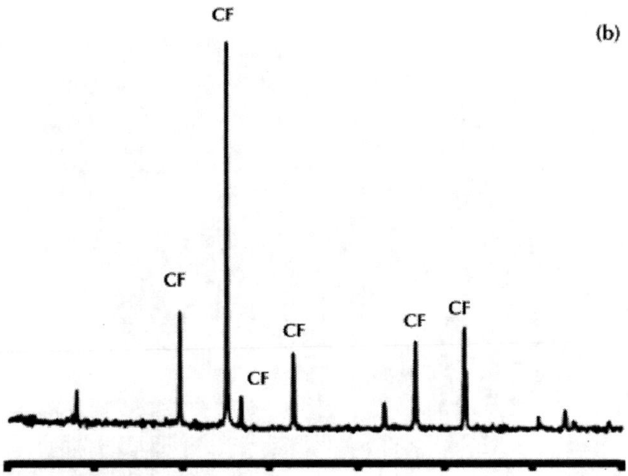

FIGURE 1 XRD patterns as obtained for the mixed $CoFe_2O_4$-Fe_2O_3 (CFH) (a) and pure-$CoFe_2O_4$ (b) magnetic ceramic particles.

The TEM image of a magnetic photocatalyst particle, exhibiting a "core-shell" structure, with a core of magnetic ceramic particle and the shell of a sol-gel derived nanocrystalline coating of anatase-TiO_2 particles, is shown in Figure 2(a). The SAED pattern as obtained from the core is shown as an inset in Figure 2(a), which confirms the crystalline nature of the mixed $CoFe_2O_4$-Fe_2O_3 magnetic ceramic particle. The presence of anatase-TiO_2 in the shell has also been confirmed *via* XRD analysis as reported elsewhere [27, 28]. As observed in Figure 2(b), the hydrothermal treatment results in the morphological transformation within the shell involving the conversion of nanocrystalline anatase-TiO_2 particles into the nanotubes of hydrogen titanate ($H_2Ti_3O_7$) as identified via high-magnification images of the shell presented in Figure 2(c) and 2(d), and the corresponding SAED pattern shown as an inset in Figure 2(c). The "core-shell" magnetic nanocomposite, Figure 2(b), with the core of a magnetic ceramic particle and the shell of nanotubes of $H_2Ti_3O_7$ is termed as a "magnetic dye-adsorbent catalyst", which possesses the magnetic, dye-adsorption (in the dark), and catalytic properties.

The comparison of the morphologies of magnetic photocatalyst and the magnetic dye-adsorbent catalyst is schematically shown in Figure 3. The formation mechanism of magnetic dye-adsorbent catalyst *via* hydrothermal treatment of the magnetic photocatalyst, under highly alkaline condition, can be explained using the model originally proposed for the free-standing powder [32-34], which is applied here for the similar conversion in the form of coating. When the anatase-TiO_2 particles, present in the coating form on the surface of magnetic ceramic particle, Figure 3(a), are subjected to the hydrothermal treatment under highly alkaline condition, an exfoliation of single-layer nanosheets of sodium titanates ($Na_2Ti_3O_7$) results from the bulk anatase-TiO_2 structure, which continuously undergo the dissolution and crystallization processes [34].

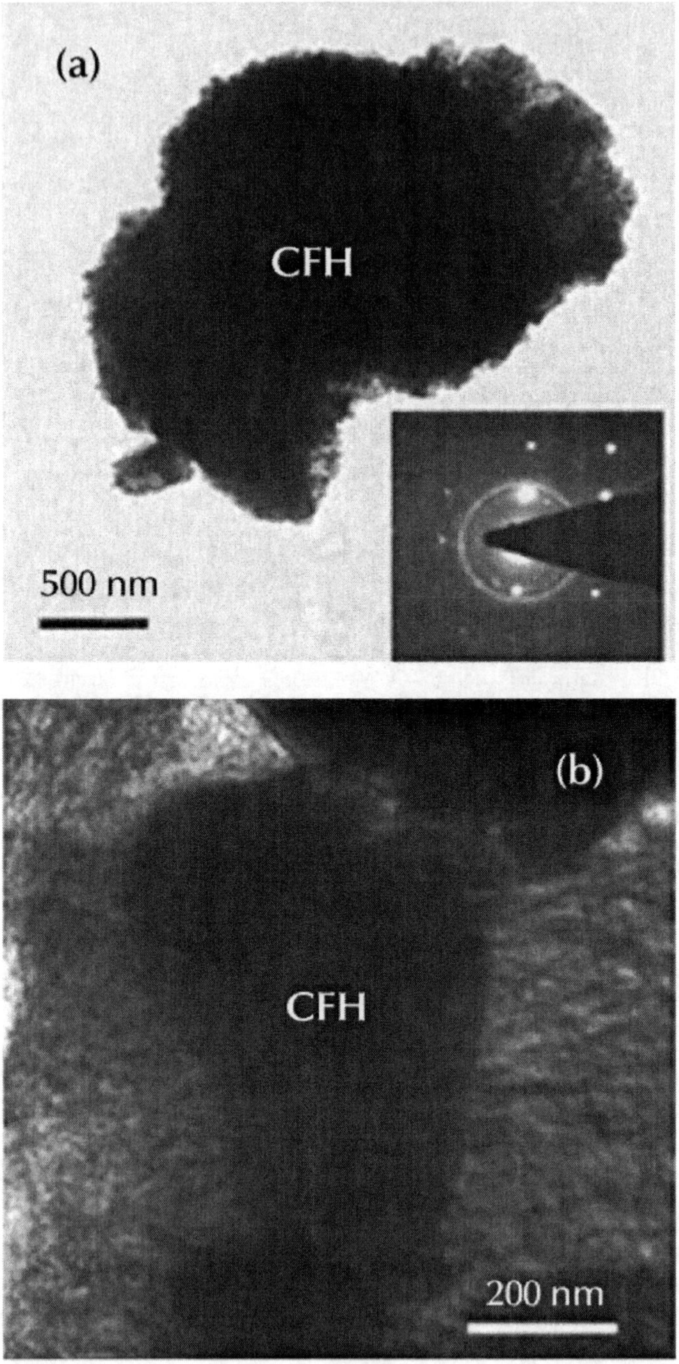

FIGURE 2 *(Continued)*

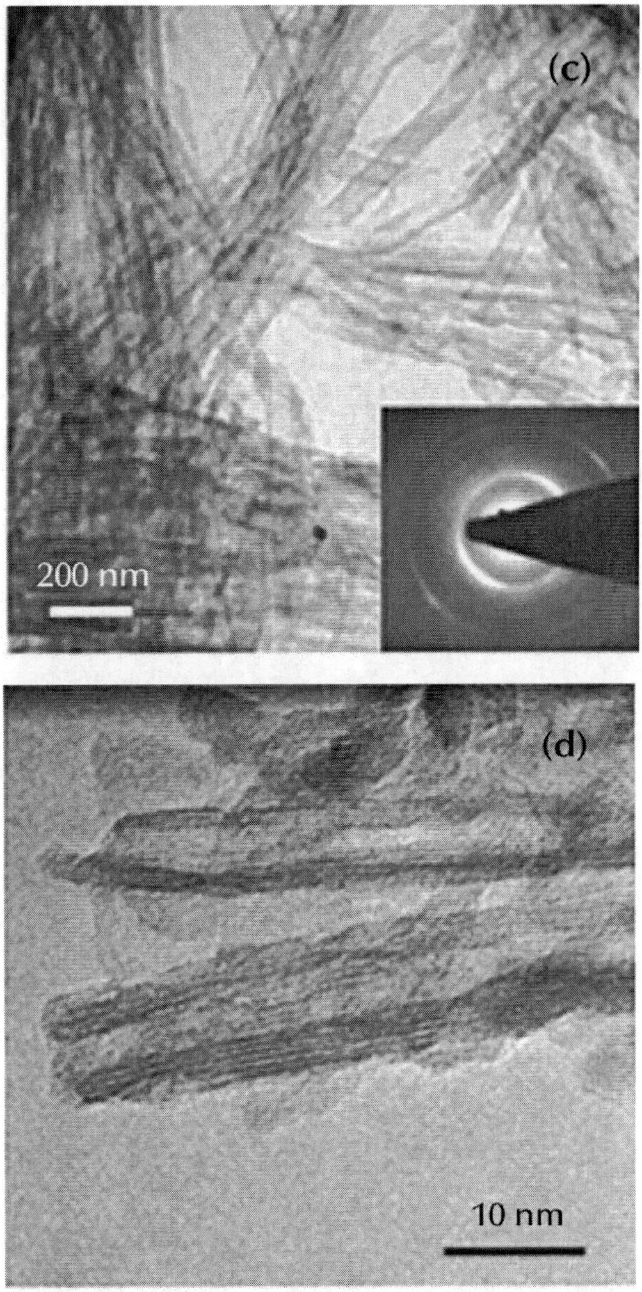

FIGURE 2 The TEM images of the magnetic photocatalyst (a) and the magnetic dye-adsorbent catalyst (b). The high-magnification images of the shell of magnetic dye-adsorbent catalyst are presented in (c) and (d). The SAED pattern corresponding to the core of magnetic photocatalyst is shown as an inset in (a); while, that corresponding to the shell of magnetic dye-adsorbent catalyst is shown as an inset in (c). The samples are processed with $R = 5$.

$$3TiO_2 + 2NaOH \rightarrow Na_2Ti_3O_7 + H_2O \tag{5}$$

$$Na_2Ti_3O_7 \longleftrightarrow 2Na^+ + Ti_3O_7^{2-} \tag{6}$$

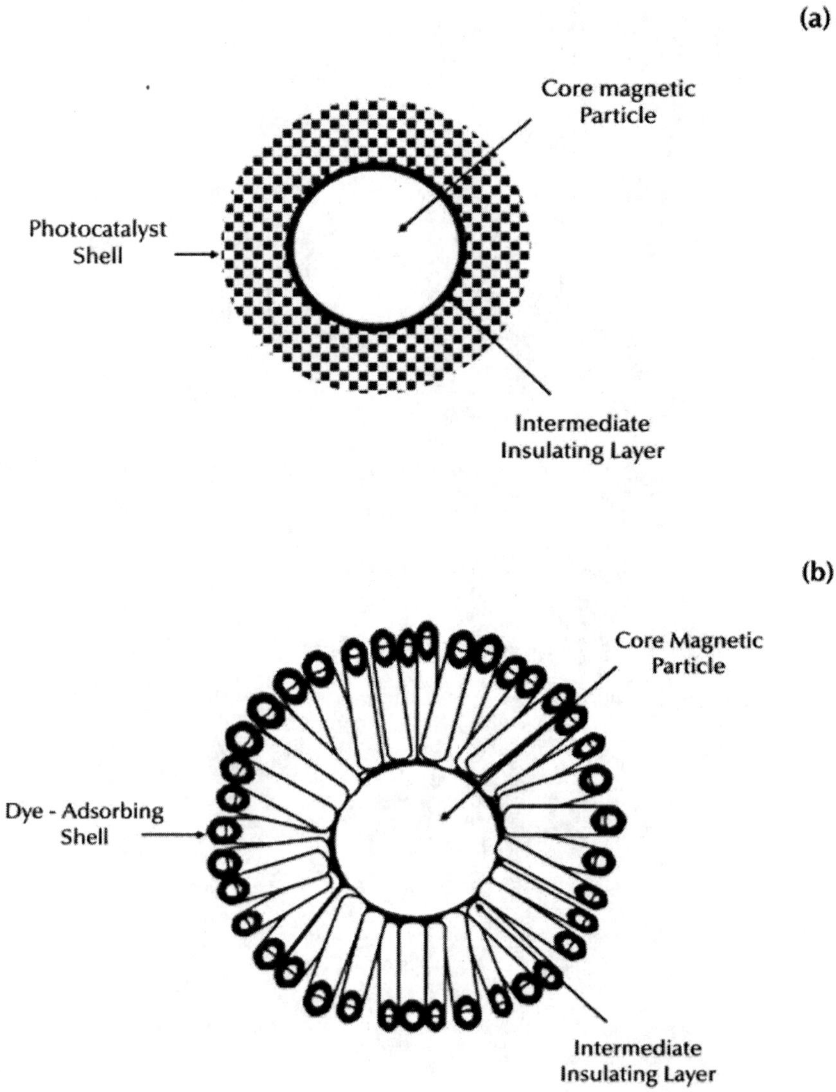

FIGURE 3 Schematic representation of the morphologies of magnetic photocatalyst (a) and magnetic dye-adsorbent catalyst (b).

Due to their higher surface area to volume ratio and the presence of dangling bonds along the two long-edges, the nanosheets of $Na_2Ti_3O_7$ have a strong drive to rollup. However, this rollup tendency is opposed by the repulsive force produced by the charge on the nanosheets created by the presence of Na^+-ions. These Na^+-ions are easily replaced *via* ion-exchange mechanism with H^+-ions in the subsequent washing steps in an acidic aqueous solution and pure-water, which reduces the repulsive force to rollup.

$$Na_2Ti_3O_7 + 2HCl \rightarrow H_2Ti_3O_7 + 2NaCl \tag{7}$$

$$Na_2Ti_3O_7 + 2H_2O \rightarrow H_2Ti_3O_7 + 2NaOH \tag{8}$$

This results in the formation of nanotubes of $H_2Ti_3O_7$ which are normally transformed to those of anatase-TiO_2 following the calcination treatment at higher temperature.

$$H_2Ti_3O_7 \xrightarrow{\Delta} 3TiO_2 + H_2O \tag{9}$$

The XRD pattern of the magnetic dye-adsorbent catalyst (calcined-sample), however, did not reveal the presence of anatase-TiO_2 on the surface [27, 28]. Since the $H_2Ti_3O_7$-to-anatase TiO_2 transformation has been observed earlier in the powder form [35], it appears that, such transformation is possibly retarded in the coating form due to the substrate effect.

The M-H graphs, as obtained for the magnetic photocatalyst and the magnetic dye-adsorbent catalyst (calcined-sample), are presented in Figure 4. The presence of a hysteresis loop is noted, which suggests the ferromagnetic nature of these composite particles. For the magnetic photocatalyst, the obtained values of saturation magnetization, remenance magnetization, and coercivity are 59 emu g^{-1}, 24 emu g^{-1}, 1410 Oe; while, those for the magnetic dye-adsorbent catalyst (calcined-sample) are 45 emu g^{-1}, 15 emu g^{-1}, 578 Oe. It is noted that, the magnetic dye-adsorbent catalyst (calcined-sample) shows reduced saturation magnetization, remenance magnetization, and coercivity relative to those observed for the magnetic photocatalyst. This has been attributed to the combined effect of decrease in the volume fraction and increase in the average particle size and crystallinity of the core magnetic ceramic particle following the hydrothermal, drying, and calcination treatments [28]. Nevertheless, the hydrothermal product (calcined-sample) also possesses the magnetic property and is suitable for its separation, after the photocatalytic and dye-adsorption experiments, using an external magnetic field.

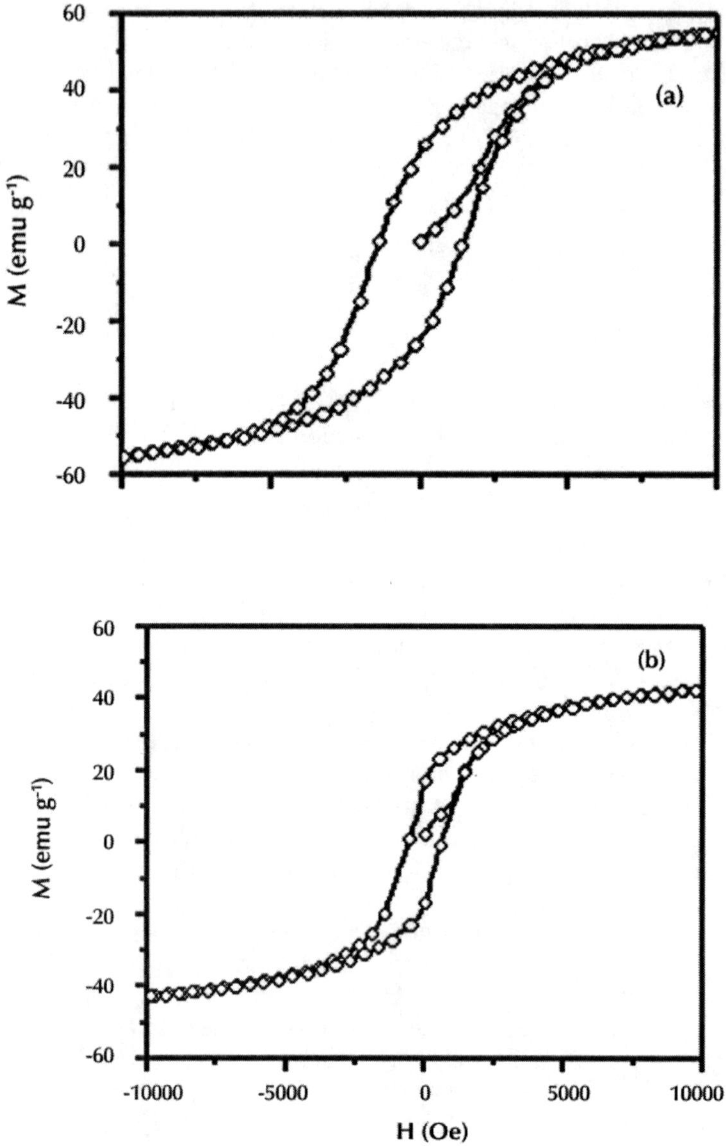

FIGURE 4 Variation in the induced magnetization (M) as a function of external magnetic field strength (H) as obtained for the magnetic photocatalyst (a) and magnetic dye-adsorbent catalyst (calcined-sample) (b). The samples are processed with $R = 5$. Copyright 2010 John Wiley and Sons.

The variation in the normalized residual MB dye concentration as a function of stirring time, under the UV-radiation exposure, as obtained for the pure and Pd-deposited [36] magnetic photocatalyst is shown in Figure 5(a). It has been demonstrated earlier that, the sol-gel derived pure nanocrystalline anatase-TiO_2 particles completely remove the MB dye via photocatalytic degradation mechanism under similar test-conditions within the UV-radiation exposure time of 1 hr [37]. It, hence, appears that the magnetic photocatalyst possesses very slow MB dye degradation kinetics. The comparison of the kinetics of MB dye removal via surface adsorption mechanism, under the dark-condition, using the magnetic photocatalyst and the magnetic dye-adsorbent catalyst is presented in Figure 5(b). Due to its higher specific surface area, the magnetic dye-adsorbent catalyst (dried-sample) exhibits > 99% of dye-adsorption, which is larger than that (~90%) of the calcined sample. This is attributed here to some loss in the specific surface area of the magnetic dye-adsorbent catalyst as a result of the calcination treatment. Nevertheless, the magnetic dye-adsorbent catalysts (dried and calcined samples) show significantly higher amount of dye-adsorption on the surface, under the dark condition, relative to that (~50–55%) shown by the magnetic photocatalyst. It is, thus, successfully demonstrated that under the dark condition, the magnetic dye-adsorbent catalyst removes an organic-dye from an aqueous solution predominantly *via* surface adsorption mechanism. On the other hand, the magnetic photocatalyst cannot completely remove the MB dye neither *via* photocatalytic degradation mechanism (under the UV-radiation exposure) nor *via* surface-adsorption mechanism (under the dark-condition). This is further supported by the comparison of the qualitative variation in the color of the MB dye solution as a function of stirring time, as observed for the magnetic photocatalyst and the magnetic dye-adsorbent catalyst (dried-sample), Figure 6(a). In Figure 6(b), the qualitative variation in the color of the $H_2Ti_3O_7$ nanotubes (without the core magnetic ceramic particle) is presented under different conditions. It is noted that, the initial white-color of the nanotubes changes to blue after the surface-adsorption of the MB dye under the dark condition. After a typical surface-cleaning treatment [27], the surface-adsorbed MB dye can be decomposed, which is suggested by the disappearance of the blue color. This produces surface cleaned $H_2Ti_3O_7$ nanotubes which can be recycled for the next cycle of dye-adsorption under the dark condition with the dye-adsorption capacity comparable with the original powder. (Note: Since the surface cleaning treatment is effective with pure-$H_2Ti_3O_7$ nanotubes, it would also be effective in the presence of the core magnetic ceramic particle). As a result, it seems that, the magnetic dye-adsorbent can be recycled and used as a catalyst for the dye-removal application.

The increased surface adsorption of the MB dye, under the dark condition, as observed earlier in Figure 5(b) for $R = 5$ following the hydrothermal treatment of the magnetic photocatalyst, is also shown by those processed with larger R-values (10 and 20), Figure 7(a) and 7(b). Interestingly, comparison of Figure 8 and Figure 7(b) shows that, at higher solution-pH in the basic region (pH = 10), both the magnetic photocatalyst as well as the magnetic dye-adsorbent catalyst exhibit high and comparable surface adsorption of the MB dye under the dark condition. However, when the same catalyst is used for the repetitive dye-adsorption cycles, at higher solution pH in the basic region (pH = 10), the magnetic photocatalyst rapidly loses its maximum dye-adsorption capacity, Figure 9(a), during each successive cycles. On the other hand, under similar test-conditions, the magnetic dye-adsorbent catalyst (calcined-sample) retains its maximum dye-adsorption capacity, Figure 9(b).

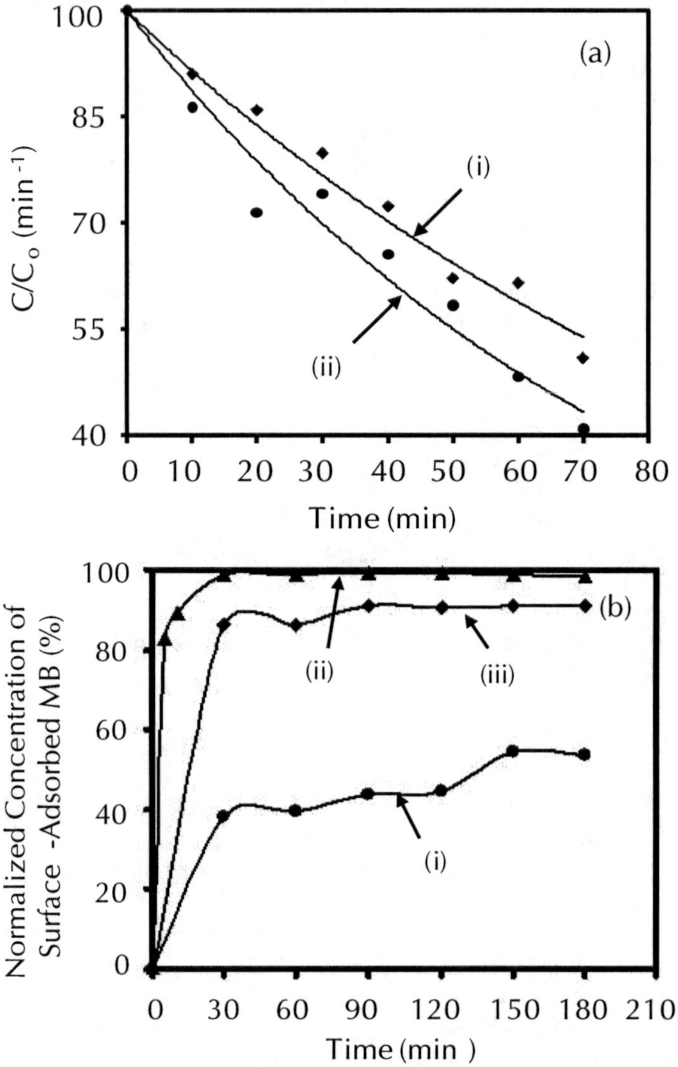

FIGURE 5 (a) Variation in the normalized residual MB dye concentration as a function of stirring time under the UV-radiation exposure as obtained for the pure (i) and palladium (Pd)-deposited (ii) magnetic photocatalyst. (b) Variation in the normalized concentration of surface-adsorbed MB as a function of stirring time under the dark-condition as obtained for the magnetic photocatalyst (i) and magnetic dye-adsorbent catalyst, dried (ii) and calcined (iii) samples. The samples are processed with $R = 5$. Copyright 2010 John Wiley and Sons (b).

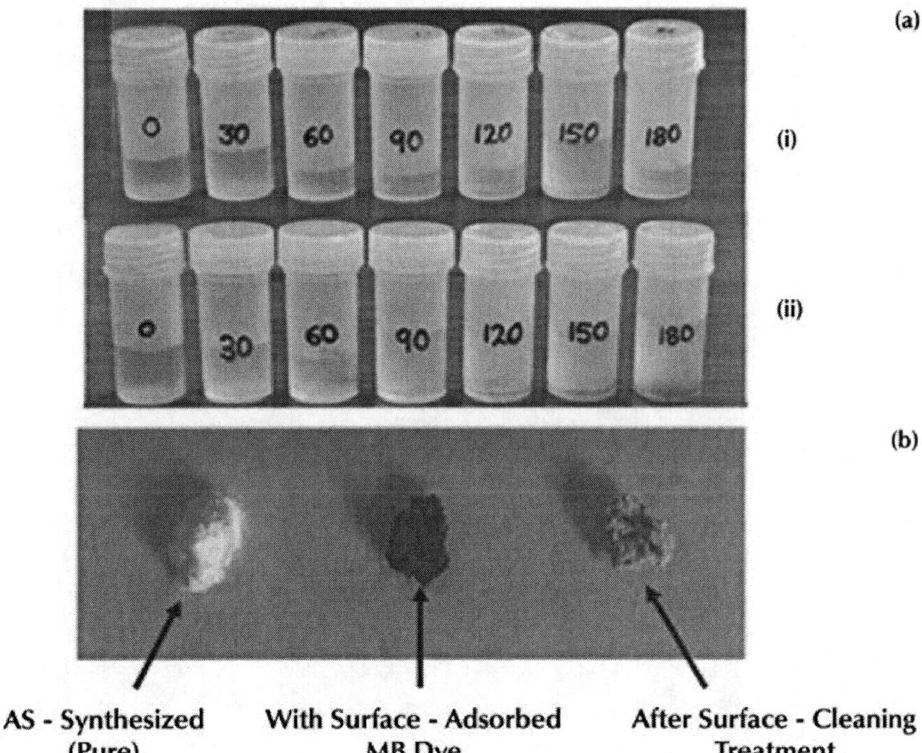

FIGURE 6 (a) Qualitative variation in the color of the MB dye solution as a function of stirring time under the dark-condition as obtained for the magnetic photocatalyst (i) and magnetic dye-adsorbent catalyst (dried-sample) (ii). The samples are processed with $R = 5$. (b) Qualitative variation in the color of the $H_2Ti_3O_7$ nanotubes powder (without the core magnetic ceramic particle) under different conditions.

As a result, overall it appears that, the transformation of magnetic photocatalyst to the magnetic dye-adsorbent catalyst is accompanied by a concurrent change in the dye-removal mechanism from the photocatalytic degradation under the UV-radiation exposure to the surface adsorption under the dark condition.

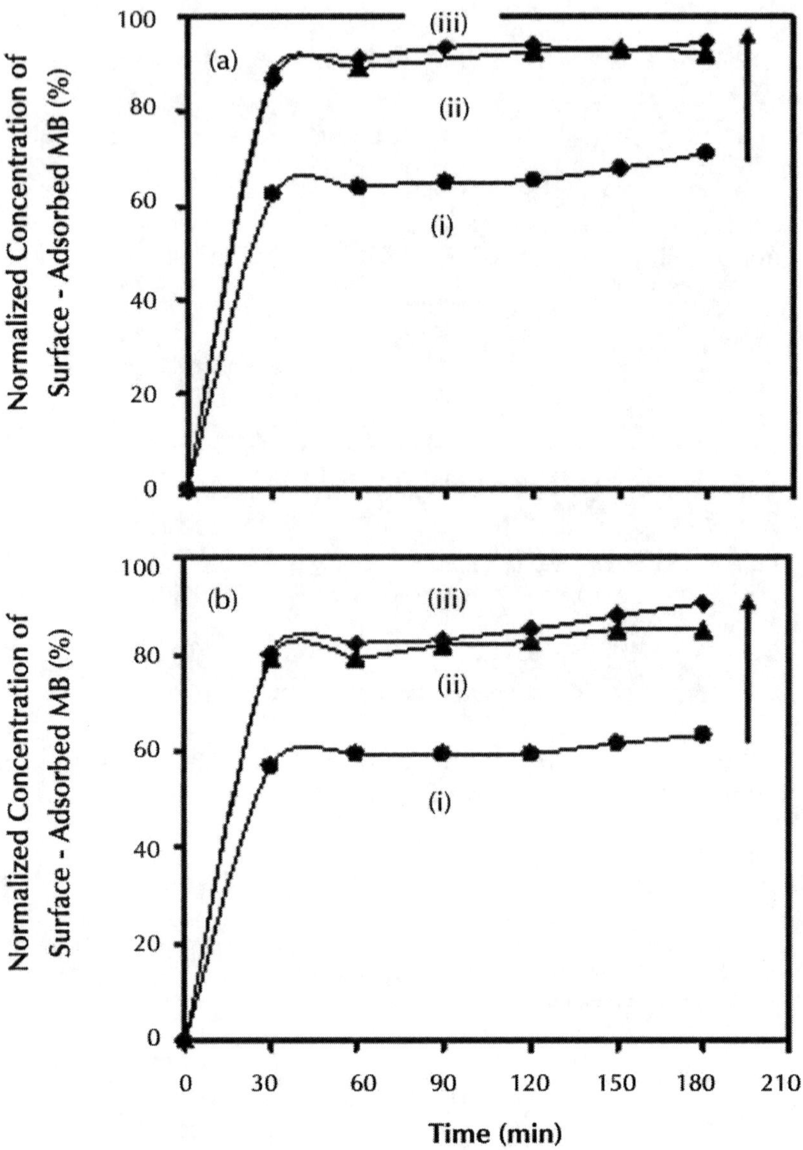

FIGURE 7 Variation in the normalized concentration of surface-adsorbed MB as a function of stirring time under the dark-condition as obtained for different samples: magnetic photocatalyst (i) and magnetic dye-adsorbent catalyst, dried (ii) and calcined (iii) samples. The samples are processed with R = 10 (a) and *R* = 20 (b), and the dye-adsorption measurements are conducted at the neutral solution-pH (6.4).

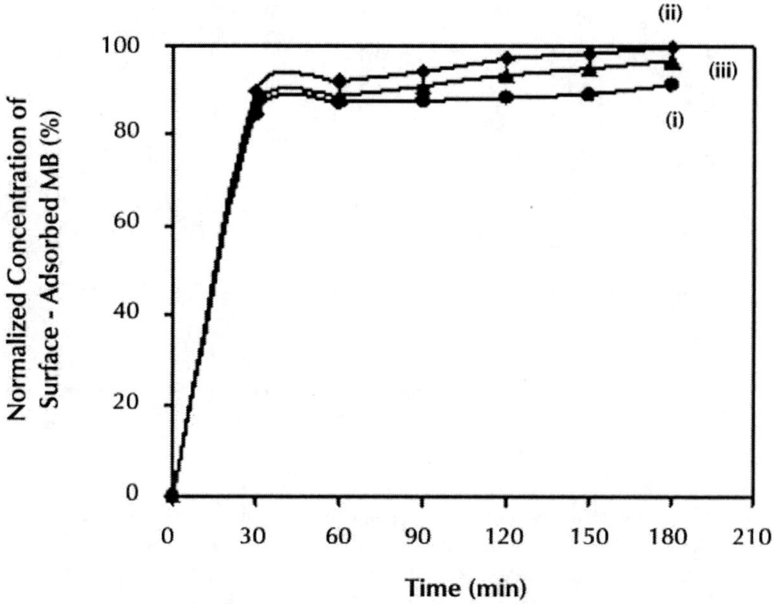

FIGURE 8 Variation in the normalized concentration of surface-adsorbed MB as a function of stirring time under the dark-condition as obtained for different samples: magnetic photocatalyst (i) and magnetic dye-adsorbent catalyst, dried (ii) and calcined (iii) samples. The samples are processed with $R = 20$ and the dye-adsorption measurements are conducted at the basic solution-pH (10).

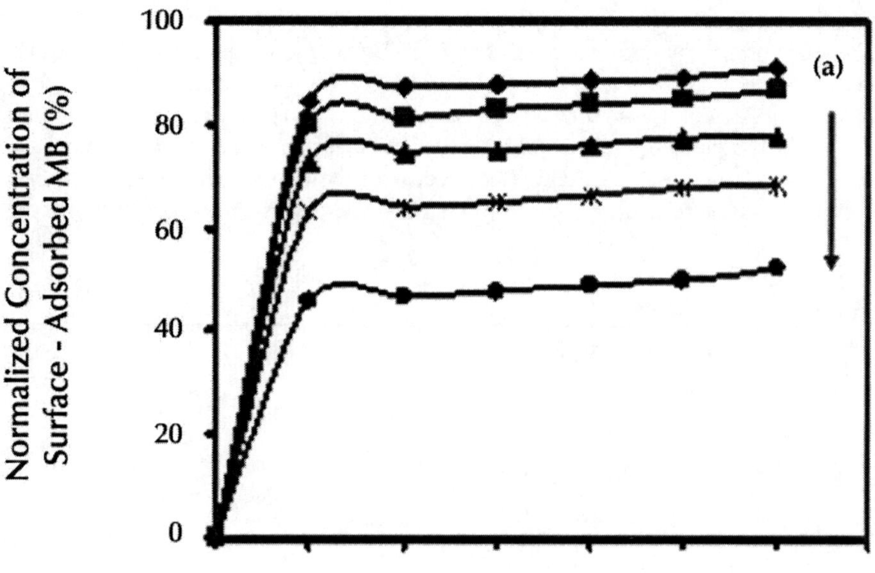

FIGURE 9 *(Continued)*

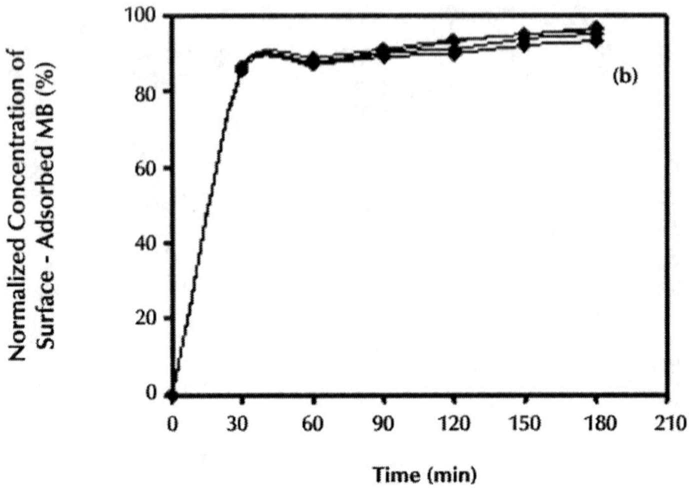

FIGURE 9 Variation in the normalized concentration of surface-adsorbed MB as a function of stirring time under the dark-condition, for the successive dye-adsorption cycles, as obtained for different samples – magnetic photocatalyst (a) and magnetic dye-adsorbent catalyst (calcined-sample) (b). The samples are processed with $R = 20$ and the dye-adsorption measurements are conducted at the basic solution-pH (10).

9.4 CONCLUSION

The magnetic photocatalyst has been successfully converted to the magnetic dye-adsorbent catalyst *via* hydrothermal followed by typical washing and thermal treatments. The latter consists of a core shell nanocomposite with the core of a magnetic ceramic particle, such as mixed $CoFe_2O_4$-Fe_2O_3 and pure-$CoFe_2O_4$, and the shell of nanotubes of dye-adsorbing material such as $H_2Ti_3O_7$. The magnetic dye-adsorbent catalyst has been successfully utilized for the removal of organic dye from an aqueous solution, under the dark condition, *via* surface adsorption mechanism. Due to their magnetic nature, the magnetic dye-adsorbent catalyst can be separated from the treated solution using an external magnetic field. The successful removal of previously adsorbed-dye from the surface of nanotubes, *via* typical surface-treatment, suggests that the magnetic dye-adsorbent catalyst can be recycled.

KEYWORDS

- **Dye-adsorbent catalyst**
- **Methylene blue**
- **Photocatalyst**
- **Vibrating sample magnetometer**
- **Waste waters containing**

ACKNOWLEDGMENT

Author thanks CSIR, India for funding the photocatalysis and nanotechnology research at NIIST-CSIR, India through the Projects # NWP0010 and P81113.

REFERENCES

1. Forgacsa, E., Cserhati, T. and Oros, G. *Environ. International*, **30**, 953–971 (2004).
2. Gupta, G. S., Shukla, S., Prasad, G., and Singh, V. N. *Environ. Technol.*, **13**, 925–936 (1992).
3. Shukla, S. P. and Gupta, G. S. *Ecotoxicol. Environ. Saf.*, 24,155–163 (1992).
4. Sokolowska-Gajda, J., Freeman, H. S., and Reife, A. *Dyes Pigments*, **30**, 1–20 (1996).
5. Robinson, T., McMullan, G., Marchant, R., and Nigam, P. *Bioresource Technol.*, **77**, 247–255 (2001).
6. Carp, O., Huisman, C. L., and Reller, A. *Prog. Solid State Ch.*, **32**, 33–177 (2004).
7. Tachikawa, T., Fujitsuka, M., and Majima, T. *J. Phys. Chem. C*, **111**, 5259–5275 (2007).
8. Chen, X. and Mao, S. S. *Chem. Rev.*, **107**, 2891–2959 (2007).
9. Fujishima, A., Zhang, X., and Tryk, D. A. *Surf. Sci. Rep.*, **63**, 515–582 (2008).
10. Marto, J., Marcos, P. S., Trindade, T., and Labrincha, J. A. *J. Haz. Mat.*, **163**, 36–42 (2009).
11. Pan, S. S., Shen, Y. D., Teng, X. M., Zhang, Y. X., Li, L., Chu, Z. Q., Zhang, J. P., Li,G. H., and Hub, X. *Mater. Res. Bull.*, **44**, 2092–2098 (2009).
12. Feng, S., Zhao, J., and Zhu, Z. *Mater. Sci. Eng. B*, **150**, 116–120 (2008).
13. Datta, A., Priyama, A., Bhattacharyya, S. N., Mukherjee, K. K., and Saha, A. *J. Colloid Interf. Sci.*, **322**, 128–135 (2008).
14. Shaul, G. M., Holdsworth, T. J., Dempsey, C. R., and Dostal, K. A. *Chemosphere*, **22**, 107–119 (1991).
15. Gupta, V. K. and Suhas J. *Environ. Manage.*, **90**, 2313–2342 (2009).
16. Crini, G. *Bioresource Technol.*, **97**, 1061–1085 (2006).
17. Beydoun, D., Amal, R., Scott, J., Low, G., Evoy, S. M. *Chem. Eng. Technol.*, **24**, 745–748 (2001).
18. Song, X. and Gao, L. *J. Am. Ceram. Soc.*, **90**, 4015–4019 (2007).
19. Gao, Y., Chen, B., Li, H., and Ma, Y. *Mater. Chem. Phys.*, **80**, 348–355 (2003).
20. Xiao, H. M., Liu, X. M., and Fu, S. Y. *Compos. Sci. Technol.*, **66**, 2003–2008 (2006).
21. Rana, S., Rawat, J., Sorensson, M. M., and Misra, R. D. K. *Acta Biomater.*, **2**, 421–432 (2006).
22. Lee, S. W., Drwiega, J., Mazyck, D., Wu, C. Y., and Sigmund, W. M. *Mater. Chem. Phys.*, **96**, 483–488 (2006).
23. Fu, W., Yang, H., Li, M., Li, M., Yang, N., and Zou, G. *Mater. Lett.*, **59**, 3530–3534 (2005).
24. Jiang, J., Gao, Q., Chen, Z., Hu, J., and Wu, C. *Mater. Lett.*, **60**, 3803–3808 (2006).
25. Beydoun, D., Amal, R., Low, G., and McEvoy, S. *J. Mol. Catal. A*, **180**, 193–200 (2002).
26. Siddiquey, I. A., Furusawa, T., Sato, M., and Suzuki, N. *Mater. Res. Bull.*, **43**, 3416–3424 (2008).
27. Shukla, S., Warrier, K. G. K., Varma, M. R., Lajina, M. T., Harsha, N., and Reshmi, C. P. PCT Application No. PCT/IN2010/000198.
28. Thazhe, L., Shereef, A., Shukla, S., Pattelath, R., Varma, M. R., Suresh, K. G., Patil, K., and Warrier, K. G. K. *J. Am. Ceram. Soc.*, **93**, 3642–3650 (2010).
29. Shukla, S., Varma, M. R., Suresh, K. G., and Warrier, K. G. K. "*Magnetic dye-adsorbent catalyst: a "core-shell" nanocomposite*", In Proceedings of NanoTech conference and expo, TechConnect world summit conferences and expo 2010, Anaheim, California, USA,, Vol. 1, pp. 830–833 (2010).
30. Varma, P. C. R., Manna, R. S., Banerjee, D. Varma, M. R., Suresh, K. G., and Nigam, A. K. *J. Alloy Compd.*, **453**, 298–303 (2008).
31. Lee, S. W., Drwiega, J., Mazyck, D., Wu, C. Y., and Sigmund, W. M. *Mater. Chem. Phys.* , **96**, 483–488 (2006).
32. Kasuga, T., Hiramatsu, M., Hoson, A., Sekino, T., and Niihara, K. *Langmuir*, **14**, 3160–3163 (1998).

33. Kasuga, T., Hiramatsu, M., Hoson, A., Sekino, T., and Niihara, K. *Adv. Mater.*, **11**, 1307–1311 (1999).
34. Bavykin, D. V., Parmon, V. N., Lapkina, A. A., and Walsh, F. C. *J. Mater. Chem.*, **14**, 3370–3377 (2004).
35. Sun, X. and Li, Y. *Chem. Eur. J.*, **9**, 2229–2238 (2003).
36. Harsha, N., Ranya, K. R., Shukla, S., Biju, S., Reddy, M. L. P., and Warrier, K. G. K. *J. Nanosci. Nanotech.*, **11**, 2440–2449 (2011).
37. Baiju, K. V., Shukla, S., Sandhya, K. S., James, J., and Warrier, K. G. K. *J. Phys. Chem. C*, **111**, 7612–7622 (2007).

10 Synthesis of Metal Nanoclusters in Dielectric Matrices by Ion Implantation: Plasmonic Aspects

P. Mazzoldi, G. Mattei, G. Pellegrini, and V. Bello

CONTENTS

10.1 Introduction 167
10.2 Nucleation and Growth of Metal Nanoparticles 169
10.3 Optical Properties of Metal NCS 172
10.4 Ion Beam Direct Synthesis: Bimetallic Clusters 175
10.5 Optical Sensing Properties of Nanocomposite Films 178
10.6 Core-Satellite Systems 189
10.7 Plasmonic Nanostructures 190
10.8 Conclusion 193
Keywords 193
Acknowledgment 194
References 194

10.1 INTRODUCTION

Composite materials formed by metal nanoclusters (NCs) embedded in dielectric films (Metal Nanocluster Composite Film, MNCF) are the object of several studies owing to their peculiar properties suitable for application in several fields, such as nonlinear optics, photoluminescence, catalysis, or magnetism [1-4]. Recently, also sensitizing effects for rare earth ions luminescence have been reported as due to energy transfer between metallic NCs and Er ions in silica glass [5]. Updated review chapter dealing with MNCFs are currently published, each one covering one or more particular aspects, ranging from preparation techniques to properties and characterization. In particular, by ion implantation, very large doping concentration values can be obtained in the ion irradiated region, with a modification of chemical and physical material properties. A proper choice of implantation energies and fluences allows to predetermining the composition, the depth and the spatial shape of the modified layer. A fundamental feature of ion implantation is that the implantation process does not take place under thermodynamic equilibrium and consequently the usual solubility limits of the implanted ions in the host can be largely overcome, achieving impurity local concentrations inaccessible by conventional synthesis routes.

Two operating modes of ion implantation may be used: the first is related to the direct synthesis of NCs as resulting from the precipitation of the supersaturated solid solution produced by the implanted ions, the second is related to the modification induced by the ion beam on already formed nanostructures and is therefore referred to as indirect synthesis. Technological applications of the different kind of NCs by both approaches are obtained. The ion distribution is controlled by the experimental parameters of the implantation process (i.e., energy, current, and fluence), but also by the diffusion coefficients of the different species (implanted ions and displaced matrix atoms). The change of diffusion rates into the solid, enhanced by the production of defects due to irradiation can favor either the aggregation of the dopant or their diffusion inside the target. Depending on the choice of the implanted atom and the dielectric target, implantation of "metal" ions in dielectric substrates gives rise to the formation of new compounds and/or metallic nanoparticles. The processes governing the chemical and physical interaction between the implanted ions and the host matrix atoms very crucial for the final system configuration are still not completely understood, in particular in terms of the relative roles of electronic and nuclear energy release. Moreover, postimplantation thermal treatments are normally performed on the as-implanted systems with a twofold meaning: (i) annealing of the implantation-induced defects, (ii) growth of the nucleated embryos by means of suitable combination of annealing atmosphere, temperature, and time. In silica, for instance, annealing of the implantation damage requires temperature near or above 600°C: at these temperatures the thermal diffusion of the implanted species can be quite effective in modifying the postimplantation dopant distribution, promoting either redistribution of implanted species or clustering around nucleated embryos. Sequential double implantation of noble or transition metals in glass has been explored as a suitable method in the formation of binary metal clusters with composition and crystalline structure sometimes different from those predicted by thermodynamics considerations due to the nonequilibrium nature of the process. Focusing on the optical properties of MNCFs, conventional Surface Plasmon Resonance (SPR) measurements performed onto Au or Ag-polyimide nanocomposite films, obtained by ion implantation or by chemical methods, including ion-exchange process, showed an important sensing activity [6-11]. Optical Gas-Sensing properties are also evidenced by nanocomposites containing cookie-like Au/NiO nanoparticles [12].

The use of ion implantation as a tool for synthesizing and processing dielectric thin films doped with metal nanoparticles is addressed with particular emphasis on the nanoparticle nucleation and growth. The plasmonic aspects of nanoparticles and functional application will be discussed.

10.2 NUCLEATION AND GROWTH OF METAL NANOPARTICLES

The precipitation processes that occur during either implantation or annealing of ion implanted materials may be schematically divided in three steps not necessarily strictly separated: (i) nucleation, (ii) noncompetitive or diffusion-limited growth, and (iii) competitive growth (i.e. coarsening or Ostwald ripening) regime. The cluster evolution has been studied in Au implanted silica, as function of thermal annealing.

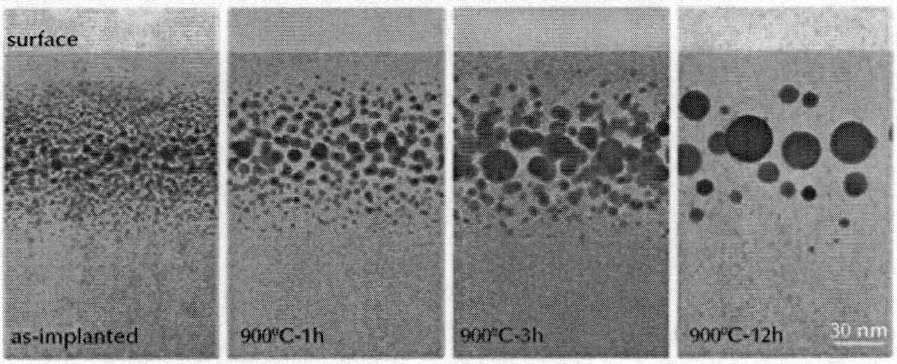

FIGURE 1 The TEM cross-sectional bright field micrographs for the Au-implanted samples as a function of the annealing time from the as-implanted to samples annealed at 900°C for 1, 3, and 12 hr in air.

The use of gold is particularly suitable when attempting to decouple as much as possible the chemical from the physical phenomena triggering the precipitation of a metallic species during ion implantation. Indeed, gold has a reduced chemical interaction with the elements constituting the matrix (Si and O, in the considered case) and a low diffusivity in comparison with other noble metals like Ag for example: this minimizes the role of diffusion-controlled processes, which occur in postimplantation thermal annealing, allowing to vary the NPs size and density distribution by means of the atmosphere in which postimplantation annealings take place. The main results of the investigation. temperature dependence of the gold cluster radius under isochronal annealing (1 hr) may be summarized as follows: (i) annealing in air is more effective in promoting cluster aggregation with respect to reducing or neutral atmosphere, (ii) the squared average cluster radius in an Arrhenius plot shows two different regimes upon air annealing which can be explained by a general model for gold atom diffusion interacting with excess oxygen coming from the external ambient. The results of transmission electron microscopy (TEM) analysis (Figure 1) for 190 keV Au ions implanted in silica at 3×10^{16} ions/cm^2 give for the average cluster diameter in as implanted sample $D = 1.6 \pm 0.8$ nm (similar mean diameter value is obtained for the sample annealed at 400°C). After annealing at 900°C in air the average nanocluster size increases at $D_{900°C}$

= 5.3 ± 3.9 nm. On the other hand, samples annealed in Ar and H_2-Ar atmospheres at 900°C exhibit diameter values ($D_{900°C}$ = 2.1 ± 0.9 nm and $D_{900°C}$ = 2.5 ± 1.2 nm, respectively) comparable to that in as-implanted. To better understand these results, we have analyzed them as a function of the temperature. Figure 2 reports the Arrhenius plot of the squared average cluster radius after annealing in air or Ar, at fixed time, 1 hr.

We observe that below 700–800°C the cluster radius increases very slowly with the annealing temperature, at constant time, independently from the atmosphere composition, due to low gold diffusivity, suggesting a diffusion mechanism controlled by radiation damage. The measured activation energy of 1.17 eV/atom for gold diffusion in silica during annealing in air the temperature range from 750 to 900°C, is very different from the literature value of 2.14 eV/atom.

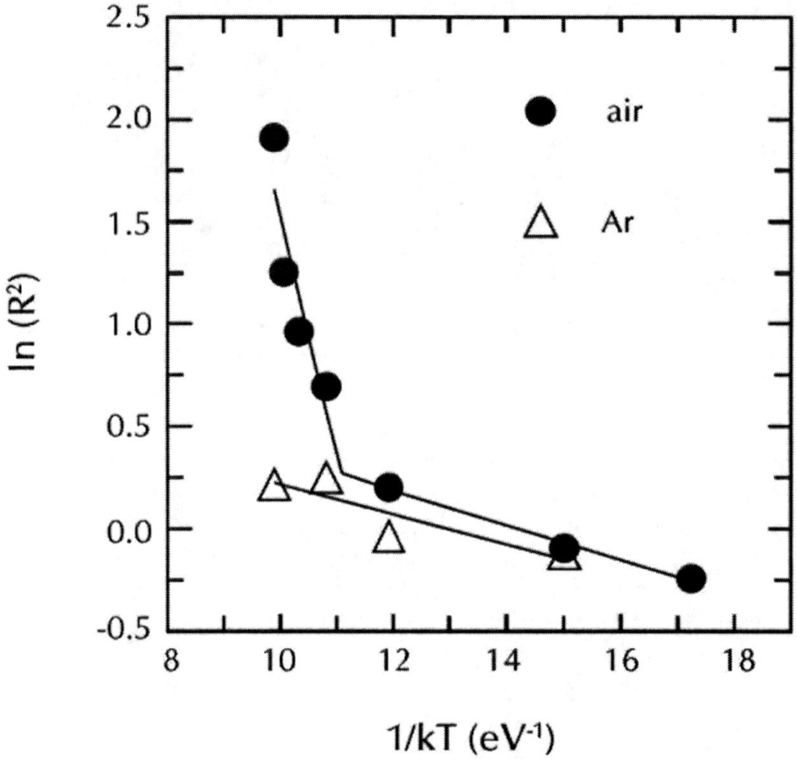

FIGURE 2 Arrhenius plot of the squared average cluster radius R^2 after 1 hr annealing in air (filled circles) or argon (empty triangles). Solid lines are linear fit to the experimental data.

Note however that the activation energy for the molecular oxygen diffusion in silica through an interstitial mechanism is in the range from 1.1 to 1.3 eV/atom. Considering that the clustering process is associated with gold diffusion, in order to explain the role of oxygen in promoting gold diffusivity, we interpreted the results by assuming a thermodynamic interaction between oxygen and gold, following the model Kelvin-Onsager [13], and consequently a correlated diffusion, which is absent in the inert or reducing annealing.

The analysis of the gold clustering problem for annealing in air has been extended at annealing time intervals exceeding 1 hr, when coarsening becomes most probably the relevant cluster growth mechanism. As the annealing time interval increases, larger spherical Au clusters are formed and a corresponding broadening of the distribution is evident.

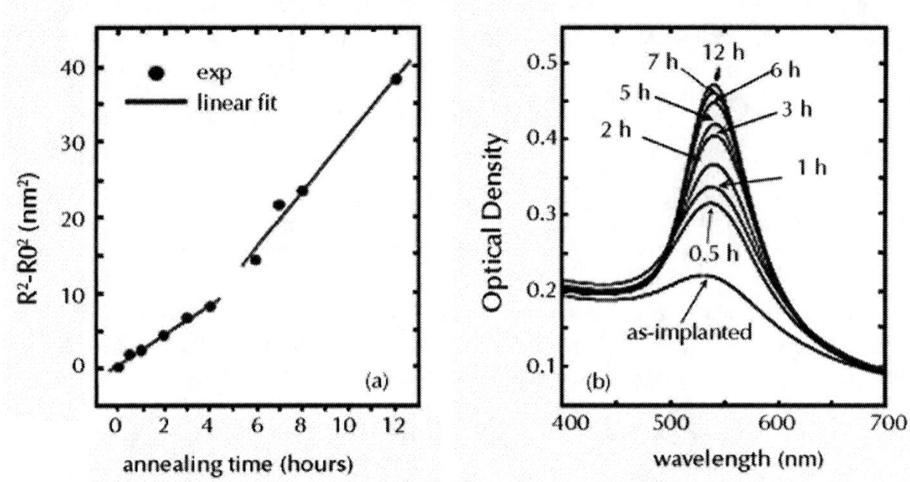

FIGURE 3 (a) $R^2(t)$ evolution in Au-implanted silica samples annealed in air at 900°C for different time intervals, (b) corresponding evolution of the optical absorption spectra.

The results of TEM analysis are shown in Figure 3(a), which reports R^2 (the square of the average radius of the growing particles) as a function of t. R_0 is the value of R at $t = 0$ (which accounts for the radius of the already formed precipitates by implantation). We observe a linear relation between the two quantities with a change of the slope in the range between 4 and 5 hr. This linearity is expected when the cluster growth is only due to the precipitation process of a supersaturated solution (i.e., diffusion-limited aggregation stage). The change of the slope in the plot suggests that the kinetics of cluster growth is modified. The discontinuity may be explained by the onset of the coarsening regime. Thus, we distinguish two different kinetic regimes of NP growth: (i) diffusion (occurring at the earlier stage of growth) characterized by a time dependency of NP radius scaling as $(Dt)^{1/2}$, where D is the diffusion coefficient and t the diffusion time, (ii) a coarsening regime (occurring at longer annealing times) with a radius scaling as $(Dt)^{1/3}$. In this second process, the mass transfer from the matrix to the NP is controlled by the Gibbs-Thomson equation [14]: since smaller NPs have a higher solute concentration than larger ones, the diffusional balance promotes matter transfer from smaller to larger precipitates. Consequently, the average size R increases as $R^3(t) = R^3_0 + K_2Dt$ [14-15]. From the slopes of the linear fit we estimated a gold cluster surface tension value of 1.5×10^{-4} J/cm^2, consistent with the measured gold surface tension value of a "free surface".

10.3 OPTICAL PROPERTIES OF METAL NCS

One of the fingerprints of noble metal cluster formation is the development of a well-defined absorption band in the visible or near UV spectrum which is called the SPR absorption. The SPR is typical of s-type metals like noble and alkali metals and it is associated with the collective oscillations of the conduction electrons of a metallic nanoparticle in the interaction with an electromagnetic wave [15]. The theory developed by G. Mie in 1908 [15], for spherical non-interacting nanoparticles of radius R embedded in a non-absorbing medium with dielectric constant ε_m (i.e. with a refractive index $n = \varepsilon_m^{1/2}$) gives the extinction cross section $\sigma(\omega,R)$ in the dipolar approximation as:

$$\sigma(\omega, R) = 9\frac{\omega}{c}\varepsilon_m^{3/2} V_0 \frac{\varepsilon_2(\omega, R)}{(\varepsilon_1(\omega, R)+2\varepsilon_m)^2 + \varepsilon_2^2(\omega, R))} \tag{1}$$

where V_0 is the cluster volume, c is the speed of light in vacuum and $e(\omega, R) \equiv e_1(\omega,R) + i\, e_2(\omega,R)$ is the size dependent complex dielectric function of the cluster. The SPR resonance holds when the denominator is vanishingly small. This approximation is valid for isolated clusters when the radius is much less than the wavelength $\lambda = c(2\pi/\omega)$: when this is not the case, retardation effects must be taken into account and the general Mie formula, including high order multipolar contributions, has to be employed [16].

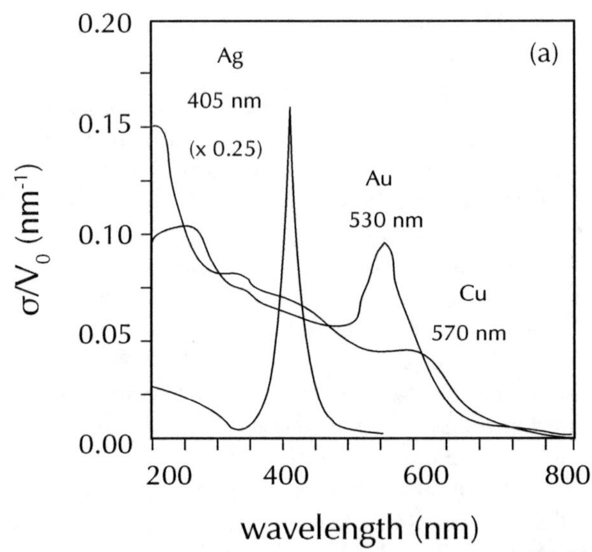

FIGURE 4 *(Continued)*

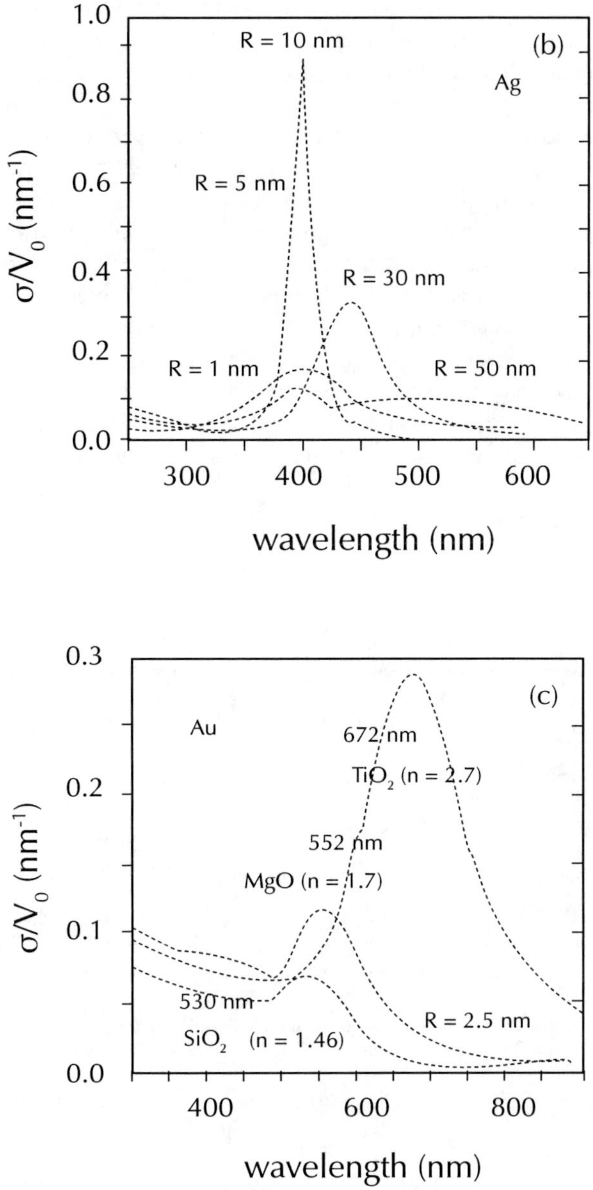

FIGURE 4 Absorption spectra of spherical non-interacting NPs embedded in non-absorbing matrices (Mie calculations): (a) effect of the NP composition for R = 5 nm NPs (Ag, Au, Cu) in silica (n = 1.46); (b) effect of the size for Ag NPs in silica; (c) effect of the matrix for R = 2.5 nm Au NPs (the refractive index $n = \varepsilon_m^{1/2}$ and the position of the surface plasma resonance are reported for each considered matrix).

One relevant point concerns the size dependence of the cluster dielectric function. Following [16], the size correction is obtained by modifying the Drude-like part due to delocalized free s-electrons introducing a size dependent damping frequency which accounts for the increased scattering at the cluster surface. To better illustrate the main factors influencing the position and the shape of the SPR absorption band, we report in Figure 4(a) the effect of the cluster composition (Ag , Au, and Cu) in silica (refractive index, n = 1.46), in Figure 4(b) of the Ag NP radius when the size is not longer negligible with respect to the wavelength there is a red shift of the SPR position with the appearance of multipolar peaks in the UV-blue region and in Figure 4(c) of the matrix, related to the change of refractive index, considering TiO_2, MgO, and SiO_2 substrates. The higher the matrix dielectric constant, the more red shift is the SPR wavelength.

The optical absorption spectra of Au-implanted silica samples annealed in air or Ar for 1 hr at different temperatures are shown in Figures 5(a) and 5(b) respectively. The above formalism has been applied (Figure 5) to extract the size distribution of Au clusters upon thermal annealing from the optical density (OD) absorption measurements to complement the TEM analysis, as reported in Figures 3 and 4. The good level of agreement is shown in Figure 6.

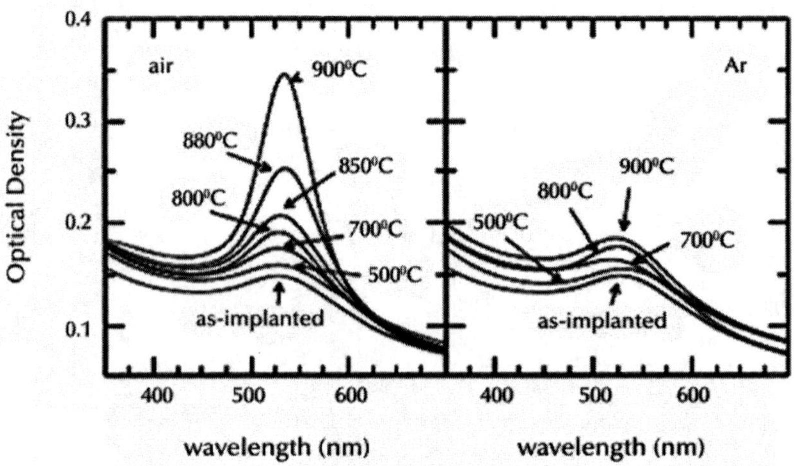

FIGURE 5 Optical absorption spectra of Au-implanted silica samples annealed in air (a) or Ar (b) for 1 hr at different temperatures.

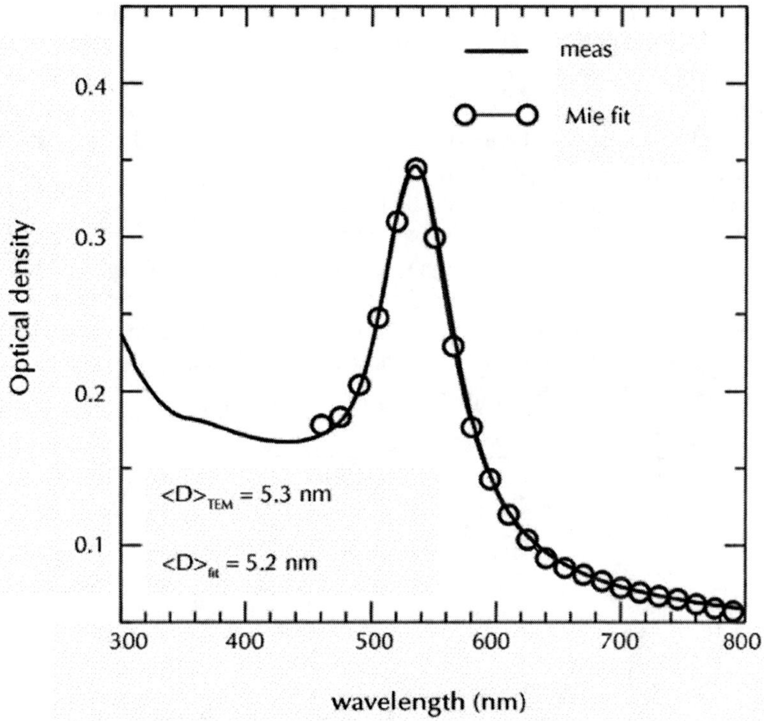

FIGURE 6 Nonlinear fit (empty circles) to the optical absorption spectrum of the sample annealed at 900°C in air, from which the average cluster diameter <D> fit is obtained and compared to the TEM measured one, $<D>_{TEM}$.

10.4 ION BEAM DIRECT SYNTHESIS: BIMETALLIC CLUSTERS

Sequential ion implantation of two different metal species at suitable energies and fluences to maximize the overlap between the implanted species and control their local relative concentration may give rise to different nanocluster structures, with the possible presence of separated families of pure metal clusters, crystalline alloy clusters, or core-shell structures. The formation of clusters of a certain nature depends critically on the implantation parameters, implants sequence and the temperature at which the process is realized. Moreover, postimplantation treatments such as annealing in controlled atmosphere and/or ion or laser irradiation have demonstrated to be effective in driving the system towards different stable clusters structures. As a general rule, the criterion valid for bulk systems of miscibility of the two elements as a constraint for alloy formation is not so stringent in the case of NCs. This is due to the incomplete onset of the bulk properties triggered by the large number of atoms at the surface that makes a cluster more similar to a molecular than to a massive system [17]. This leads to new possible alloy phases, which may be thermodynamically unfavored in the bulk. In the case of noble-metal-based systems (Au-Cu, Au-Ag, Pd-Ag, and Pd-Cu) perfect miscibility is expected from the bulk phase diagrams and in fact sequentially as-implanted samples exhibit direct alloying [1, 18-19]. On the contrary, systems like

Co-Cu or Au-Fe which are not miscible in the bulk showed nanoalloy formation after sequentially implantation in silica. Binary nanoparticles can be present as alloys but also as core-shell structure that can have new properties due to the interaction or combination of properties of the two parts of the nanoparticle. The nanostructure of the binary nanoparticles depends both on the chemical reactivity of the implanted species [20-21] and on the alloy formation heat. In the case of noble metal clusters the SPR in the visible range is a clear fingerprint of nanoparticles formation [22]. Similarly, for noble metals alloy the SPR resonance is located in between those of the pure elements, and is triggered by the complex interplay between the modified free electrons and interband absorptions. This can be seen in Figure 7, which shows a comparison between optical absorption, OD, spectra either simulated with the Mie theory [16, 22] for 3 nm clusters of pure Au, Ag, and $Au_{0.4}Ag_{0.6}$ alloy in silica (Figure 7(a)), or measured for analogous systems [23] in ion implanted silica (Figure 7(b)).

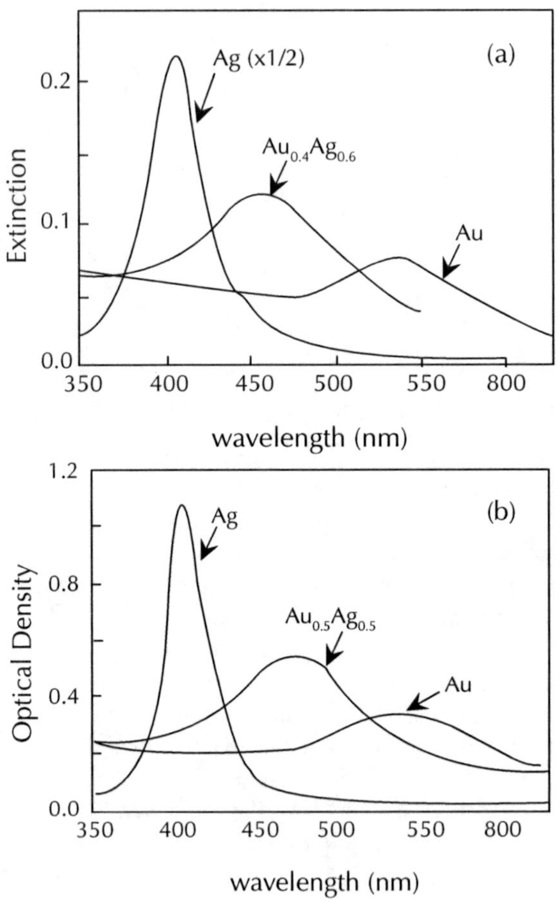

FIGURE 7 Comparison between a simulations based on the Mie theory of the optical absorption in the UV-vis range for 3 nm NPs of pure Au, Ag, and $Au_{0.4}Ag_{0.6}$ alloy in silica (a), with the experimental optical density of the same systems in ion implanted silica (b).

Linear absorption measurements can therefore give the first indication of possible alloy formation. Nevertheless, in systems containing transition metals (Pd-Ag, Co-Ni, ...) such a simple technique is no longer effective as interband transitions completely mask the SPR peak, resulting in a structurless absorption, which hinders any unambiguous identification of the alloy. In such cases, one has to rely on structural techniques like TEM (selected-area electron diffraction, selected area electron diffraction (SAED) and energy-dispersive X-ray spectroscopy (EDS) or EXAFS (Extended X-ray Absorption Fine Structure) to establish alloy formation.

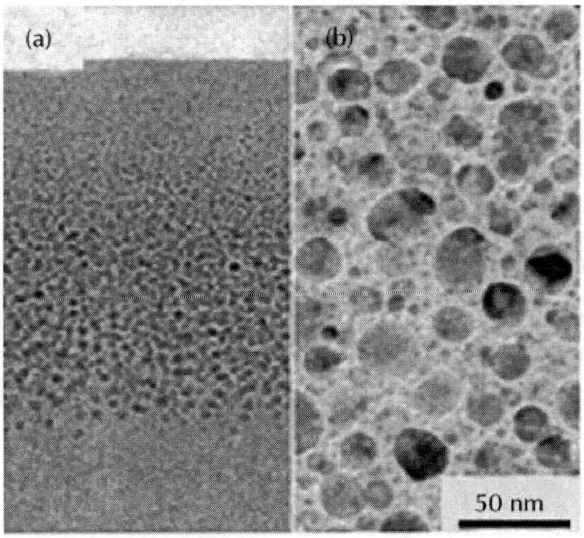

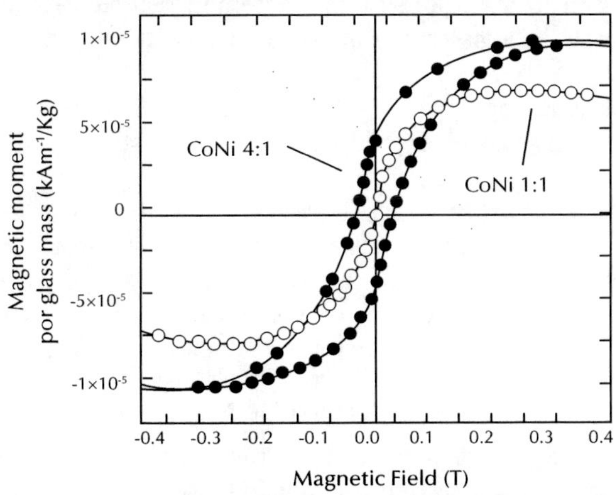

FIGURE 8 (a) Cross-sectional bright field TEM image of sequentially ion implanted silica with Co and Ni with a concentration ratio of 4:1 for a total fluence 3 × 1017 ions/cm^{2,} (b) planar bright field TEM image of sequentially ion implanted silica with Co and Ni with a concentration ratio of 1:1 for a total fluence 4 × 1017 ions/cm^2, (c) comparison between the room temperature hysteresis loop of f.c.c. Co1Ni1 and h.c.p. Co4Ni1 samples.

The Padova Group obtained some relevant results on binary alloy NCs in silica by using ion implantation, without subsequent annealing. In particular the formation of alloy has been observed for Au-Cu, Au-Ag, Au-Fe (metastable magnetic alloy), Fe-Al, Pd-Ag, Pd-Fe, Pd-Cu, superparamagnetic nanoparticles Cu-Ni Ni-Co Co-Cu (h.c.p. and f.c.c. phase coexistence), and Ag-S(core-shell) [3, 24-32]. As an example of binary clusters for magnetic applications, we present the study performed on Ni-Co alloy [31]. The Co-Ni phase diagram was investigated by performing sequential ion implantation in silica of Co and Ni at the same energy but with different fluences in order to have a constant total Co + Ni fluence. All the samples investigated exhibit Co_xNi_{1-x} alloy NCs. The SAED analysis was able to monitor a phase transition from f.c.c. to h.c.p. as the Co content in the system is greater than 70%, in agreement with bulk Co-Ni alloy. It is interesting to note that similar Co-Ni alloy system obtained by our group with the sol-gel route, exhibited at all the Co/Ni ratios the f.c.c. structure [33]. As the lattice parameters of the f.c.c. phases of Co and Ni differ by a quantity that is at the limit of SAED quantification for nanoclustered systems (mostly due to the size dependent broadening of the diffraction peaks), we performed also EDS compositional analysis with a sub-nanometer electron probe on single clusters which demonstrated the presence of both Co and Ni. Figure 8(a) shows a TEM image of a sample containing Co and Ni in the ratio 4-1 (Co4Ni1) whereas Figure 8(b) is a TEM image of the Co1Ni1 sample: the SAED pattern for Co1Ni1 sample exhibits a single alloy f.c.c. phase with lattice parameter $a = 0.3533(12)$ nm, whereas, when the Co concentration in the alloy is increased above 70% (Co4Ni1), an h.c.p. diffraction pattern is obtained. The difference in structure as a function of the Co contents in the alloy is of paramount importance for controlling the magnetic properties. In Figure 8(c) the hysteresis loop of f.c.c. Co1Ni1 and h.c.p. Co4Ni1 samples (at the same total fluence) are compared: the average cluster size in both samples is similar (about 5 nm) but the h.c.p. phase (Co4Ni1) exhibits a coercive field which is absent in the f.c.c. phase (Co1Ni1). This is due to the reduction in symmetry of the hexagonal unit cell with respect to the cubic one.

Very interesting fast nonlinear optical properties are shown by the nanocomposites containing AuAg and AuCu NCs, produced by ion implantation, with nonlinear refractive index n_2 of $(-1.6 \pm .3) \times 10^{-10}$ cm²/W and $(+6.3 \pm 1.2) \times 10^{-11}$ cm²/W, at 527 nm of wavelength [34], respectively. Just to have an idea of the increase in the nonlinear optical properties due to the NCs, the pure silica matrix has a $n_2(SiO_2) = 5 \times 10^{-16}$ cm²/W. The peculiarities of this experimental finding are the very large modulus value of n_2 and the sign change according to the cluster composition. Such high value of the fast n_2 coefficient has never been detected in metal nanocomposites up to now [35].

10.5 OPTICAL SENSING PROPERTIES OF NANOCOMPOSITE FILMS

With the aim of giving a further contribution to the understanding of nanocomposites properties, we have investigated the gold precipitation process induced by ion implantation in a polyimide matrix and application for gas sensor. We focused on gold for two main reasons: (i) its chemical inertness which allows to decoupling cluster nucleation from the chemical interaction of the implanted species with the matrix components, (ii) the large optical absorption cross section of gold NCs (SPR) in the VIS region at wavelengths larger than the optical absorption edge of the chosen polyimide. Gold ions were implanted at different fluencies on pyromellitic dianhydride-4,4' oxydianiline (PMDA-ODA) polyimide 100 ± 2 nm thin films, deposited by glow discharge vapor deposition polymerization (GDVDP) on pure silica matrix [6, 8].

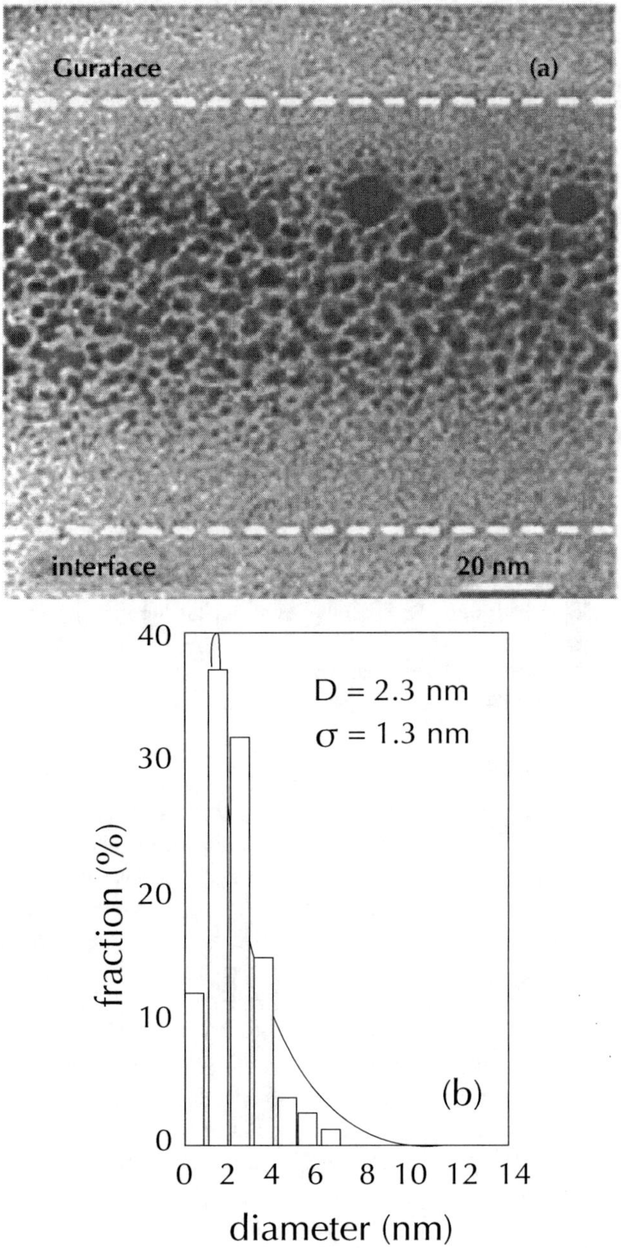

FIGURE 9 *(Continued)*

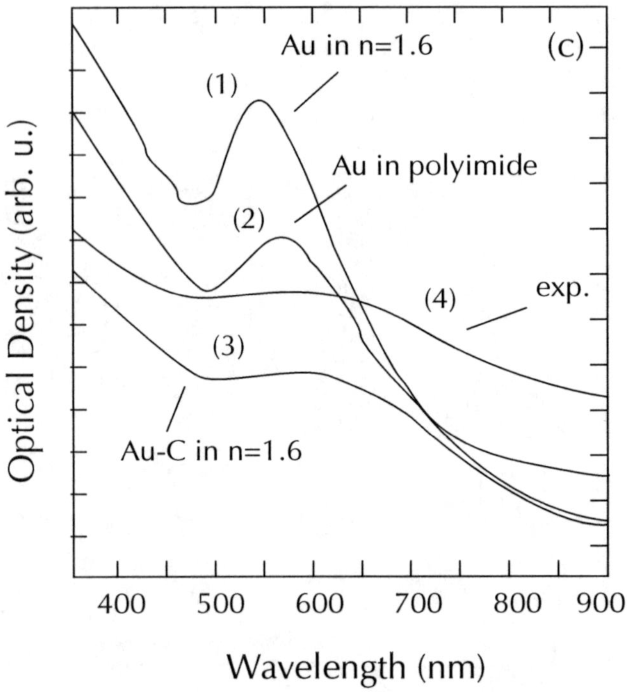

FIGURE 9 The TEM and optical simulation in the sample implanted with 5×10^{16} Au^+/cm^2: (a) TEM cross-sectional micrograph (dashed lines represent the free surface and film-substrate interface), (b) nanoparticles size distribution, (c) simulated optical spectra: (1) Au cluster in a nonabsorbing medium with n = 1.6, (2) Au cluster in polyimide (absorbing), (3) Au(core)-C(shell) cluster in a nonabsorbing medium with n = 1.6; (4) the experimental spectrum of Au-implanted polyimide sample.

The bright field TEM micrograph of the sample, in Figure 9(a), shows spherical Au nanoparticles dispersed in a 40 nm thick layer, embedded between two nanoparticle free layers: a 10 nm thick surface layer and a bottom layer of about 30 nm. The particle size distribution, reported in Figure 9(b), presents an average value of 2.3 nm and a standard deviation of 1.3 nm. The EDS analysis evidenced a C content of the surface layer higher with respect to the rest of the film, in agreement with literature results, giving a realistic representation of the polymer composition around the clusters: indeed, the C:H ratio in this region could be more similar to the nominal 2:1 value, considering that the C-rich surface layer partially compensates the measured C-excess, which refers to the whole structure. The TEM data have been used to simulate in the frame of the Mie theory and Maxwell–Garnett effective medium approximation [22] the optical absorption spectra of the gold implanted sample. The results are reported in Figure 9(c), with three different experimental assumptions. In the first analysis (spectrum 1), we used the first Mie model to describe the nanocomposite absorption properties, assuming spherical Au clusters, with size 2.3 nm, surrounded by a matrix with refractive index of about 1.6, typical of the aromatic polyimides, considering the analysis on the local polymer composition near the Au clusters. The simulation (spectrum 2), in which the experimental bulk dielectric function of gold has been size-corrected [22], gives a narrow resonance centered at 550 nm instead of the experimental large band centered at 560 nm. Even considering that the matrix could have a larger refraction index due to implantation damage, the simulation does not reproduce the measured spectrum. The SPR simply shifts to larger wavelengths without any FWHM broadening. We considered a second model in which the Au cluster is surrounded by a very thin carbon layer, forming a core shell particle dispersed in a polyimide matrix. The best agreement between experimental and simulations was obtained for a C shell thickness of about 0.8 nm, indicating that locally the composition of the matrix can be carbon enriched.

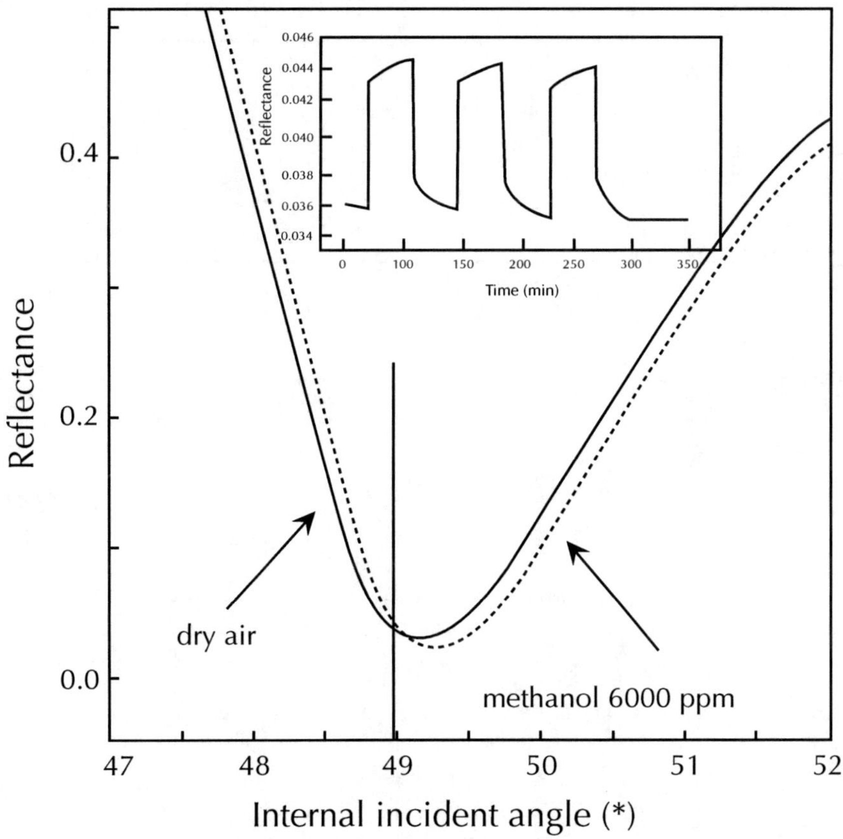

FIGURE 10 The SPR spectra of the implanted polyimide film in presence of dry and of 6,000 ppm methanol virgin vapors (dashed line) Inset: dynamic response upon repeated exposure to saturated methanol vapor.

The optical gas sensing features of this system has been studied, by using optical absorption and SPR transduction techniques, in the presence of mixtures containing vapors of methanol, ethanol, and gases, like NO_2 and NH_3, in dry air. A specific experimental setup was realized to acquire simultaneously the array of optical responses to gases and/or vapors in terms of the absorption curves variations in the UV–vis spectral range (300–700 nm). All the measurements were carried out at room temperature and at normal incidence of the light beam. In Figure 10 we report the optical absorption spectra of the sample as implanted and in presence of methanol vapor (6,000 ppm). As can be observed, the presence of the vapor produces an increase of the optical absorption in the entire spectral region larger than the optical absorption edge of the nanocomposite. It can be observed that the change consists of a shift in the SPR angle. Similar effect is observed in presence of ethanol vapor, but the experiments performed with NO_2 and NH_3 gases do not produce significant changes in the optical spectrum of the nanocomposite.

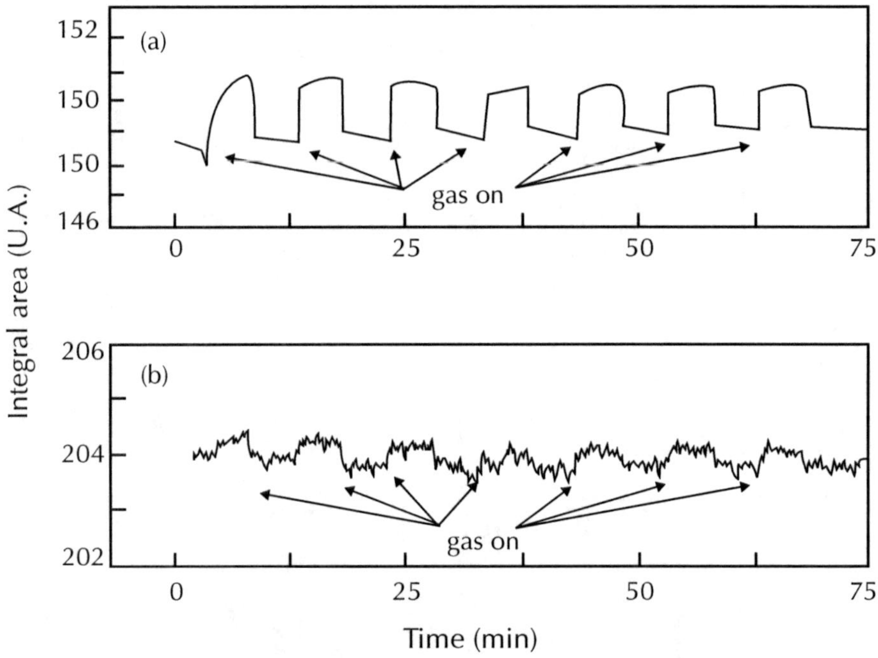

FIGURE 11 Dynamical optical absorption response for (a) the polymide film implanted with 5×10^{16} Au^+/cm^2 and for (b) the film obtained upon different exposures to methanol vapor.

Figure 11 shows the dynamic response curves relative to the virgin polyimide film (b) and 5×10^{16} Au^+/cm^2 implanted film (a) in the presence of a mixture of dry air containing methanol vapors at a concentration of about 6,000 ppm. The response curves have been obtained by monitoring time *versus* the integral area calculated under the absorption curve in the 350–800 nm spectral range in the case of polyimide film and 450–650 nm spectral range in the case of implanted polyimide film. In the last case, we have analyzed the spectral region centered around the typical plasmon peak of the gold implanted nanoparticles. As one can see, the virgin polyimide film does not present any variation in the absorption curves in the presence of alcohol vapors. A response appears in the case of gold implanted polyimide film with an increase of the integral area around the region of the plasmon peak that is not limited at this region but involved all the mixed structure gold nanoparticles/polymer as evidenced from the absorption spectra carried out in dry air and in alcohol vapors and reported in Figure 10. Similar experiments realized onto samples implanted with smaller ion fluences showed that there were negligible changes in their optical properties in presence of all the analytes. In conclusion, only implanted samples at high fluence can be used for vapor sensing.

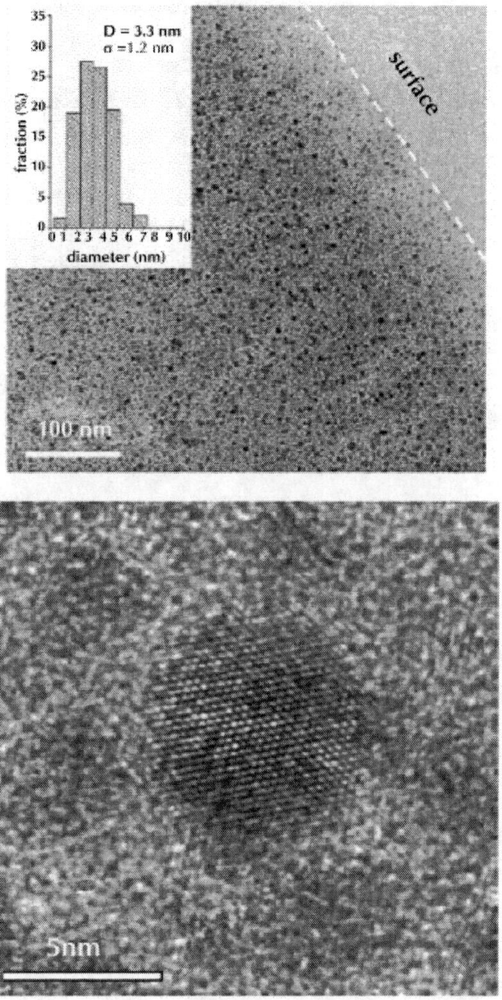

FIGURE 12 Cross sectional TEM images at different magnifications of 6FDA-DAD polyimide treated for 10 min in KOH, for 20 min in $AgNO_3$ solution and annealed at 300°C in reducing atmosphere. In the inset of the left image, the size histogram is reported.

Moreover the formation of silver nanoparticles in transparent fluorinated polyimides has been obtained, as a consequence of K-assisted Ag doping and thermal reduction in hydrogen atmosphere [10]. The first stage of the process, leading to the formation of the potassium polyamate salt, and the following K^+-Ag^+ ion exchange process have been correlated to the starting polyimide chemical structure. The K penetration depth is a linear function of the treatment time in KOH, thereby proving that the chemical reaction between the nucleophile and the polyimide is in both cases a first order process. Moreover, the K penetration rate depends on the polyimide structure: both the chemical nature of the substituents along the polymeric chain and the network permeability to aqueous solution are considered to affect the K penetration rate. The ion exchange step in $AgNO_3$ solutions affords complete substitution of K by Ag ions in the polyimides. The Ag nanocrystals precipitation in reducing atmosphere has been correlated to the silver reduction process, which in turns is related to the silver (I) polyamate complex stability in the polyimides. The subsequent step of clusters growth as a function of the annealing temperature has been investigated by X-ray Diffraction (XRD) and TEM analyses (Figure 12). Similar clusters dimensions have been estimated in different polyimides, ranging from 2 to 3.5 nm. The optical sensing capabilities of silver-polyimide nanocomposite materials have been tested upon exposure to vapors of water, ethanol and acetone. In order to enhance the porosity of the polyimide films, small amounts of azodicarbonamide (ADC) have been added to the dissolved polyimide before the film casting. [9]. Analyte/film interaction produces a change of the nanoparticle plasmon absorption peak owing to a change of the average refractive index of the environing medium (Figure 13). The absorbance change ΔA is defined as: $\Delta A = A_{analyte} - A_{nitrogen}$ where $A_{analyte}$ is the steady absorbance during the exposure to vapor and $A_{nitrogen}$ is the absorbance before exposure to vapor.

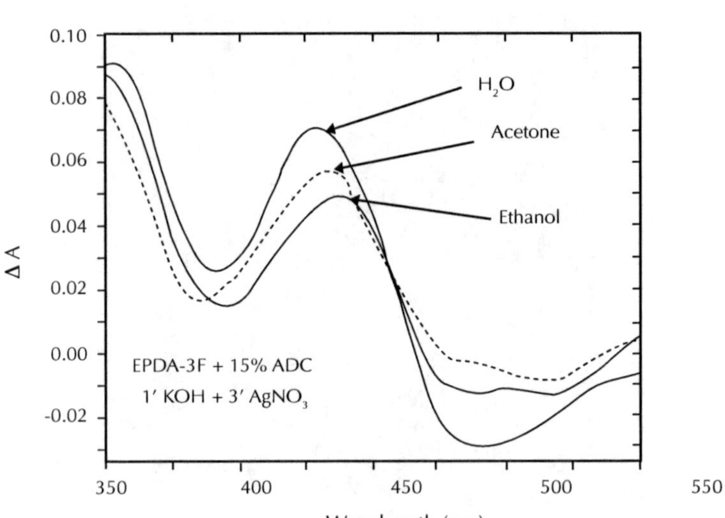

FIGURE 13 Absorbance variation ΔA in presence of saturated vapors of water, ethanol, and acetone for the polyimide thin film doped with 5% of ADC and treated for 1 and 3 min in KOH and $AgNO_3$ aqueous solutions, respectively.

As can be observed, the variation is more pronounced in the presence of water saturated vapors and decreases with acetone and ethanol. Measurements of the absorbance changes at a fixed wavelength show that the nanocomposite films respond to all the analytes, at concentrations ranging from 800 to 88,000 ppm.

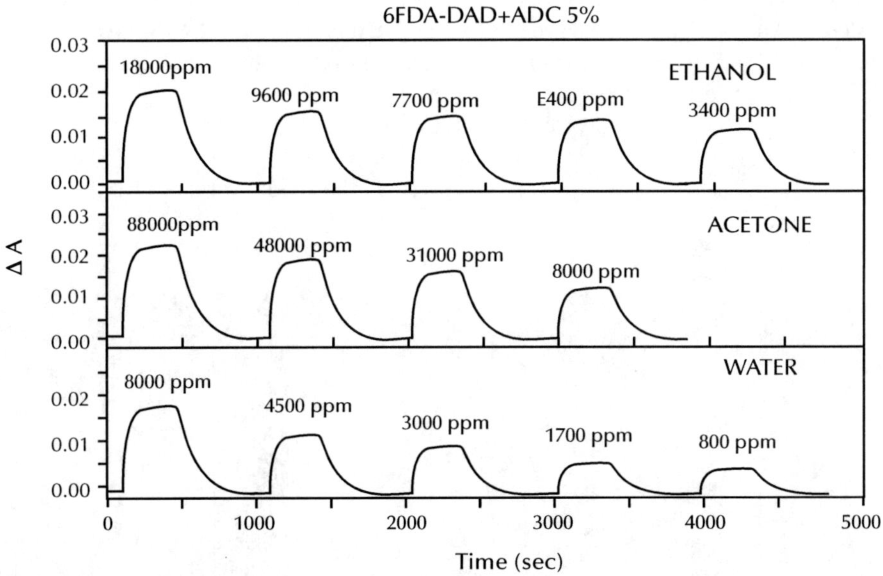

FIGURE 14 Dynamic absorbance variation in presence of vapors at different concentrations of the polyimide thin film doped with 5% of ADC.

Response times ranging from 0.7 to 2.0 min and recovery times in the order of few minutes have been obtained for all the nanocomposite films. The intensity of the optical response increases at increasing ADC content. All the nanocomposite films exhibit a stronger response towards water with respect to ethanol and acetone and this behavior is due to the easier permeation of water molecules into the nanocomposites with respect to acetone and ethanol, owing to the considerably lower molecular dimension and steric hindrance of water (Figure 14).

As third example, we report a study on the formation of cookie-like Au/NiO nanoparticles with Optical Gas Sensing [12]. Au/NiO nanoparticles have been embedded in a porous silica matrix with a high specific surface area which increases the number of active sites for gas reaction and enhances sensor functionality.

The nanocomposites were prepared by the sol-gel technique. Films, 525 nm thick, with nominal molar ratios of $NiO/SiO_2 = 2:3$ and NiO/Au = 5:1, were deposited on SiO_2 glass and heated at 700°C for 1 hr in air. Such films are still porous at this temperature [36] and the porosity of the matrix provides a path for the gas molecules to reach the functional ultrafine particles embedded in the glass matrix.

In Figure 15(a), we present a high-resolution TEM (HRTEM) image of Au/NiO cookie-like nanocluster. Cross section TEM measurements showed the presence of such clusters throughout the film thickness. Figure 15(b) shows a detail of the interface between Au and NiO at a higher magnification, from which the epitaxial growth at the interface can be clearly seen. Indeed, the FT of the HRTEM image (Figurre 15(c)) highlights the presence of two sets of parallel planes: the region with darker contrast in Figure 15(b) shows the lattice fringes of face-centered cubic (fcc) Au (111) planes (with interplanar distance $d_{(111)} = 0.236$ nm), whereas a second smaller periodicity, arising from the region with brighter contrast in Figure 12(b), can be indexed as originating from fcc NiO (200) planes ($d_{(200)} = 0.207$ nm). A closer inspection of the HRTEM image shows, in the NiO part of the cluster, the presence of small amorphous zones superimposed on the crystalline regions. This result suggests a possible island like growth of the oxide, directed by the underlying Au template.

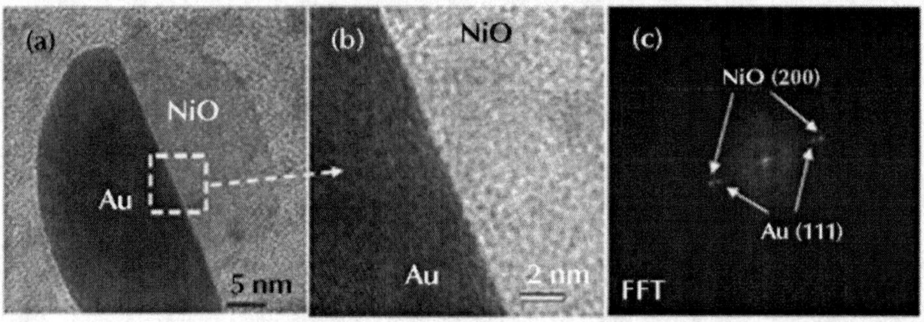

FIGURE 15 (a, b) High-resolution TEM (HRTEM) images of the twofold Au/NiO cluster topology. c) Fourier transform (FT) of the HRTEM image in (b), showing the presence of two sets of parallel planes in Au and NiO.

The cluster consists of two parts: an Au region and a Ni-rich part. The data obtained in the HRTEM–STEM analyses have been used to model the optical absorption spectrum of the sample in the visible range. The experimental absorption curve is shown in Figure 16 and exhibits two main bands, the most intense band is centered at 613 nm and the other is a shoulder at about 530 nm.

To model both absorption bands, we redefined the geometry of the twofold clusters in a way that is depicted in the inset of Figure 16, that is as a sum of spherical Au clusters (type 1 cluster) and a spherical core/shell structure (type 2 cluster) in which the Au core has the same size as the previous Au type and is surrounded by a shell of NiO. The surface free electrons of Au experience direct dielectric coupling with the silica matrix from one side (type 1 cluster) and, on the opposite side, with the silica mediated by a thick shell of NiO (type 2 cluster).

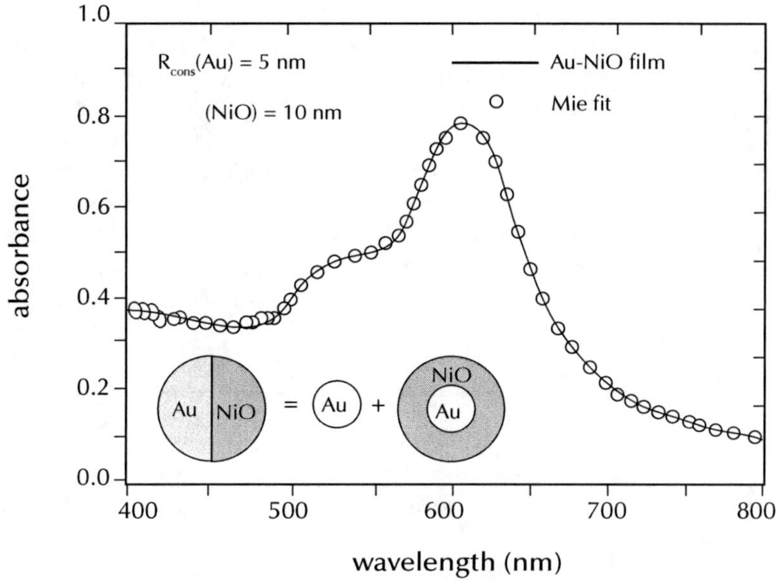

FIGURE 16 Experimental optical absorption spectrum of the sample (solid line) and corresponding effective medium fit (open circles) assuming a combination of monoelemental and core/shell NCs.

By applying Mie theory, we simulated the experimental spectrum by adding the contribution of type 1 clusters, which have a SPR band at 530 nm in silica, with that of type 2 clusters, which exhibit a red shift of the Au SPR band induced by the higher dielectric constant (e) of NiO (e_{NiO} = 5.4) compared with that of silica (e_{SiO2} = 2.13), and account for the band at 613 nm. The agreement between the simulated and the experimental spectrum is quite satisfactory, as shown in Figure 16. Comparable simulations made by assuming ellipsoidal cluster shapes did not give the same level of agreement. Optical simulation evidences that the most intense absorption band is controlled by dielectric coupling at the Au/NiO interface, which red shifts the position of the Au SPR band. This is important for explaining the results of the influence of gas absorption on the optical response of the material, as it will be presented.

The gas sensing properties of the nanocomposite films were evaluated by measuring the variation of optical transmittance of the film when exposed to CO in air. The nanocomposite Au/NiO films exhibited clear and reversible absorbance changes in the vis-NIR wavelength region when exposed to CO, as shown in Figure 17. Upon exposure to 1% CO, and with the exception of wavelengths near the main plasmon absorption band around 613 nm, there is a decrease in absorbance over the whole wavelength range (Figure 17(a)). This feature can be better appreciated when considering the data showing the absorbance in air minus absorbance in 1% CO, defined as ΔA, reported in Figure 17(b). The changes in absorbance clearly display dependence in magnitude and sign of ΔA with respect to wavelength. Thus, in the region of most intense absorption, ($585 < k < 635$ nm) there is a reversal of sign of the absorbance change (ΔA) in comparison with other wavelengths. The present data is thus the first that provides evidence of negative ΔA in the plasmon wavelength region, which accompanies the exposure of NCs with Au/NiO interfaces to CO. In NiO films that have no noble metal component, the decrease in absorbance is close to constant in the 350–850 nm wavelength range and is ascribed to a decrease in the positive whole density of NiO during catalytic oxidation of CO [37]. For the present case with NiO/Au, the decrease in absorbance at wavelengths outside the plasmon band can be ascribed to the same mechanism. The variation in absorbance at 613 nm decreases as the CO concentration decreases (Figure 17(c)). In conclusion we reported some example on optical sensing properties of nanocomposite films, synthesized by ion implantation or chemical methods.

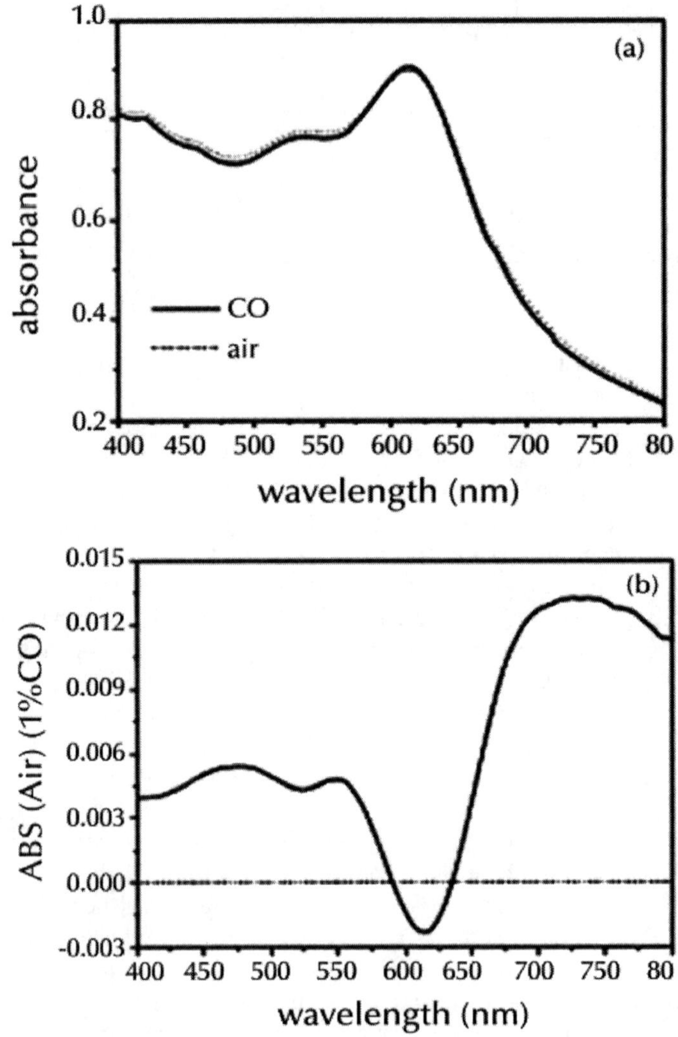

FIGURE 17 *(Continued)*

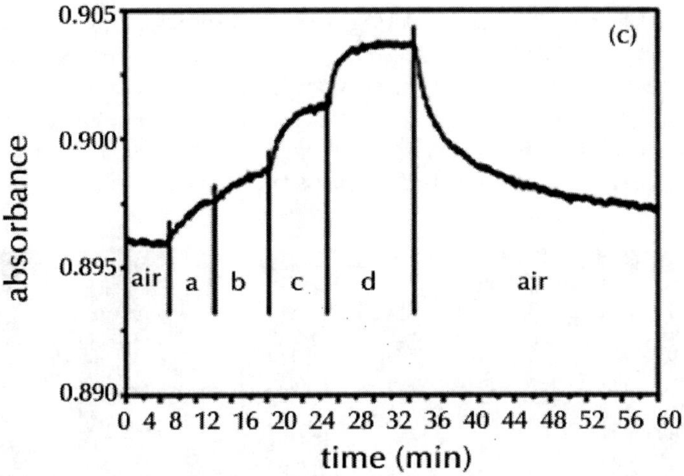

FIGURE 17 (a) Absorption spectra of Au/NiO nanocomposite film in dry air (dotted line) and after exposure to 1 vol% CO (solid line). (b) Absorbance in dry air minus absorbance in 1% CO (ΔA) of the Au/NiO nanocomposite film. (c) Absorbance measured at 613 nm and different CO concentrations (a = 10, b = 100, c = 1000, and d = 10,000 ppm) of the Au/NiO nanocomposite film.

10.6 CORE-SATELLITE SYSTEMS

An important aspect of ion implantation, referred to as indirect synthesis, is correlated to the modification induced by the ion beam on already formed nanostructures. We have seen how the metal NPs can be considered as a functional optical building block and how ion implantation can be used, in combination with thermal treatments, to control both of them. The next level of our hierarchical approach is the one in which the interaction between NPs can be used to enhance the optical performances of the materials. Ion implantation has been used to obtain nanostructures that exhibit a dimer like interaction. Such structures have been called nanoplanets and are composed by a large central NP (5-30 nm in diameter) whose surface is surrounded by a halo of smaller (1–5 nm) NPs (the satellites) extending up to 10–20 nm from the central NP. These are obtained for instance by irradiating Au_xCu_{1-x} or Au_xAg_{1-x} alloy NPs different ions like He, Ne, Ar, or Kr ions at different fluence and energy [38].

The main advantage of ion irradiation is that the nanoplanet topology can be tailored by tuning the irradiation parameters (fluence, energy, ions, and flux) [29, 39]. The irradiation effect is shown for Au_xAg_{1-x} alloy NPs (x = 0.6) in Figure 18(b) in comparison with the unirradiated reference, Figure 18(a) [40]. The most evident result is the new topology is obtained: around each original NP a set of satellite NPs of about 1-2 nm are formed with an average distance of about 3–5 nm from the NP surface. Their size and density can be increased by increasing the nuclear fraction of the energy loss by using different ions. The EDS compositional analysis with a focused 2 nm electron beam of the FEG-TEM in the central part of mother cluster on the Ar-irradiated AuAg sample, gives an Au/Ag ratio (measured at AuL and AgL edges) of 1.4 ± 0.1, whereas the same ratio measured on the satellite clusters is 2.3 ± 0.8. Similar ratios have been found from EDS analysis on AuAg sample irradiated with He, Ne, or Kr ions. The EDS analysis reveals therefore a preferential extraction of Au atoms from the original cluster and this selective dealloying process is independent of the particular system investigated (we obtained similar results for Ne-irradiated AuCu cluster).

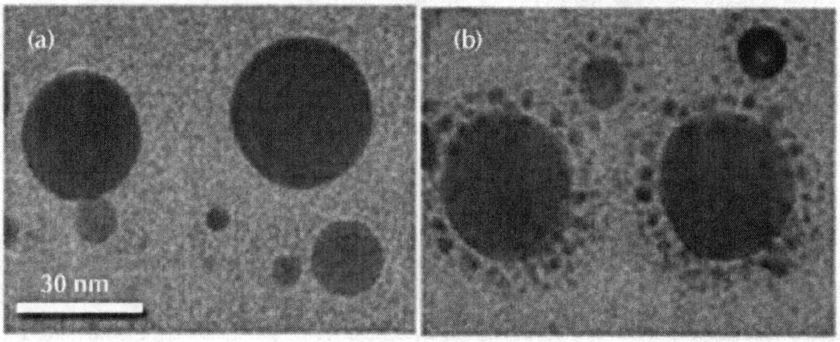

FIGURE 18 Bright field TEM cross sectional micrographs of AuxAg1-x alloy NPs (x = 0.6, annealed in air at 800°C for 1 hr) before (a) and after (b) irradiation at room temperature with 190 keV Ar^+, at a current density 2.5×10^{16} ions/cm^2 and fluence of 0.84 μA/cm^2.

The peculiar topology of the core-satellite NPs allows controlling the optical properties in the ion irradiated samples by producing a red shift of the SPR absorbance, as we will discuss.

10.7 PLASMONIC NANOSTRUCTURES

The interaction of light with metal NCs in insulating matrices has received increasing attention in the last decade. Glass embedded noble metal NCs exhibit strong surface plasmon absorption in the visible spectrum [22], and can increase the third order susceptibility of the matrix by several orders of magnitude [41-42]. In the case of spherical isolated metal NCs, the plasmon resonance frequency and electromagnetic field configuration depend on the cluster size and on the metal and matrix dielectric functions. In ensembles of interacting clusters the plasmon peak position and the local field are influenced by the interparticle electromagnetic coupling. Parameters like particle size, number, and relative position as well as incident light polarization state influence the extinction spectrum and the local field enhancement [42-45]. If the system is formed by metal alloy NCs, alloy composition is one additional parameter to play with for the plasmon tuning [38, 46]. Strongly coupled clusters attracted much interest in the field of single molecule sensing applications, such as surface enhanced Raman scattering and molecular plasmon rulers, due to their far and local field properties, or as miniaturized nonlinear optical elements and polarization sensitive photonic devices [47-49].

The nanoplanet configuration described exhibits a red shift of the SPR absorption band, with respect to the simple non-interacting sphere case, due to a strong coupling between the core and the satellite NPs, as evidenced in Figure 19(a), which reports optical extinction spectra for bare and irradiated (He^+, Ne^+, and Kr^{++}) AuAg alloy NCs [50]. The unirradiated sample presents one single extinction band located at 478 nm, between the silver (410 nm) and gold (530 nm) plasmon resonance in silica (refractive index 1.45), with the irradiated ones showing a marked red shift. In the case of Kr^{++}, the extinction spectrum shows two clearly distinct features: a shoulder at about 495 nm and a principal peak at 538 nm, beyond the pure gold plasmon peak in silica.

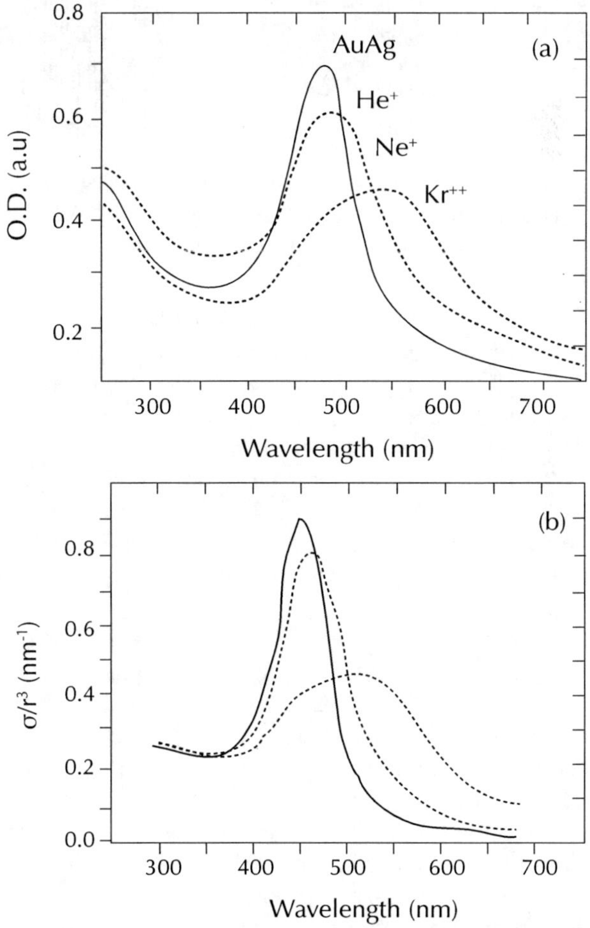

FIGURE 19 (a) Experimental optical extinctions for unirradiated and irradiated AuAg NPs, with different ions, He, Ne, and Kr, samples measured with unpolarized light. (b) Calculated extinction spectra for a single AuAg particle of 12 nm of radius (black line) and for model targets created on the base of TEM structural and compositional data as reported in Figure 18 (dashed, dot-dashed, and short dashed lines refer to He and Ne, or Kr ions), following GMM approach. Empty and filled circles correspond to spectra calculated following MMG approach.

In order to calculate the optical response of strongly interacting spherical NPs, statistically representative nanoplanet models have been created on the base of TEM structural and compositional data. Their optical response has been evaluated by using the Generalized Multiparticle Mie (GMM) and hybrid Mie-Maxwell-Garnett (MMG) [51] simulations, that strongly corroborate the experimental data interpretation and likewise revel large, strongly localized local-field enhancement in the satellite cluster corona.

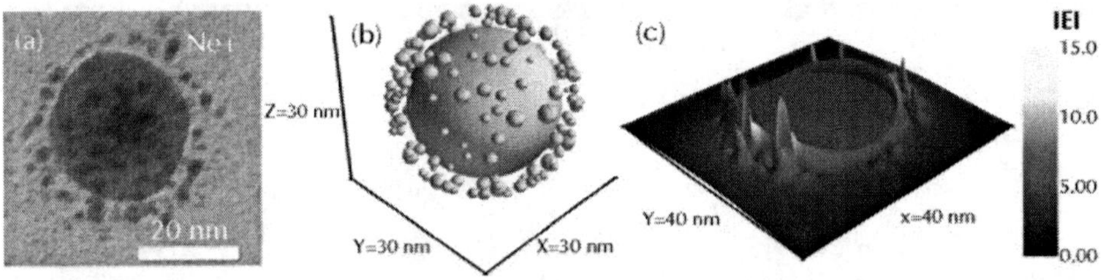

FIGURE 20 (a) The experimental core-satellite nanostructure obtained upon Ne irradiation on AuAg NP, (b) its corresponding model, (c) the computed modulus of the local field on a section in the equatorial plane of the system shown in (b) and calculated at the SPR position of the system (460 nm).

The nanoplanet configuration determines particular local field properties as the presence of hot spots. In Figure 20, we show a GMM calculation of the field distribution around the central NP due to the coupling with its satellites. The experimental core-satellite nanostructure obtained after Ne irradiation on AuAg NP is modeled as in Figure 20(b). The modulus of the local field, calculated on a equatorial section at the SPR of the system (460 nm), shows field enhancement factor as high as 15, which could be exploited for driving, as a nanoantenna, the external field to suitable emitters located in close proximity to the nanoplanet, similarly to surface enhanced Raman spectroscopy [52].

The confinement of the electromagnetic field, due to near field plasmonic coupling between NPs, has been demonstrated by photon scanning tunneling microscopy in linear chains of $100 \times 100 \times 40$ nm^3 Au NPs made by EBL [47]. The possibility of guiding an electromagnetic field along chains of Ag metal NPs has been suggested [53] and demonstrated [54] giving strong impulse to the research in this field. The NP size and distance in the chains are the two main constrains: (i) the NPs size should be larger than the electronic mean free path in order to reduce the scattering losses at the NP surface, responsible of resistive heating and (ii) the interparticle separation should be not so large to produce dephasing in the wave confinement and reduction of the group velocity. By proper optimization, with 50 nm Ag NPs displaced by 75 nm along a linear chain, a propagation group velocity of 0.1c with about 5 dB/μm losses can be obtained [54]. Although such losses per length are quite high, they can be feasible for new photonic devices whose dimensions are less than or of the order of 1 μm, that is a size comparable to those of the functional units in electronic devices.

Ion implantation can be a possible alternative to lithography especially when extended plasmonic structures have to be synthesized, for example sensing applications. Nanopatterning of a substrate may be obtained by ion implantation through some mask which is able to stop incident ions in prescribed regions. Self assembled structures can be deposited on the surface of the substrate to be implanted and then chemically removed after the implantation. Subsequent additional thermal treatments for promoting cluster growth will be performed. Low energy (20–50 keV) ion implantation may be performed through the mask, so that most of the ions are stopped within the nanospheres and the remaining ones are buried just beneath the substrate surface, where they can grow under post-implantation treatments. Additional control on the exposed area can be obtained by suitable thermal treatments of the masks, promoting partial sintering and reduction of the mask "holes". Recently, we combined ion implantation and nanosphere lithography to regularly dope, by a mask-assisted process, a SiO_2 substrate with rare earth ions (Er) by ion implantation and to fabricate by sputtering a plasmonic 2D periodic array of Au nanostructures on the silica surface spatially coupled to the implanted Er^{3+} ions.

10.8 CONCLUSION

The aim of this work was to study how Er^{3+} emission at 1.5 μm can be affected by the interaction with a plasmonic nanostructure. A variation of the radiative lifetime of the Er^{3+} emission and a change from single exponential to biexponential of the luminescence intensity decay has been observed. Obviously the synthesis of such kind of optimized nanostructures will require new strategies if ion implantation techniques have to be used. Although competing techniques like colloidal chemistry or lithography are at present more widely diffused for their better control over size and spatial resolution, respectively, ion implantation based techniques also coupled in some hybrid schemes to self assembly strategies, can be fruitfully applied to obtain a large class of optically interesting materials and devices, embedded in transparent matrices, which can preserve their structure and functionality, allowing chemical, thermal, and mechanical stability. Moreover, ion implantation, coupling its highly desirable intrinsic feasibility for large area substrate patterning to its compatibility with contemporary silicon nanotechnology will still be one of the reference techniques in the next future for nanophotonics and nanoplasmonic devices.

KEYWORDS

- **Azodicarbonamide**
- **Diffusion coefficients**
- **Luminescence**
- **Nanoclusters**
- **Surface plasmon resonance**
- **Transmission electron microscopy**

ACKNOWLEDGMENT

The present review is dedicated to Professor Viswhas Kulkarni, Director of IIT Research Centre at Kanpur, India, a special friend of us, who died on July 24th, 2010, and who gave fundamental contributions to the field of ion beam assisted processing of materials.

REFERENCES

1. Mazzoldi, P. and Mattei, G. *Rivista del Nuovo Cimento*, **26**(7), 1 (2005).
2. Mazzoldi, P. and Mattei, G. *Physica Status Solidi A.*, **204,** 621–623 (2007).
3. Mazzoldi, P. and Mattei, G. *Metal nanoclusters in catalysis and material Science in Part II.* B. Corain, G. Schmid and N. Toshiba (Eds.). Elsevier, p. 269 (2008).
4. Mattei, G., Mazzoldi, P., and Bernas, H. Metal nanoclusters for optical properties, *Material Science with Ion beams*, Topics in Applied Physics. H. Bernas (Ed.) Springer -Verlag Heidelberg, **116**, (2010).
5. Trave, E., Mattei, G., Mazzoldi, P., Pellegrini, G., Scian, C., Maurizio, C., and Battaglin, G. *Applied Phys. Letters*, **89,** 151121-1/151121-3 (2006).
6. Aggioni, G., Vomiero, A., Carturan, S., Scian, C., Mattei, G., Bazzan, M., Fernandez, de Julian C., Mazzoldi, P., Quaranta, A., and Della Mea, G. *Applied Phys. Letters*, **85**(23), 5712–5714(2004).
7. Fernandez, de Julian C., Manera, M. G., Spadavecchia, J., Maggioni, G., Quaranta, A., Mattei, G., Bazzan, M., Cattaruzza, E., Bonafini, M., Negro, E., Vomiero, A., Carturan, S., Scian, C., Della Mea, G., Rella, R., Vasanelli, L., and Mazzoldi, P. *Sensors and Actuators B*, **111-112**, 225–229 (2005).
8. Quaranta, A., Carturan, S., Bonafini, M., Maggioni, G., Tonezzer, M., Mattei, G., Fernandez, de Julian C., Della Mea, G., and Mazzoldi, P. *Sensors and Actuators B*, **118**, 418–424 (2006).
9. Carturan, S., Quaranta, A., Bonafini, M., Vomiero, A., Maggioni, G., Mattei, G., Fernandez, de Julian C., Bersani, M., Mazzoldi, P., and Della Mea, G. *Eur. Phys. J. D*, **42**, 243–251 (2007).
10. Manera, M., Fernandez, de Julian C., Maggioni, G., Mattei, G., Carturan, S., Quaranta, A., Della Mea, G., Rella, R., Vasanelli, L., and Mazzoldi, P. *Sensors and Actuators B*, **120**, 712 718 (2007).
11. Mattei, G., Mazzoldi, P., Post, L. M., Buso, D., Guglielmi, M., and Martucci, A. *Advanced Materials*, **00930**, 1–5 (2006).
12. Miotello, A., Marchi, G. De, Mattei, G., Mazzoldi, P., and Sada, C. *Phys. Rev. B* **63**, 075409 (2001).
13. Wagner, C. Z. *Elektrochem.*, **65**, 581 (1961).
14. Lifshitz, I. and Slezof, V. J. *Phys. Chem. Solids*, **19**, 3–5 (1961).
15. Mie, G. *Ann. Phys., Leipzig*, **25**, **377** (1908).
16. Yasuda, H. and Mori, H. *Phys. D*, **31**, 131 (1994).
17. Mazzoldi, P., Mattei, G., Maurizio, C., Cattaruzza, E., and Gonella, F. *In Engineering Thin Films and Nanostructures with Ion Beams*, Chapt. 7. Emile Knystautas (Ed.). CRC Press pp. 82–158 (2005).
18. Gonella, F. and Mazzoldi, P. Metal Nanocluster Composite Glasses, *in Handbook of Nanostructured Materials and Nanotechnology*. H. S. Nalwa (Ed.). Academic Press, S. Diego 4, 82–158 (2000).
19. Ito, T., Kitakami, O., Shimada, Y., Kamo, Y., and Kikuchi, S. *J. Magn. Mater.*, **235**, 165 (2000).
20. Cattaruzza, E. *Nucl. Instrum. Methods B*, **169**, 141 (2000).
21. Kreibig, U. and Vollmer, M. *Optical Properties of Metal Clusters*, Springer-Verlag, Berlin Heidelberg (1995).
22. Battaglin, G., Cattaruzza, E., Gonella, F., Mattei, G., Mazzoldi, P., Sada, C., and Zhang, X. *Nucl. Instrum. Methods B*, **166–167**, 857 (2000).

23. Battaglin, G., Catalano, M., Cattaruzza, E., D'Acapito, F., Fernández, de Julián, C., De Marchi, G., Gonella, F., Mattei, G., Maurizio, C., Mazzoldi, P., Miotello, A., and Sada, C. *Nucl. Instrum. Methods B*, **178**, 176 (2001).
24. Mattei, G., Fernández, de Julián C., Battaglin, G., Maurizio, C., Mazzoldi, P., and Scian, C. *Nucl. Instrum. Methods B*, **250**, 225 (2006).
25. Fernández, de Julián C., Tagliente, M., Mattei, G., Sada, C., Bello, V., Maurizio, C., Battaglin, G., Sangregorio, C., Gatteschi, D., Tapfer, L., and Mazzoldi, P. *Nucl. Instrum. Methods B*, **216**, 245–250 (2004).
26. Mattei, G., Battaglin, G., Bello, V., Cattaruzza, E., Fernández, de Julián C., De Marchi, G., Maurizio, C., Mazzoldi, P., Parolin, M., and Sada, C. *Nucl. Instrum. Methods B*, **218**, 433 (2004).
27. Fernández, de Julián C., Mattei, G., Sangregorio, C., Tagliente, M., Bello, V., Battaglin, G., Sada, C., Tapfer, L., Gatteschi, D., and Mazzoldi, P. *J. Non-Cryst. Solids*, 345–346, 682 (2004).
28. Mattei, G., Maurizio, C., Mazzoldi, P., D'Acapito, F., Battaglin, G., Cattaruzza, E., Fernández, de Julián C., and Sada, C. *Phys. Rev. B*, **71**, 195–418 (2005).
29. Cattaruzza, E., D'Acapito, F., Gonella, F., Longo, A., Martorana, A., Mattei, G., Maurizio, C., and Thiaudiere, D. *J. Appl. Cryst.*, **33**, 740 (2000).
30. Fernández, de Julián C., Sangregorio, C., Mattei, G., Maurizio, C., Battaglin, G., Gonella, F., Lascialfari, A., Lo Russo, S., Gatteschi, D., Mazzoldi, P. Gonzalez, J., and D'Acapito, F. *Nucl. Instrum. Methods B*, **175–177**, 479 (2001).
31. Bertoncello, R., Gross, S., Trivillin, F., Caccavale, F., Cattaruzza, E., Mazzoldi, P., Mattei, G., Battaglin, G., and Daolio, S. *J. Mater. Res.*, **14**, 2449 (1999).
32. Mattei, G., Fernández, de Julián C., Mazzoldi, P., Sada, C., De, G., Battaglin, G., Sangregorio, C., and Gatteschi, D. *Chem. Mater.*, **14**, 3440 (2002).
33. Cattaruzza, E., Battaglin, G., Gonella, F., Mattei, G., Mazzoldi, P., Polloni, R., and Scremin, B. *Appl. Surf. Sci.*, **247**, 390 (2005).
34. Cattaruzza, E., Battaglin, G., Gonella, F., Polloni, R., Mattei, G., Maurizio, C., Mazzoldi, P., Sada, C., Tosello, C., Montagna, M., and Ferrari, M. Philos. Mag. B, **82**, 735 (2002).
35. Martucci, A., Pasquale, M., Guglielmi, M., Post, M., and Pivin, J. C. *J. Am. Ceram. Soc.*, **86**, 1638 (2003).
36. Ando, M., Kobayashi, T., and Haruta, M. Catal. Today, **36**, 135 (1997).
37. Mattei, G., De Marchi, G., Mazzoldi, P., Sada, C., Bello, V., and Battaglin, G. *Phys. Rev. Lett.*, **90**, 085502/1 (2003).
38. Mattei, G., Battaglin, G., Bello, V., De Marchi, G., Maurizio, C., Mazzoldi, P., Parolin, M., and Sada, C. *J. Non-Cryst. Solids*, **322**, 17 (2003).
39. Hache, F., Ricard, D., and Flytzanis, C. *J. Opt. Soc. Am. B-Opt. Phys*. **3**, 1647 (1986).
40. Hamanaka, Y., Fukuta, K., Nakamura, A., Liz-Marzan, L. M., and Mulvaney, P. *Appl. Phys. Lett.*, **84**, 4938 (2004).
41. Maier, S. A., Brongersma, M. L., Kik, P. G., and Atwater, H. A. *Phys. Rev. B*, **65**, 193–408 (2002).
42. Penninkhof, J. J., Polman, A., Sweatlock, L. A., Maier, S. A., Atwater, H. A., Vredenberg, A. M., and Kooi, B. *J. Appl. Phys. Lett.*, **83**, 4137 (2003).
43. Zou, S. L., Janel, N., and Schatz, G. C. *J. Chem. Phys.*, **120**, 10871 (2004).
44. Sweatlock, L. A., Maier, S. A., Atwater, H. A., Penninkhof, J. J., and Polman, A. *Phys. Rev. B*, **71**, 235408 (2005).
45. Gaudry, M., Lermé, J., Cottancin, E., Pellarin, M., Vialle, J. L., Broyer, M., Prével, B., Treilleux, M., and Mélinon, P. *Phys. Rev. B*. **64**, 085407 (2001).
46. Krenn, J. R., Dereux, A., Weeber, J. C., Bourillot, E., Lacroute, Y., Goudonnet, J. P., Schider, G., Gotschy, W., Leitner, A., Aussenegg, F. R. et al. *Phys. Rev. Lett.*, **82**, 2590 (1999).
47. Li, K. R., Stockman, M. I., and Bergman, D. *J. Phys. Rev. Lett.*, **91**, 227402 (2003).
48. Talley, C. E., Jackson, J. B., Oubre, C., Grady, N. K., Hollars, C. W., Lane, S. M., Huser, T. R., Nordlander, P., and Halas, N. *J. Nano Lett.*, **5**, 1569 (2005).
49. Pellegrini, G., Bello, V., Mattei, G., and Mazzoldi, P. *Optics Express*, **15**, 10097–10102 (2007).
50. Pellegrini, G., Mattei, G., Bello, V., and Mazzoldi, P. *Mat. Sci. Eng. C*, **27**, 1347 (2007).

51. Talley, C. E., Jackson, J. B., Oubre, C., Grady, N. K., Hollars, C. W., Lane, S. M., and Huser, T.R., Nordlander, P. and Halas, N. *J. Nano Lett.*, **5**, 1569 (2005).
52. Brongersma, M. L., Hartman, J. W., and Atwater, H. A. *Phys. Rev. B*, **62** (24), R16356 (2000).
53. Maier, S. A., Kik, P. G., Atwater, H. A., Meltzer, S., Harel, E., Koel, B. E., and Requicha, A. A. G. *Nature Mater.*, **2**, 229 (2003).
54. Perotto, G., Bello, V., Cesca, T., Mattei, G., Mazzoldi, P., Pellegrini, G., and Scian, C. *Nuclear Instruments and Methods B*, **268**, 3211–3214 (2010).

11 From Single Atom to Nanocomposites

Beata Kalska-Szostko

CONTENTS

11.1 Introduction 197
11.2 Nanoparticle Preparation 198
 11.2.1 Neighborhood Modification 200
 11.2.2 Core Modification 203
 11.2.3 Spontaneous Core-shell Structures 208
 11.2.4 Purposely Core-shell Structures 215
11.3 Nanowires Preparation 223
 11.3.1 Aao Matrix 224
 11.3.2 Single Phase Nanowires Deposition 227
 11.3.3 Core-shell Nanowires 227
11.4 Surface Functionalization 231
11.5 Composites 232
11.6 Conclusion 233
Keywords 234
Acknowledgment 234
References 234

11.1 INTRODUCTION

The last decade of the 20th century was very fruitful for nanotechnology. Many different materials were designed and their properties were measured. New materials with novel properties are discovered. Many scientists are working on the studies to develop certain needed physical and chemical properties of nanomaterials. Nowadays everybody can find products with some nanotechnological achievements [1, 2]. Nevertheless, there are still many possibilities to study novel materials with unique properties and find proper application for them.

There is a very weak border between nanomaterials and bulk materials in defining size [3, 4]. Thin films possess macroscopic dimension in two directions and only one is of the order of few nanometers. Nanowires have also length in micrometers but here the diameter is in nanometers [5]. This is enough for the fact that the properties change drastically when even only one of the dimensions goes below 10^{-7} m and it does not have to the structure built up of only few atoms. For example, 10 nm Co or Au nanoparticles can have between 35,000–45,000 atoms (depending on the crystal structure) but the properties which they possess are much different from the bulk one [6]. For some, such structures are small when we consider specific application and must deal with manipulation of the objects. For others, they can be huge when one studies the adsorption of separated atoms on the surface. It is very similar in case of the definition of nanopores below 100 nm. For some, they are called micropores, for others, quasi-zero dimensional structures [7]. Every specific society has established their own language and a set of definitions for specific materials and applications.

Nanomaterials can be produced in two different approaches by self assembly of single atoms in bigger structures, which was firstly observed in biology and chemistry. This way was called a bottom-up approach [8]. On the other hand, nanostructures can be obtained by purposeful degradation of a bigger one by physical or chemical means until single structures will be obtained in nanosize. This approach is called a top-down [9].

In this work I would like to present two different approaches which lead to obtain nanocomposites.

11.2 NANOPARTICLE PREPARATION

Nanoparticles are the objects which have been known since the 60s of the 20th century [10]. At the beginning they were not very popular due to few facts. At first the homogeneity of such powders was not good enough. On the other hand, there was no proper equipment for nano studies. The imaging of the structures was very problematic and scientists had no experience how to deal with materials at such a small scale. The number of publications was increasing slowly reaching a very high rate at the end of the last century after assembling atomic force microscopes and developing transmission and scanning electron microscopes [11] for material science. The last decade was then dedicated mainly to nanoparticle research. Many various publications appear every year, which is presented in Figure 1 on the graph. It can be seen there that every five years a number of publications increases five times (2010 only until July). This is a huge number but much more research is needed to fully understand the problems connected with nanoscience.

Concerning nanoparticles, one should think about objects and their environment as one. This causes that the complexity of the problems present in the systems in connection with nanoparticles and nanopowders is huge. In many cases the properties of fine particles depend on the way of particles preparation, their crystal structure, size, organic material which is attached to the surface, solution and so on [12]. This is due to the fact that the particles have a relatively large surface to volume ratio, even for the size of 20–30 nm it can be of the order 30% depending on the structure and the surface area [13]. The broken bonds at the surfaces can strongly modify physical and chemical properties of the materials. The same can happen by introducing interfaces inside the particles. The origin of it can be due to the presence of a layered structure in nanoparticles or due to polycrystalinity. In both cases the modifications of the properties are expected but their nature can be totally different. In such instance one can see that the nanoparticle system contains: core, interface core-shell, shell, interface shell-neighborhood, and neighborhood (Figure 2) [14]. Such scenario stresses once again that even small particle can be described and considered as an extremely huge source of problems to be resolved by scientists. Each variation of any of the mentioned parameter can totally change properties and this is a large area for studies for scientists: chemists, physicists, and technologists.

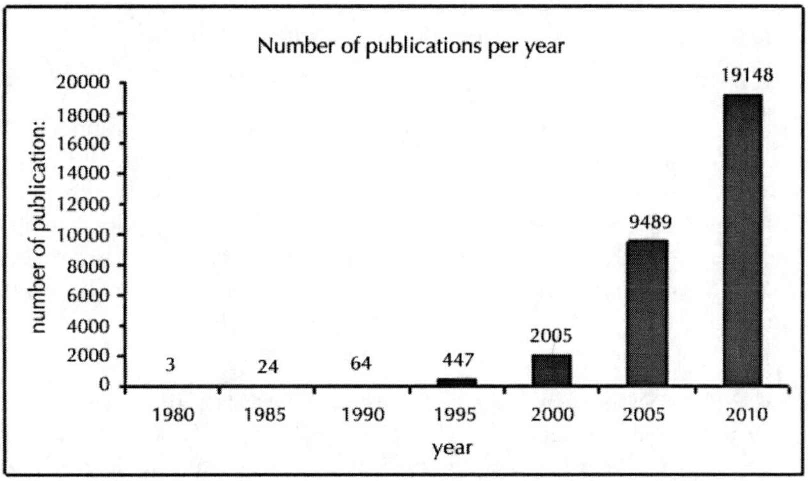

FIGURE 1 Number of publications appearing every five years (based on selected databases).

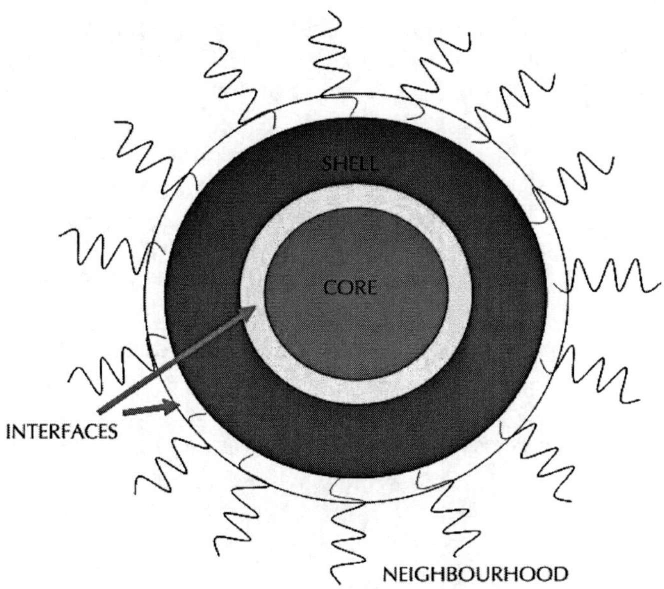

FIGURE 2 Schematic drawing of the particles inert layered structure.

In this study, I will focus on some structural and also magnetic properties seen mainly by Mössbauer spectroscopy as well as their modifications due to the variation of different nature in the nanoparticle system. Any structural changes in the particles cause changes in inert crystallographical structures of the objects and therefore in the properties too. In principle, the crystal structure can be changed or relaxation of a large number of atoms due to close vicinity of the surface can appear [15]. Both influence drastically on physical properties including the magnetic ones.

Magnetism of the particles in nanoscale is very useful not only in technology (ferrofluids, IT industry and so on) but also in medicine and biology where such particles can be used as diagnostic contrasts or drug delivery media [16].

11.2.1 Neighborhood Modification

The wet chemical route to prepare nanoparticles is an alternative to preparation by physical methods like lithography and sputtering, among others. For this study, Massart's [17] method of nanoparticles preparation was adapted with some modifications [18]. It was observed that the conditions of the sample preparation had a significant influence on the properties of the final particles as evidenced, for example, by the difference in Mössbauer spectra [19]. This fact suggests an influence of the chemical environment and size on the magnetic properties of the particles.

Characterization

The X-ray diffraction measurements to determine the crystal structure of the particles were carried out. The shape and position of main diffraction peaks allow to conclude that the obtained material is crystalline and chemical structure is magnetite or maghemite. The position and relative intensity of all diffraction peaks match well with Fe_3O_4 nanoparticles published [20].

To describe the shape of obtained nanomaterials the example transmission electron microscopic (TEM) images of the particles synthesized in water and in alcohol are shown in Figures 3(a) and (b), respectively. Both pictures were taken with the same magnification.

From the images can be seen that by using different solvents, keeping other conditions the same, a change of the final size of the particles is observed [18]. The reactions taking place in water result in bigger particles compared to in alcohol ones. From the analysis of the TEM images was found that the average size of particles in water is about 13 nm, and in alcohol case 8 nm diameters is obtained [18]. Error bar the same for both cases.

Normally to prepare samples for Mössbauer spectroscopy the initial solutions were dried out to obtain powder. This powder was mixed with BN and formed in tablets which contain about 12 mgFe/cm^2. The room temperature Mössbauer spectra were measured in all cases by the spectrometer working in the constant acceleration mode with ^{57}Co*Rh* as a source. All spectra analyses were done with NORMOS fitting package.

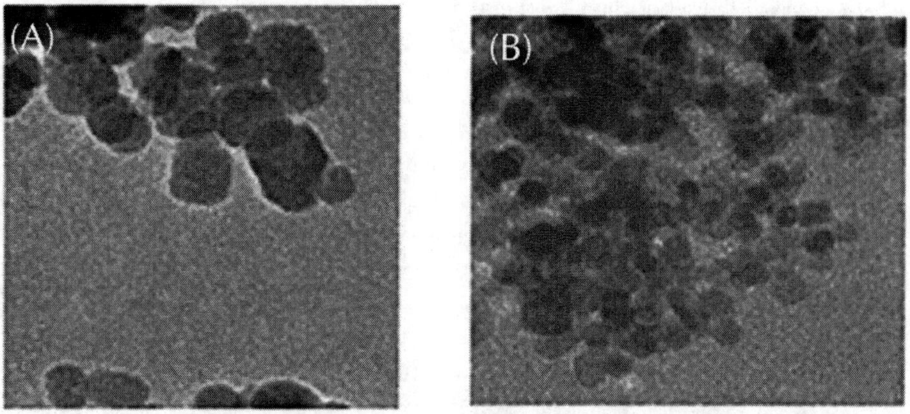

FIGURE 3 The TEM (100 × 100 nm) image of the magnetite nanoparticles: (a) synthesis in water; (b) in alcohol (M. Giersig FU Berlin).

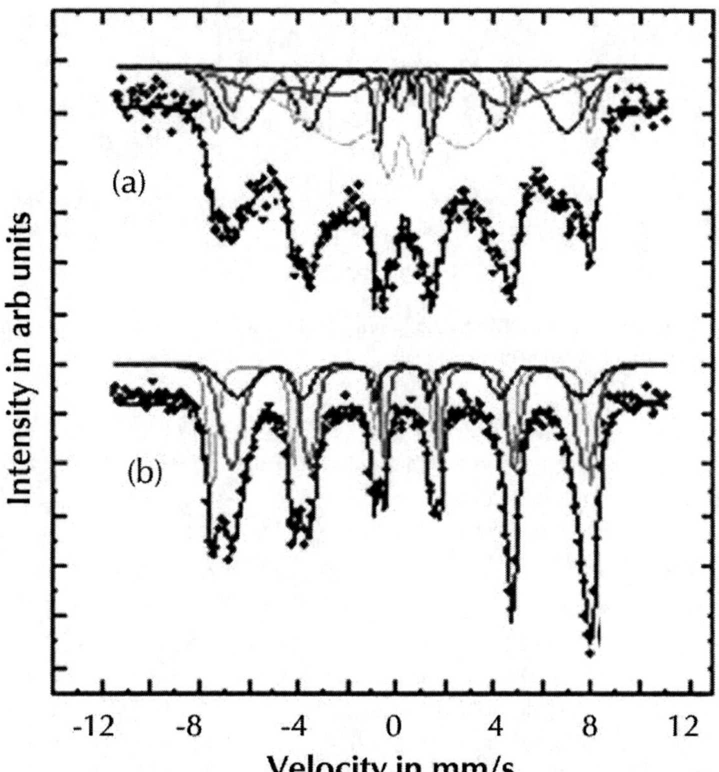

FIGURE 4 Room temperature Mössbauer spectra collected for synthesis where alcohol was as a solvent (a) and water was as a solvent (b) Synthesis time 3 hr.

In Figure 4 the set of room temperature Mössbauer spectra are presented. Spectra (a) from in alcohol and (b) from in water 3 hr lasting reactions are shown. The spectra measured for nanoparticles prepared in water are well resolved while those prepared in alcohol are broader with large superparamagnetic central part. The presence of the low hyperfine field component is due to the presence in the sample particles with the size for which the relaxation processes start below room temperature. These results are well correlated with one obtained from the TEM analysis. There presented images shows that in alcohol smaller particles compare to water condition have been obtained [18].

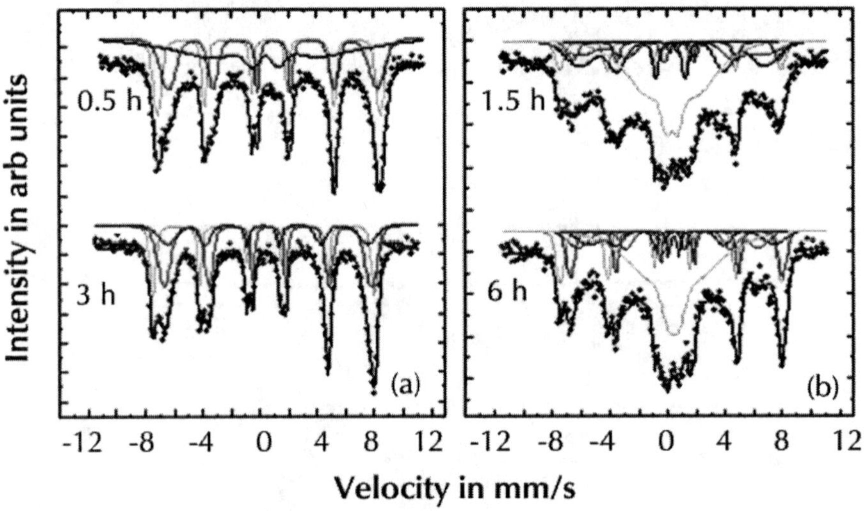

FIGURE 5 Room temperature Mössbauer spectra collected for in water (a) and in alcohol (b) synthesis with changed reaction time.

The time evolution of Mössbauer spectra is depicted in Figure 5. Here we see that prolongation of the reaction time influences shape of the spectra which become closer to the bulk magnetite one where clear two subspectra (two sextets) suppose to be seen. This process, however, is quite slow and especially for alcoholic reaction it must be prolonged significantly [18].

Discussion

To evaluate spectra have been applied a model with three subspectra which were fitted to the data for nanoparticles in water solutions (Figure 4(b) easier case). The two components with largest hyperfine field were assigned to A and B site of Fe in the magnetite crystal structure. The obtained values of hyperfine parameters such as isomer shift and quadrupole splitting were very close to typical bulk parameters [21]. The fitted hyperfine magnetic fields were slightly lower than in bulk, which is consistent with the data published [18, 19, 22, 23]. The third sextet has hyperfine magnetic field, isomer shift and quadrupole splitting belonging to neither A or B position. It was assigned to:

the surface part of the particles where the hyperfine parameters are modified [19], the existence of relaxation process [24], and oxidation process [25].

The reaction time influences the shape of the measured spectra, which can be seen from Figure 5(a). The longer the reaction time, the better I(Fe(B))/I(Fe(A)) ratio reaching almost 1.7 for longest reaction in water [18]. Obtained values were, however, smaller than 1.94 observed for stoichiometric bulk magnetite [22]. In such instance it was concluded that longer reaction time improves slightly the stoichiometry of the obtained nanoparticles. And is almost equal value obtained in first studies on nano-magnetite [26].

In Figure 5(b) time evolution of the Mössbauer spectra for samples prepared with the use of alcohol as a solvent are presented. The model to be fitted to those spectra was slightly more complicated and needed to add two more components. Here also have been seen that the increase of the reaction time led to gradually better resolved spectra particularly in the outermost part. The obtained I(Fe(B))/I(Fe(A)) ratios, however, were much lower than in water case (for the longest reaction time is less than 0.85) [18].

With the increase of the reaction time, the fraction in the middle part of the spectra (collapsed sextet) was decreasing from 82 to 68% [18]. This observation can be explained as a gradual increase of the particles size with increasing reaction time as well as improvement of stoichiometry of the core. The discussed part of the spectrum comes from the particles which are close to the transition from superparamagnetic to ferrimagnetic state. It means that for round in shape magnetite nanoparticles covered with TBAOH the critical size to become ferromagnetic is slightly below 8 nm. It has been reported that the paramagnetic doublet can be observed for the magnetite particles smaller than 10 nm embedded in polymer matrix [27]. The difference in the magnetic behavior of the particles can have origin in the strength of dipole interaction due to the interaction nature and the distance between nanomaterials in different matrixes. It shows the importance of environment chemistry for nanoscale objects.

To conclude this part it was proven that: aqueous and alcoholic synthesis leads to various sizes of the particles for all other synthesis parameters kept the same, the prolongation of the reaction time improves the stoichiometry of nanoparticles, the transition to superparamagnetic state at room temperature occurs for particles below 8 nm of diameter in case of magnetite nanoparticles covered with TBAOH.

11.2.2 Core Modification

Alloy like Composition

Magnetite is an oxide present in nature in inverse spinel structure with Fe^{3+} in the tetrahedral A-site and Fe^{3+} and Fe^{2+} in octahedral B-site [28, 29]. A lot of fabrication methods of iron oxides have been studied in details [30-32] but the formation of nanoferrites or their derivatives by incorporation of the substitution elements into their structure is still under discussion [33]. Stochiometric ferrites $MeFe_2O_3$ can have very complicated magnetic structure such as: non collinear spin structures, spin-glass-like structures or radial reorientation due to unique nanoparticles properties [34-37]. Non-stoichimetric ferrites $Me_{(1-x)}Fe_{(2+x)}O_3$ can have Fe in both valence states or only one

Fe(III) depending on the preparation procedure [35, 38]. In general consideration the substitution elements can occupy the A Fe sites and/or B Fe sites [38-41].

There will be presented changes in magnetic properties of the ferrite nanoparticles of the size in the range 13±2 nm by changing the Me/Fe^{2+} ratio in the magnetite based materials. The used substitution materials are magnetic elements (Mn, Co, Ni) and also nonmagnetic (Zn, Ca, Au).

The Mössbauer measurements were carried out in order to study the influence of the substitution transition metals Me/Fe^{2+} (where Me = Co, Ni, Mn) 10:90, and 90:10 on magnetic properties of the ferrite nanoparticles [42] as well as non magnetic metals as Ca, Al, Zn for composition 50:50.

Characterization

The magnetite nanoparticles were synthesized following the same originally Massart's methods as in the paragraph where the chlorides of the respective metals were used [17]. The substitution elements were added instead of part of Fe^{2+} chlorides. The description 90:10 means 90% Fe^{2+} 10 % Me^{2+} (where Me are Co, Ni, Mn, Zn, Al, Ca). The Mössbauer studies are made on the powder samples after drying the solution containing the fabricated particles. The Mössbauer equipment with constant acceleration mode was used with α-Fe foil as a reference and the Co*Rh* radioactive source.

Discussion

In Figure 6 the example of the TEM images of pure magnetite and substituted nonstochiometric ferrites are depicted. The reference material (magnetite) contains spherical particles with an average size of the order of 13±2nm [42].

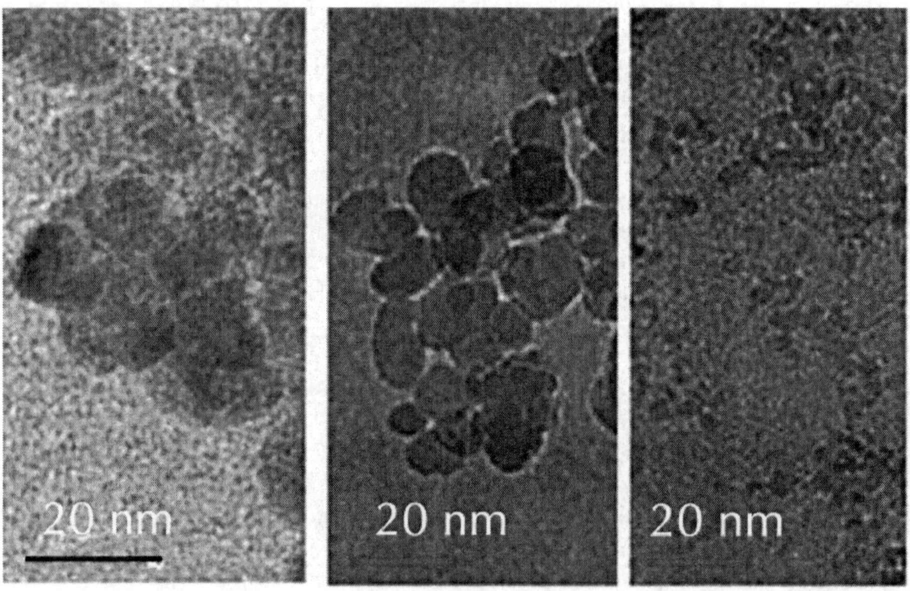

FIGURE 6 The TEM images of: (a) magnetite; (b) Ni (90:10) and (c) Ni (10:90) particles (M. Giersig FU Berlin).

The Mössbauer spectra were fitted with the simplest model in which the three sextets and one doublet were taken into account. The same as on page 4 of this paragraph. The doublet describes nonmagnetic and superparamagnetic part of the spectra [42].

(a) Substitution of Magnetic Elements Co, Ni, Mn

In Figure 7 the set of Mössbauer spectra is presented. As one can see the top most spectrums shows the pure magnetite composition where the sextets belonging to Fe in octahedral and tetrahedral crystallographic positions are partly separated. The 10% substitution Fe^{2+} by Co drastically influences the Mössbauer spectra. The change of the relative intensity between Fe(A) and Fe(B) sextets is clearly seen. Further substitution causes presence of the doublet in central part of the spectra. For sample (10:90) the doublet is taking more than 50% of the total spectral intensity, which suggests that the blocking temperature for this sample falls far below the room temperature. As a result the magnetic part of these spectra is represented only by a very broad sextet [42].

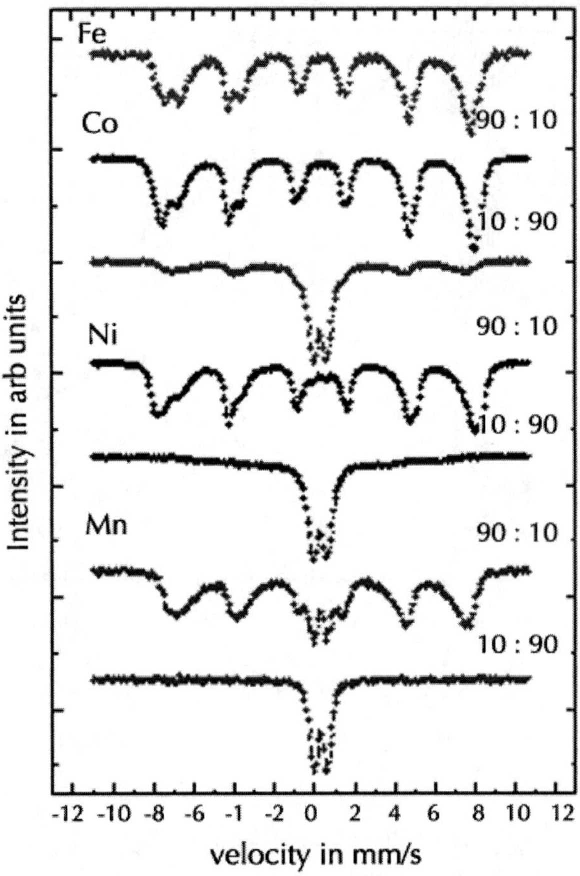

FIGURE 7 Mössbauer spectra of magnetite substituted by magnetic elements in two different ratio Me^{2+}/Fe^{2+} 10:90 and 90:10. In the figure the reference magnetite spectra are included.

In Figure 7 the ferrites samples with composition 10% and 90% substituted Fe^{2+} by Ni are also depicted. In this case similar to the previous one the coexistence of sextets and a doublet is seen. However, the presence of a doublet is remarkable already in the first sample (case with substituted 10% of Fe). Further addition of Ni causes great changes of the room temperature Mössbauer spectra transition of the sextet into doublet. The last composition containing 90% of Ni shows almost only paramagnetic doublet, which is different to the case showed.

In the same Figure 7 the samples containing the Mn as the doping material are presented. The clear change of the spectra is seen after 10% substitution Fe by Mn compared to magnetite reference. The significant doublet in the middle part of the spectra is seen already for first sample. The broadening of the sextet is the same as in case of the transition from magnetic to paramagnetic state.

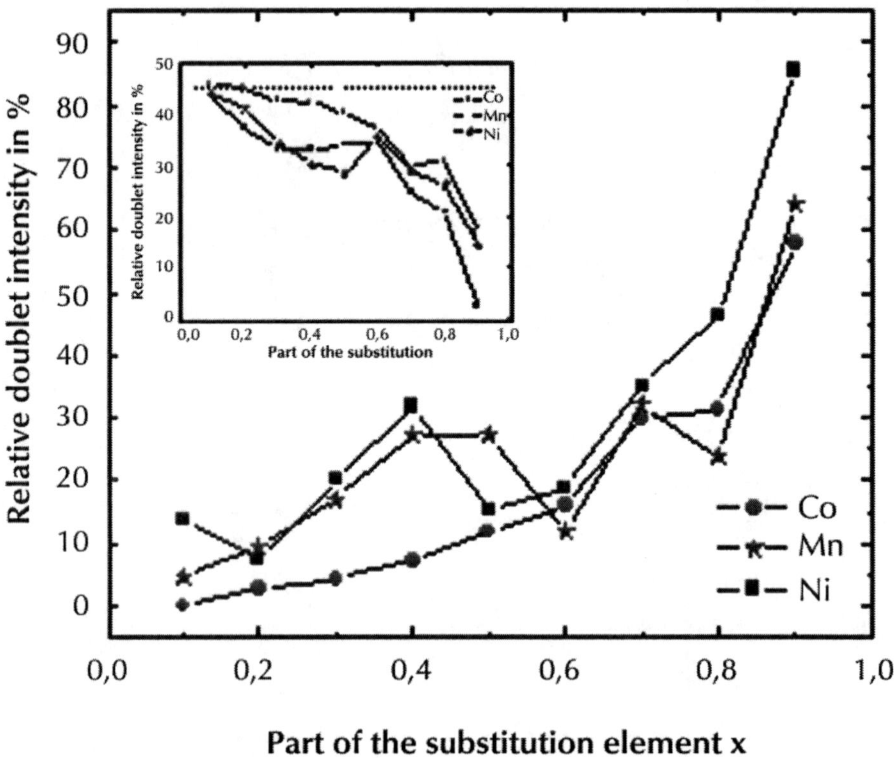

FIGURE 8 Plot of the doublet intensity *versus* the Co, Ni and Mn concentration (inset shows the change of average hyperfine field).

In Figure 8 the changes of doublet intensities for all three substituents are plotted. The depicted graph shows the gradual increasing of the doublet intensity with the amount of Co added to the sample [42]. The hyperfine parameters extracted from the fitting procedure are following: quadrupole splitting = (0.67±0.03)mm/s and isomer shift = (0.34±0.02)mm/s, which falls well with the values obtained for superparamagnetic $CoFe_2O_4$ [35, 43], but disagrees with these reported by others [30, 44, 45].

The observed doublet intensity as a function of the concentration for Ni (Figure 8), can be seen that the increase is not as monotonic as in Co case. A small reduction of the doublet intensity around 50 : 50 after which strong increase is observed. The fitted doublet hyperfine parameters has exactly the same values as in Co case, which does not agree with the data reported for superparamagnetic $NiFe_2O_4$ (IS = 0.21(6) mm/s and QS = 0.40(1) mm/s) [46].

On the other hand, Figure 8 shows also that the increase of the doublet intensity for Mn is not as monotonic above 50% as below that concentration. But in Mn high concentration part much larger error bars have to be taken into account due to overlapping of the broad background sextet with the paramagnetic doublet. Here the observed doublet possess also the same hyperfine parameters as in Co and Ni cases.

According to our observation, from ferrites derivatives study of nonstochiometric magnetite substituted by 3d elements a similar doublet was observed for all samples [42] and it can be recognized as a mixture magnetite/maghemite at the surface of nanoparticles, which agrees with some literature data [27]. The existence of only one of the type of Fe (one valence state II or III) in the sample can be excluded due to the disagreement of the hyperfine parameters obtained by us with those reported in the literature [40, 45, 47].

The inset in Figure 8 shows changes of an average value of the hyperfine field with the concentration of a substitution element. Co substitution shows almost monotonic decrease of the total average hyperfine field where Ni and Mn curves show two different slopes for the parts below and above 50%. The most significant changes in the value of the average hyperfine fields are observed in Ni case where the field decreases from the typical magnetite value to 0, which is also seen on Mössbauer spectra.

In the literature [45] the appearance of the doublet without reduction of the value of hyperfine field regarding magnetic part of the sample was explained by the reduction of the average particle size. At the same time the increase of the hyperfine field in the core due to increasing surface effects was expected.

General reduction of the total average hyperfine field does not confirm the theory proposed in the literature [45]. The small increase of the hyperfine field can be only marked for low Co concentration.

The successive appearance of the doublet with changes of the amount of the substitution element can suggest the reduction in the blocking superparamagnetic temperature due to changes in particle size and composition.

(b) Nonmagnetic Elements Substitution

In Figure 9 Mössbauer spectra for 50:50 ratio Me/Fe^{2+} are depicted. There both magnetic and nonmagnetic elements are included. The presence of nonmagnetic ions Al, Zn and Ca in the structure causes more drastic changes in Mössbauer spectra compared to 3d magnetic elements (Co, Ni and Mn). In the fitting procedure used model has been modified and applied. The observed modification of the spectra is other nature compare to studied case. For these ions the successive studies have not been done yet and gradual dependence of subspectra intensity for this particular case is not plotted. Similarly to magnetic dopping elements the obtained doublet in the middle part of the spectra has the same hyperfine parameters in all cases. This observation strength idea about surface magnetite/maghemite layer. In every case Mössbauer spectra with total nonmagnetic case for Al element was much more modified.

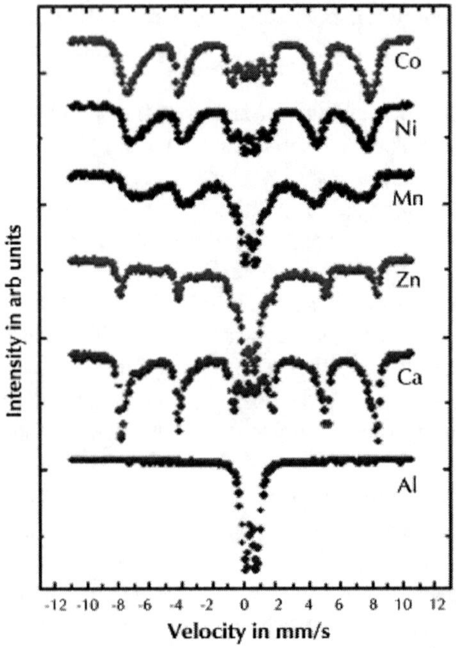

FIGURE 9 Mössbauer spectra of magnetite substituted with different element with ratio Fe^{2+}/Me^{2+} equal 50:50.

To conclude this part, it can be said that magnetic properties of the particles have changed after substitution both magnetic and non magnetic elements. The Mössbauer spectra of reference magnetite samples show well resolved magnetically split pattern connected with octahedral (A Fe positions) and tetrahedral (B Fe positions). The increase of the substitutions transition metal concentration caused broadening of the magnetic part of the spectra and the doublet appearance in the central part of the spectra for all presented samples. It hyperfine parameters is in all studied cases. The doublet develops much slower in case of Co compared to Mn and Ni. Co substitution shows monotonic increase of the doublet intensity when Ni and Mn show rather two step-like behavior. The incorporation of the substitution material into the structure leads to the decrease of the hyperfine field in all cases with the mildest reduction for Co element compared to Ni and Mn. Incorporation of nonmagnetic elements Ca, Zn, and Al changes Mössbauer spectra more drastically and in other manner compared to magnetic ions. In the case of the nominally 90% substitution by Ni pure superparamagnetic state is observed, while for Co and Mn some of the particles are still below blocking temperature. For non magnetic elements the substitution of Al only superparamagnetic state is observed. Ca causes the increase of the magnetic part in the total spectrum. Zn is in the transition state.

11.2.3 Spontaneous Core-shell Structures

It is very challenging to synthesize the particles because it was observed that the conditions of the sample preparation have large influence on its final properties [48, 49]. Quite often nominally the same samples show different Mössbauer spectra which is explained by the influence of chemical environment of the particles on the magnetic properties and their inert structure. In case of ferrite nanoparticles the microstructure of each particle can be composed of Fe^{3+} oxides (hematite, maghemite) and mixed valence oxides $Fe^{3+,2+}$ (magnetite) whose properties differ a lot especially in nanosize.

It is desired to extend the knowledge of the ferrites nanoparticles focusing on stoichiometry and the role of the surface layer.

In perfect bulk Fe_3O_4 component the ratio between Fe^{3+}/Fe^{2+} is equal to 2 [50]. The Mössbauer spectra show in this case two subcomponents with a relative ratio of intensity almost 2/1 (1.94) [22]. However, a different situation is observed in case of nanoparticles. Here the intensity ratio never reaches the value 2, which depends on the composition and size of the system [19, 24, 26.]. The surface ions have smaller hyperfine fields than the interior ones and it influences the line shape which becomes asymmetric [19]. Finally, the size distribution and/or the distribution of the blocking temperature of such superparamagnetic system changes the width and the line shape significantly, which modified the ratio between A and B subpatterns. Keeping all these points in mind, the interpretation of the spectra is not so straightforward because of the difficulties in the separation A and B.

Working with nano ferrites, it is very difficult to obtain a pure one phase system. This is because of very easy oxidation process going from magnetite (Fe_3O_4) to γ-Fe_2O_3 in the presence of oxygen [22, 51]. In the macroscopic system further oxidation to α-Fe_2O_3 is inhibited at room temperature because of the changes in the crystal structure and only the elevation of temperature to around 600–700K can lead to the phase transformation. However, in the nanoparticle systems such transformation is observed even around room temperature. The creation of α-Fe_2O_3 and γ-Fe_2O_3 is possible also by other route from α- and γ-FeOOH by water reduction from the system [52-54.].

Because the oxidation of magnetite can take place any time and the crystal structures of Fe_3O_4 and γ-Fe_2O_3 are the same, it is not easy to distinguish between these two using standard structural characterization (e.g. XRD). In case of the Mössbauer spectroscopy the situation is different. Here the spectra for bulk like materials can be easily separated. But in case of nanoparticles broadening and overlapping of the lines did not show a clear answer. In most cases one obtains the mixture between magnetite and γ-Fe_2O_3 because these two are isostructural materials. Because of that such ferrite can be named nonstechiometric magnetite or/and nonstechiometric γ-Fe_2O_3 and oxidized magnetite. All three names mean the same that is we do not obtain strictly magnetite particle but some vacancies are present. The more vacancies, the more γ-Fe_2O_3 like the spectrum is. But in general the particle has a core-shell structure with different thickness of each kind of the oxide layer.

Characterization

The magnetite nanoparticles were also synthesized following main steps from Massart's methods. [17] In all cases the reaction conditions were kept the same to be sure that changes present in the spectra are connected with modification of Fe^{3+}/Fe^{2+} ratio and not with other parameters [18, 55].

The microstructure of the selected samples is shown in the TEM in Figure 10. The particles are roughly spherical and rather agglomerated due to the strong magnetic interaction between them. The average size of the particles falls in the same range for all samples (13±2) nm [55]. The high resolution picture presents clearly that the particles core are crystalline and single crystal. The shell of the particles is not separated. The electron diffraction patterns show typical rings originating from the Fe_3O_4 and γ-Fe_2O_3 structures since the position of the diffraction patterns are almost the same in both of them [20, 56]. The obtained ring pattern does not show any additional reflexes which can be associated with the thick layer or separated particles of any other oxides.

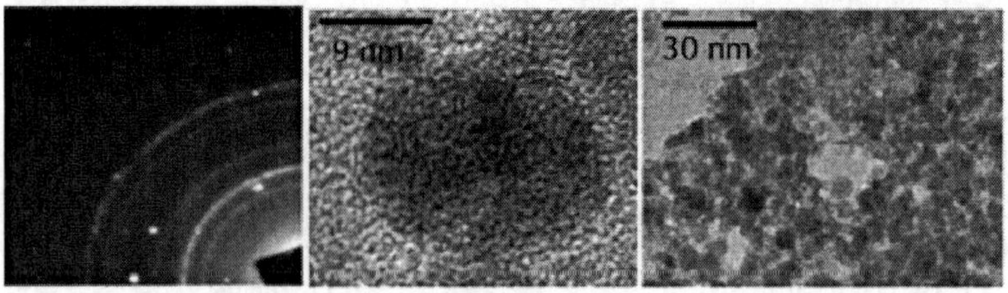

FIGURE 10 The TEM images and electron diffraction patterns of the sample $Fe^{3+}/Fe^{2+} = 2$.

Due to the fact that HR-TEM did not show any interface inside the particles, the considered α-Fe_2O_3 is present most probably at the surface of the particles with thickness less than one crystallographic unit cell.

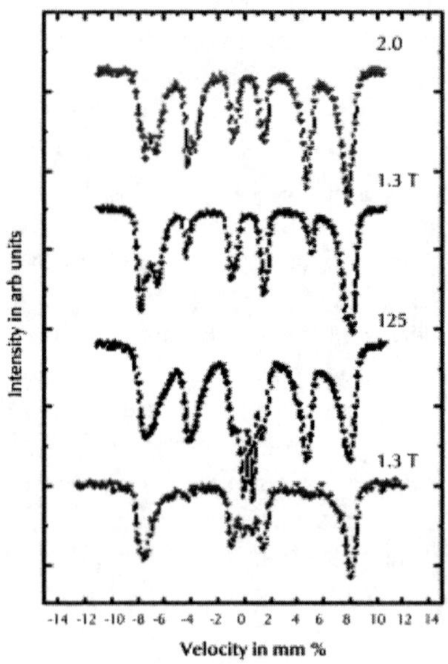

FIGURE 11 The room temperature Mössbauer spectra of the magnetite nanoparticles with different relative amount of Fe^{3+} to Fe^{2+} (1.25 and 2), respectively without and in external magnetic field of 1.3T.

In Figure 11 the set of Mössbauer spectra obtained for the particles with different ratio Fe^{3+}/Fe^{2+} in and without external magnetic field are depicted. As first the spectra collected without external magnetic field will be discuss. As can be seen the spectrum 2.0 shows quite well separated magnetic components. The 1.25 composition shows different spectra in comparison. Here the magnetic part is much overlapped due to relaxation processes and the exact hyperfine parameters cannot be extracted. In the central part of the spectrum the doublet is developed of the relative intensity of ~ 20% [55]. The fitted values of the hyperfine parameters of this doublet are equal to (isomer shift = 0.36mm/s, quadrupole splitting = 0.64mm/s) [25].

In the same figure (Figure 11) room temperature spectra collected in the external magnetic field of 1.3T parallel to γ-ray are also presented. For compositions 1.25 clear influence of the external magnetic field on the line shape of magnetic part is seen. The magnetic part of the spectra becomes much sharper. However, the intensity of the doublet was not changed. It is clearly seen that the intensity of lines 2, 5 are reduced almost to 0 and the width of the magnetic part of the spectra is significantly reduced. In case of composition 1.25 the sextet belonging to position B almost disappears and the presence of α-Fe_2O_3 or any other oxide is not so clear (which is in accordance with XRD and TEM results). In the presence of the external magnetic field spectrum 2.0 behaves differently. Here, this one shows clearly non vanishing lines 2, 5 and the reduction of the magnetic part's width is not so pronounced. From the fitting procedure have been obtained that the non vanishing lines 2, 5 are due to presence of α-Fe_2O_3 in the sample. Such conclusion was obtained on base that only the presence of a sextet with typical hyperfine parameters for α-Fe_2O_3 allows us to fit well the collected spectra.

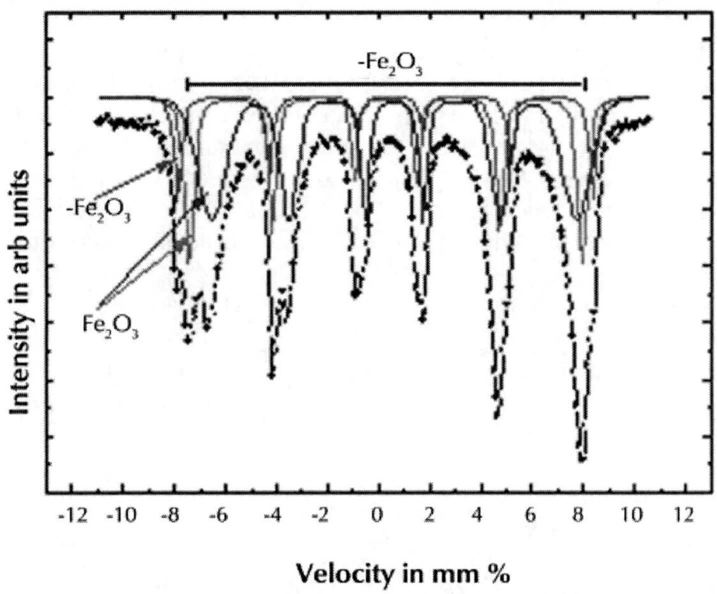

FIGURE 12 The fitting model of spectrum with marked position of subcomponents $Fe^{3+}/Fe^{2+} = 2$.

In order to describe the spectra not only magnetite and maghemite subcomponents are needed but also some additional subspectra with hyperfine parameters typical for hematite. Finally, spectrum *x=2.0* was fitted with 5 subcomponents, which were assigned to α-Fe_2O_3 (1 sextet), magnetite (2 sextets), maghemite (1sextet), surface Fe atoms, size distribution and relaxation process (1 broad sextet). For $x = 1.25$ instead of a broad sextet the doublet was needed to deconvolute the spectra [25].

Discussion

Collective magnetic excitations and relaxation processes have great influence on the shape of the spectra. The size distribution causes the asymmetry in the line shape. Surface ions have smaller hyperfine field, which leads to the asymmetric shape of the line. Therefore, small particles showed completely different spectra compared to large crystals, which can be explained as a deviation from the spinel structure, large surface effect, and changes in electron hopping. At the same time the hyperfine parameters can be changed compared to bulk-like materials. Symmetry breaking at the surface can produce electric field gradient and quadrupole moment, which causes non zero QS.

The estimation shows that for particles of around 13 nm in diameter made of magnetite the rough number of Fe atoms is about ~52000, and if we consider that one crystallographic cell is affected by the environment, this gives us the number of ~17000 of Fe atoms, which is around 30% of the total [25].

Oxygen absorption on the surface initiates electronic exchange with Fe^{2+}. This generates Fe^{3+} and cation vacancies at the surface. They diffuse into the particle and Fe^{2+} counterdiffuses from the inner layer into the surface. In such manner the gradient of oxygen in the particle creates more magnetite-like and maghemite-like layers and natural separation of these two phases and the core-shell structure can be considered [25].

The doublet hyperfine parameters are very astonishing since nonmagnetic (or superparamegnetic) α phase of Fe_2O_3 should have higher IS, for pure γ phase higher QS is expected, mixture magnetite/γ- Fe_2O_3 has hyperfine parameters quite close to our results (Table I). The presence of other oxides or rather hydroxides is also quite possible since in the preparation process they are presented as an intermediate step of the condensation process. However, the parameters did not match well with any of them, which suggest the mixture of few or the modification of parameters due to nanoscale size.

TABLE 1 Hyperfine parameters of possible Fe oxides and hydroxides, IS – isomer shift, QS – quadrupole splitting.

	β-FeOOH	*γ-FeOOH*	*Fe_5HO_x $4H_2O$*	*Particles*			
				Fe_3O_4		*γ-Fe_2O_3*	*α- Fe_2O_3*
IS in mm/s (RT)	0.38	0.37	0.35	0.36_{ref}	0.37_{exp}	0.55_{ref}	0.42_{ref}
	0.37		0.35				
QS in mm/s (RT)	0.55	0.53	0.62	0.75_{ref}	0.64_{exp}	0.75_{ref}	0.67_{ref}
	0.95		0.78				
T_N, T_C in K(bulk)	299	77	115/25				
	[57]	[57]	[57]	[27]		[58]	[59]

The relative ratio Fe^{3+}/Fe^{2+} between sextets originating in Fe type B and Fe type A describe the stoichiometry of the magnetite [22, 24, 26]. At the same time changes in the size of the particles can influence the value of the relative intensities[19, 22, 24, 26]. It was shown that such changes depend on temperature [19]. The exact evaluation of the ratio Fe^{3+}/Fe^{2+} is not easy in our opinion however, we made some attempts. On the other hand, to obtain separate intensities of subcomponents the measurements taken at room temperature in and without magnetic field should be compared. As it can be seen in Figure 11, the changes in the line shape are clearly seen. However, one should remember what happened between these two measurements. In case of subsextet described as α-Fe_2O_3 (antiferromagnetic) the use of the external magnetic field of 1.3T almost does not influence the value of the hyperfine field. The magnetite is a ferrimagnetic and sublattices A and B are opposite to each other and thus the sextets will be separated even more since A will have higher hyperfine field and B smaller, by about the value of the external magnetic field. This is not so pronounced in the figure due to overlapping between sextets. The use of the external magnetic field gives us much better separation between A and B positions of the magnetite but much more overlapping between lines A and α-Fe_2O_3. From the measurements taken without external magnetic field the more exact value of α-Fe_2O_3 can be evaluated. The obtained values of Fe^{3+}/Fe^{2+} and estimated γ-Fe_2O_3 content for extended series are plotted in Figure 13.

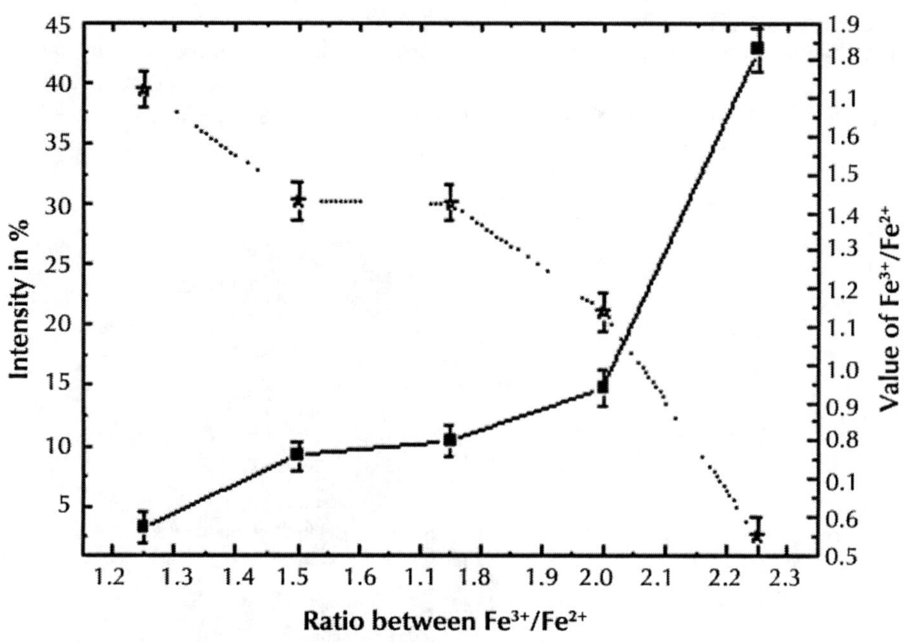

FIGURE 13 Plot of the dependence of ratio between Fe^{3+} and Fe^{2+} on β and the amount of γ-Fe_2O_3.

The values are calculated with quite large errors but the general trend is seen. It is worth mentioning that the values obtained for amount of γ-Fe_2O_3 with use of two different approaches are in good agreement. From both lines the extracted information is the same.

In principle, there are few scenarios which can be considered for describing the particle morphology but most likely we have the particle with magnetite core and γ-Fe_2O_3 and α-Fe_2O_3 in the shell layers. It is observed that the existence of thick enough α-Fe_2O_3 part stabilizes the magnetic moments of the particle and the middle doublet is absent. The lack or too thin layer of hematite causes that the magnetic moments of separated particles fluctuate and we observe the doublet. The different morphology of the particle surface due to different chemical composition changes the surface anisotropy and the magnetization of the particle.

The different magnetic behavior in particles system is explained by the change in anisotropy constant, which is much larger than in bulk crystals. The anisotropy increases with the decrease of the particles size. The surface effect contributes to the anisotropy energy and the relaxation time depends strongly on the chemisorbed molecules and leads to changes in the hyperfine parameters of Fe [19, 60].

It is known that antiferromagnetic hematite has the Morin transition at $T_M = 263K$. Above that temperature the magnetic moments are slightly canted and show a weak ferromagnetic state. According to the literature, the T_M can be significantly lowered up to 4K in the nanoparticles cases due to lattice expansion, strains and defects [61]. In the presence of the external magnetic field the perpendicular contribution of α-Fe_2O_3 to the Mössbauer spectra can be explained by the minimalization of energy and changes of magnetic anisotropy, which strongly depends on the particles volume. Energy gain due to canting the moments to the perpendicular direction is favorable. The schematic magnetic moment orientation in each part is sketched in Figure 14. This phenomenon can explain the observed orientation of α-Fe_2O_3.

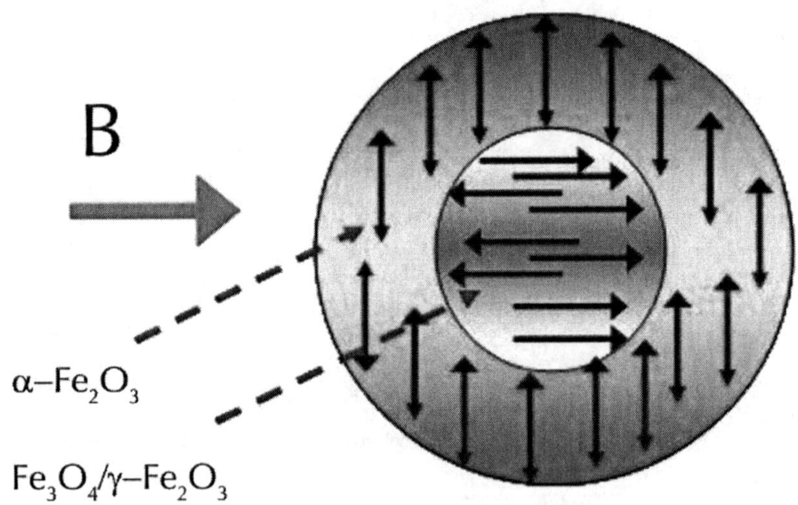

FIGURE 14 The orientation of magnetic moments in the particle.

To conclude this part: spontaneous magnetite core-shell nanoparticle synthesis with maghemite or/and hematite shell layer was obtained, the samples which contain even small amount (more than 5%) of hematite show that magnetic moments in these parts are perpendicular to the external magnetic field and the superparamagnetic fluctuation is suppressed, surface anisotropy modified by hematite causes canting of magnetic moments at the surface; relative ratio of Fe^{3+}/Fe^{2+} in a wide range did not influence the creation of the stochiometric magnetite nanoparticles.

11.2.4 Purposely Core-shell Structures

Magnetic core-shell nanoparticles were synthesized according to modified Massart, Sun, and Wagner recipe. The particles were obtained by: hydrolysis of iron chlorides and iron nitrides in proper stochiometric ratio as in magnetite followed by condensation or by decomposition of $Fe(acac)_3$ complexes [17, 20, 62].

Characteristic

a) Magnetic Core/Nonmagnetic, Nonmetallic Shell

In the next step the silica shell was deposited by policondensation process taking place at the surface of the particles [63]. The amount of TEOS taken to the synthesis was changed gradually from 0.1 to 0.8 ml. Such particles can be tested by cell mitochondrial activity coloring the modified substrate for biological application [64].

To check the layered structure of the nanoparticles TEM images were collected - for smallest and largest TEOS amount, see Figure 15. Here two different samples are imaged and the thickness of the shell layer can be easily calculated. The obtained SiO_2 shell thickness is proportional to TEOS concentration used during synthesis ranging from 0 up to 5nm.

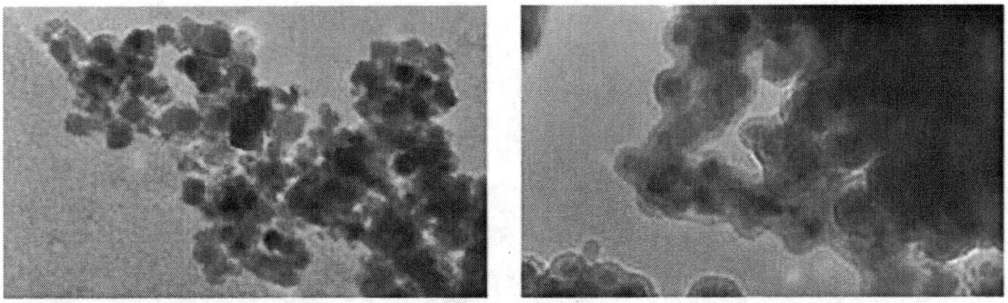

FIGURE 15 The TEM images of magnetite nanoparticles covered by the layer of SiO_2. The thickness has changed with the growing concentration of TEOS.

Quality of crystallinity of the prepared core shell particles were checked by powder XRD diffractometry.

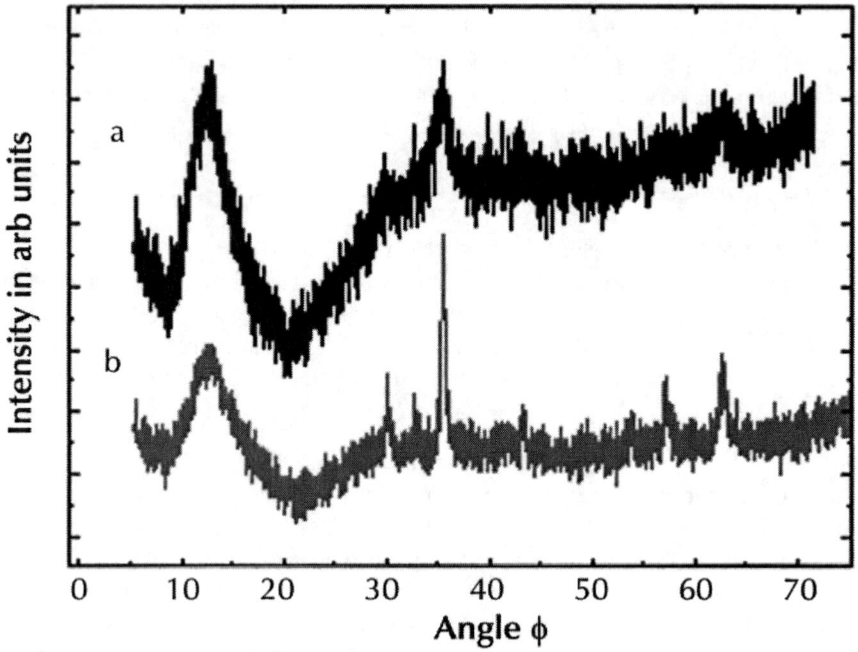

FIGURE 16 The XRD data for two different SiO_2 layers: (a) thick one, (b) thin one (K. Recko University of Bialystok).

From XRD data it can be deduced that for the sample with a thick silica layer the diffractogram is dominated by amorphous SiO_2 shell and relatively weak broad magnetite reflexes can be marked. In case of a thin shell layer the crystalline core layer dominated the diffractogram and the sharp magnetite/maghemite typical peak positions can be marked {(220), (311), (400), (422), (511), and (440)} [20].

The core thickness can be adjusted to a proper value by using the relevant amount of TEOS. In such instance also the interaction between nanoparticles, which range from a collective to individual character, can be tuned.

Discussion

In range of small enough particles below 8–10 nm the interaction between particles can be stopped very easily using a relatively thin layer of spacer (in this case SiO_2) on the surface. For larger particles the interaction between separated particles becomes so strong that the changes above the thickness of the core will result in significant changes in interparticle interaction. Our particles have a diameter of 13 ± 2 nm and the obtained shell is not thick enough to change a lot the shape of Mössbauer spectra (Figure 17). However, small changes in the resulting spectra can be marked. The most important indication about weaker dipole-dipole interaction between particles is the presence of superparamagnetic doublet in the middle part of the spectra. This doublet appears due to diminishing interparticle collective interaction which becomes more difficult (weaker) due to significant thick nonmagnetic-spacer layer.

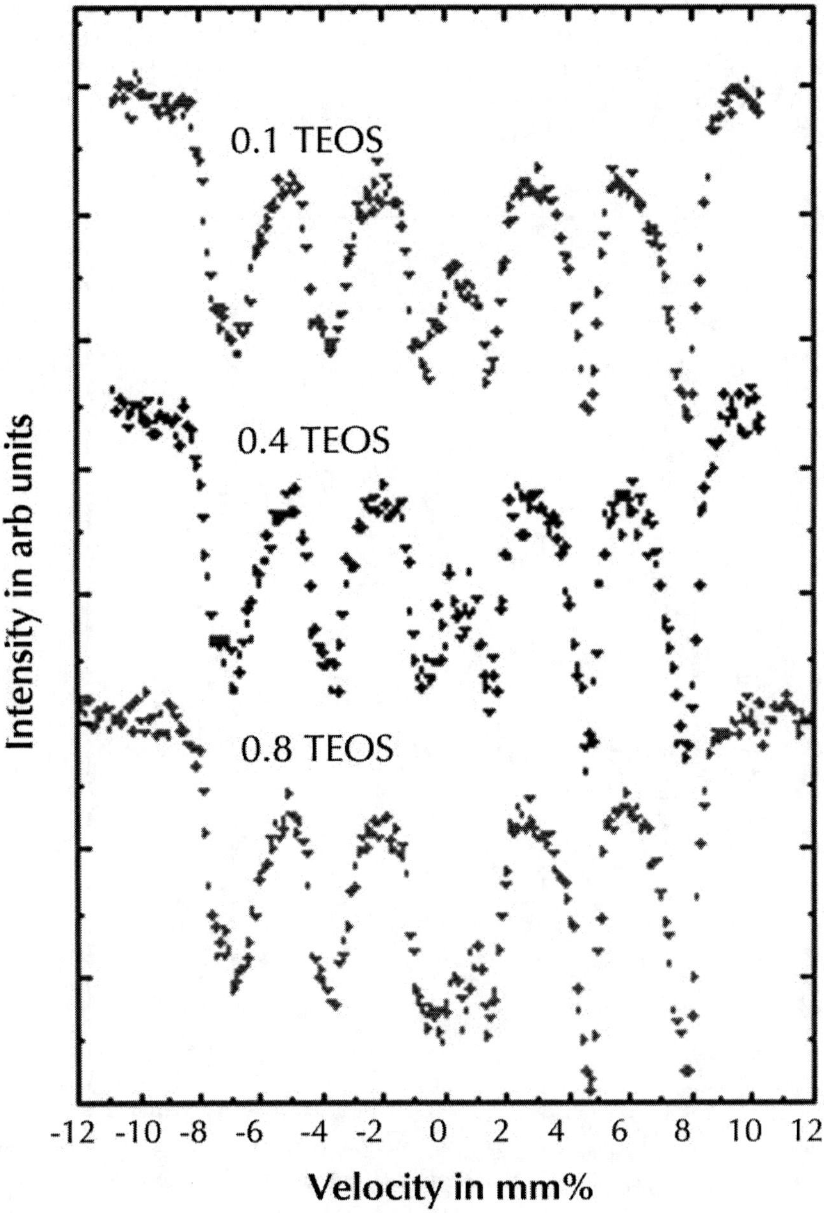

FIGURE 17 Mössbauer spectra of magnetite nanoparticles with different amount of silica shell. The growing amount of TEOS indicates thicker silica layer.

On the other hand, for the same amount of TEOS the modification in the core stoichiometry leads to quite strong changes in magnetic properties of interacting particles (Figure 18). Once (for magnetite like particles) the superparamagnetic part is rather big, which is depicted as a narrow sextet with a doublet in the middle part of the Mössbauer spectra. After modification such as oxidation by $Fe(NO_3)_3$ the spectra are more split but the core has more maghemite character [18]. *Via* the same shell thickness core interact in a different manner since the dipole-dipole interaction in two different substances has different values (strength).

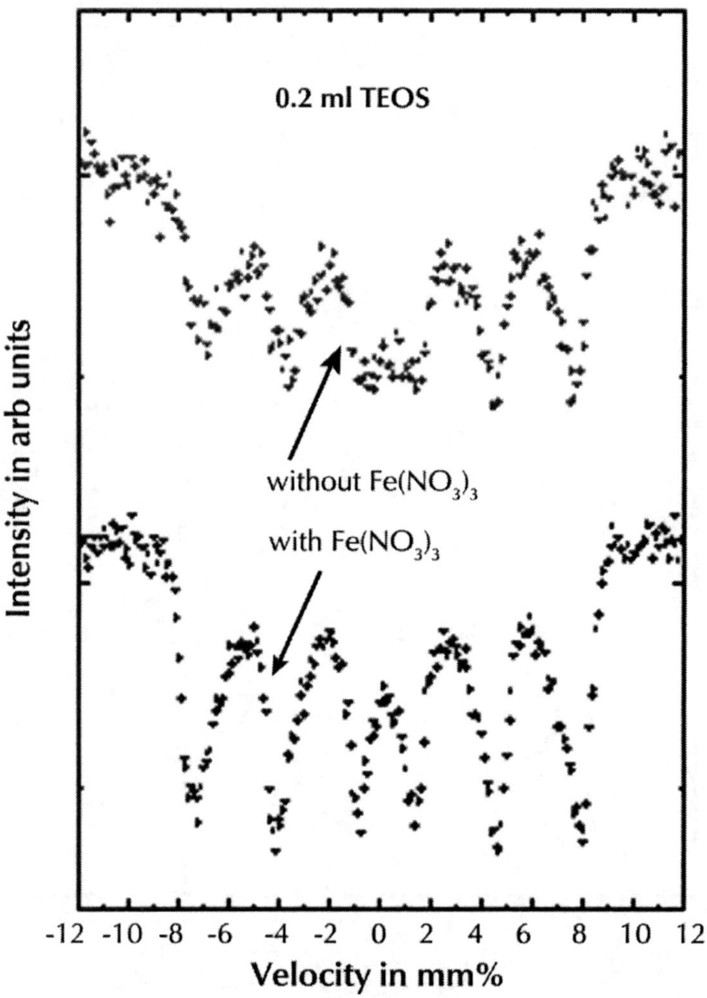

FIGURE 18 Mössbauer spectra of magnetite nanoparticles before and after oxidation by $Fe(NO_3)_3$.

b) Magnetic Core/ Magnetic Shell
Characteristic

Other type core-shell structures are one with two types of magnetic materials. Here in the set of the tested samples have been observed changes of magnetic properties in Mössbauer spectra after variation of core chemical structure with the same surface composition or after shell modification with the same core chemical structure, which schematically is presented in Figure 19.

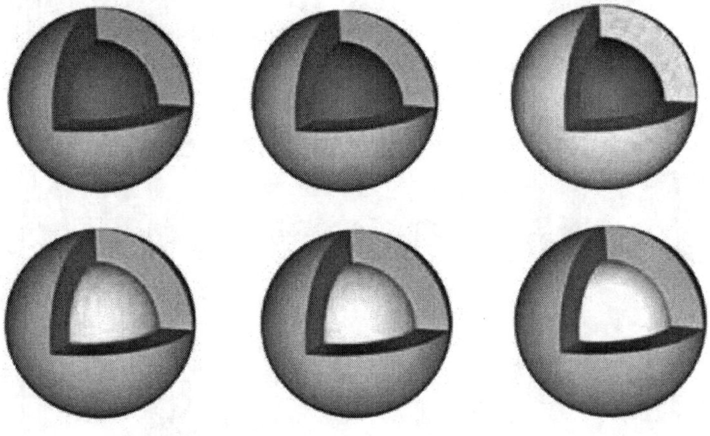

FIGURE 19 Schematic representation of core-shell nanoparticles modification. Upper row shell modification and lower row core modification.

For this study again the standard TEM characterization cannot be very useful since the particles are the same way transparent in both areas. No difference in contrast can be seen through the particle since no big differences in the crystal structure or chemical composition are present between trivalent iron oxides. The example image of the obtained core shell nanoparticles is presented in Figure 20.

FIGURE 20 TEM images of core-shell nanoparticles (M. Giersig FU Berlin).

Varying the composition of core and shell in the particle as well as their thickness the changes in Mössbauer spectra taken for nominally the same size of the particles can be observed. It can be seen that for nanoparticles not only net composition of the particle is extremely important but also the position of constituent layers in the structure is not without any role. In Figure 21, a set of Mössbauer spectra is depicted where the bottom one consists of two the same layers of maghemite on maghemite. The middle one presents maghemite on magnetite and the upper one magnetite on maghemite.

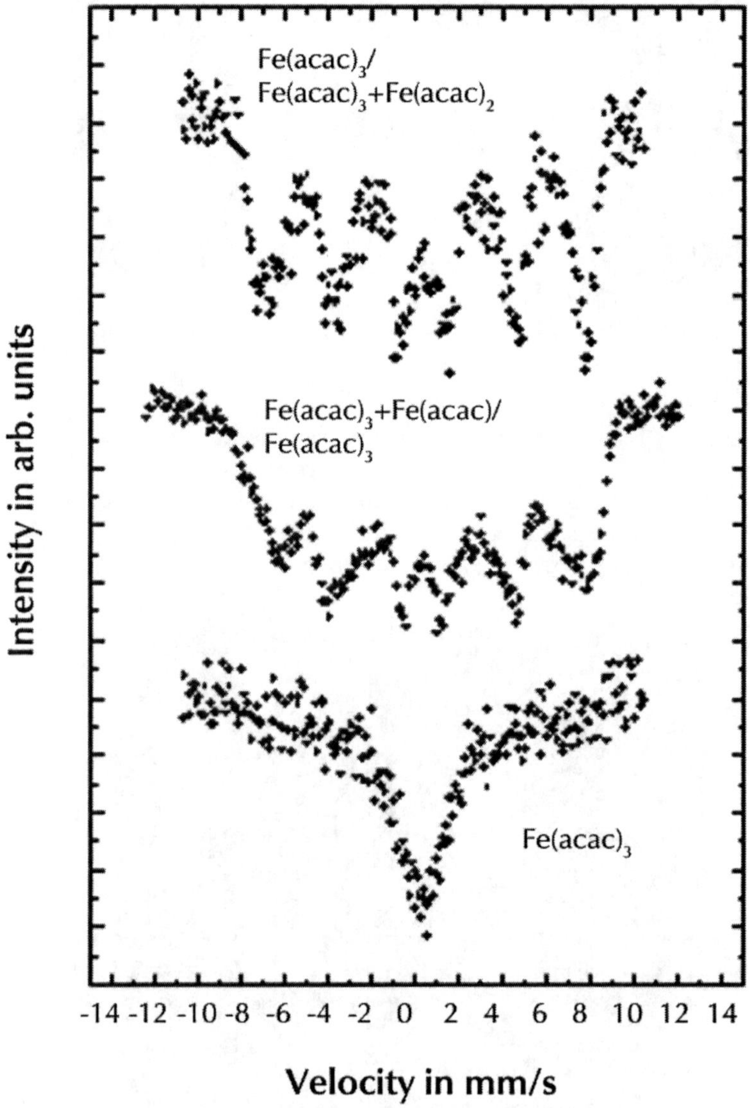

FIGURE 21 Mössbauer spectra of the nanoparticles with different layers composition.

Discussion

The changes in magnetic properties of the particles are clear. The simplest and fastest conclusion which can be taken form obtained data is that here not only dipole-dipole interparticle interactions are important but also intraparticle interlayer interaction cannot be forgotten.

The lowest spectrum is taken for nanoparticles which have a layered structure where both layers are of exactly the same composition [20]. The spectrum is not well resolved and its qualitative analysis shows that the two main subspectra should be taken into account. One narrow sextet (or doublet) and one sextet with larger hyperfine field with broad distribution must be considered. In this case the intensity of both subspectra is roughly equal. When the other layer is added, the spectrum is changed drastically. The wider sub spectrum is dominating and the positions of lines in the sextet are marked but the distribution of hyperfine field is still present. The fabrication of particles in the opposite layer order gives other Mössbauer spectra, which means different magnetic properties of the system. A reader can find more information about it in [55].

The general observation obtained from Mössbauer spectroscopy is as follows: in the system superposition of intraparticle interaction with interparticle interaction must be considered. For some particles superparamagnetic fluctuation blocking temperature decreases in comparison with other ones [65]. The collective interaction of the assembly of particles must be taken into account when magnetic properties are concerned [66]. For powdered samples both dipole-dipole and exchange interactions must be considered. Interparticle interaction must be taken into account together with surface anisotropy since both are extremely important for nanoparticle samples. Any modification of size, surface, or core composition influences drastically their values. On the other hand, local anisotropy competes with intraparticle exchange coupling between the shell and core of the particle [66]. As a result, the thickness of the core layer and the composition of such layer plays an important role in general inter and intraparticle interactions and finally, in general magnetic properties of the system. The detailed studies on the influence of the subsequent layers on the magnetic properties of the nanoparticle system will be presented elsewhere [67].

This studies shows that the particles which have different surface chemical state possess different nature and strength of inter/intraparticle interaction. The particles and their surrounding is a place where enormous competing effects of large complexity occur.

c) Magnetic Core/ Metallic Nonmagnetic Shell

To investigate the changes in magnetic properties of nanoparticles the magnetite and magnetite-silver and magnetite-gold core/shell particles were prepared [68]. The Mössbauer measurements on such set of samples were performed, and the results are depicted in Figure 22.

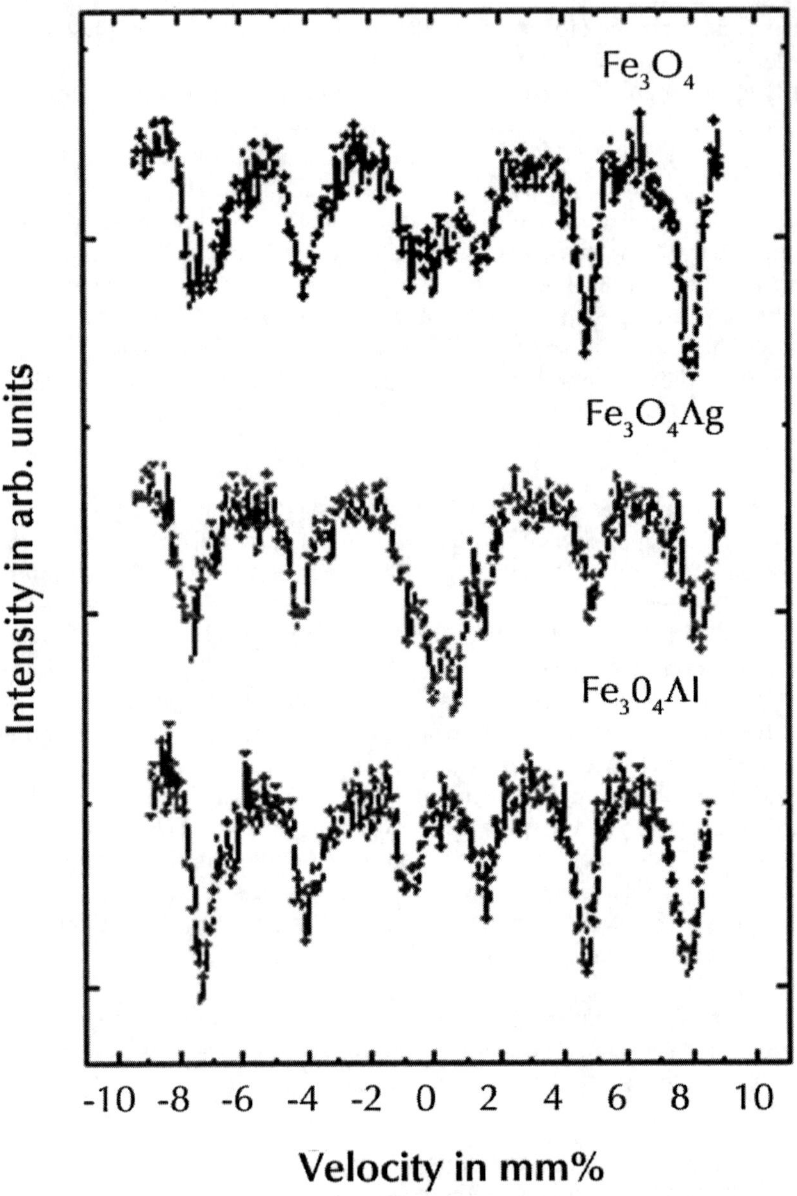

FIGURE 22 Mössbauer spectra of magnetite and silver-, gold-coated magnetite nanoparticles.

The spectra of magnetite coated with a layer of gold or silver have a different character than the one of uncoated magnetite. The spectrum of pure magnetite nanoparticles (top one) is modified compared to the one typical for bulk material but there are many parameters which influence the spectral shape [18]. There we see that besides two sextet characteristic as spectra of iron in A and B sites in magnetite crystal structure the doublet in the middle part of spectrum is visible. After covering the particles by Ag (middle spectrum) the much more significant doublet develops in the center of the spectrum. Covering the magnetite nanoparticles by Au layer causes the opposite effect–the doublet disappears and only sextets are present. In both cases the ratio between A and B subspectra is changed compared to bulk values [22]. This indicates that the covering magnetite nanoparticle with Ag or Au layer modifies magnetic properties of the magnetite core in two opposite ways. The changes suggest that Au (gold on magnetite surface) eliminate a doublet from the middle of the spectrum and influences on the value of the magnetic hyperfine field (increase) [69, 70]. In case of Ag, for some particles the value of the hyperfine field decreased [71]. Such changes can be due to changes in blocking temperature after the surface modification or *via* modification of interparticle interactions through the metallic layer [25].

To conclude this part: core-shell nanoparticles can be obtained by step-by-step decomposition of $Fe(acac)_3$ complexes, condensation from Fe chlorides in ammonia or hydroxides medium: the magnetic properties of nanoparticles are strongly modified by a number and quality of layers in the particle. The modification is caused by the competition of few parameters and can influence the properties in a complicated manner.

11.3 NANOWIRES PREPARATION

The growing interest in nanotechnology causes the development of a big variety of materials with different shape in nano scale. Magnetic dots, clusters, nanoparticles, cylinders, and wires are subjected to study due to fundamental and technological interest. The great interest in nanomagnetic materials is due to their application, for example as memories storage media, sensors, and logical devices [72]. Very important application target is also medicine where nanomaterials can play various roles from drug delivery media an contrast agent to structural material in biological composites and sensors [73, 74, 75]. Many efforts have been made to achieve multifunctional materials with simple synthetic procedure and easy processing for subsequent application. Chemical and electrochemical preparation methods are one of the options for obtaining a new class of nanomaterials. Because many methods of nanostructures fabrication need to use very expensive devices like: MBE or sputtering chambers, and so on it is worthwhile to develop methods which are less expensive and lead to similar final products. Such a method is electrochemistry which allows us to deposit materials in many different forms from big variety of solutions. Hybrid nanomaterials and nanocomposites have been recently studied because of the enhancement of their magnetic, electrical, optical, structural, and mechanical properties [76, 77, 78, 79]. The tubular or elongated structures have advantage over the round ones due to possible selective interaction with environment along and perpendicular to their main axis or due to their layered inert structure.

The nanostructures presented here were obtained by chemical and electrochemical methods [80]. Nanowires and nanotubes were deposited in anodic porous alumina template (AAO) with pore diameter range from 40 to 230 nm [80].

Electrochemical deposition is a very attractive method because the process is very effective (fast and cheap). There is a huge variety of possibly reduced ions and it has no limitation in sample shape and size. Deposition can be done in constant (DC) or accelerating current (AC) modes depending on application and wires characterization. Electrochemical deposition of nanowires is a technique which combines bottom-up and top-down approaches. This is due to the fact that the wires grown atom by atom can be obtained in the matrixes which were subjected to anodization process during which the nanoobjects are obtained from bulk material.

11.3.1 AAO Matrix

The AAO membranes can be obtained with various pores size and distance between them. The process condition such as: temperature and composition of the used solution as well as current conditions, determine matrix parameters (pore diameter and distance between each other). The dependence can be easily followed in Figure 23. There the SEM images of various matrixes obtained in different conditions are presented. The graph of the pore size dependence on voltage and temperature is plotted in Figure 24.

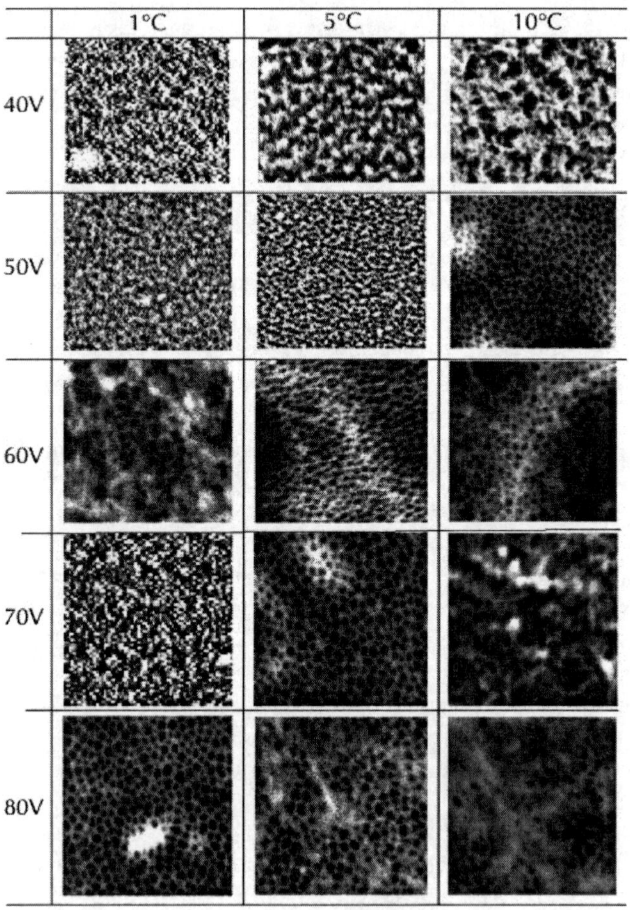

FIGURE 23 The SEM images of matrixes obtained in different current and temperature conditions.

In general, with the higher voltage the bigger particles can be obtained and the dependence is growing gradually for all temperatures (Figure 24(b)) but the dependence is the smallest for low temperature and the most rapid for highest studied temperature. The general conclusion cannot be deduced from temperature dependence since for some voltages the temperature dependence is increasing and for other decreasing. This goes from case to case with some hint that for lower voltages there is rather decreasing trend and for larger -increasing one, see Figure 24(a).

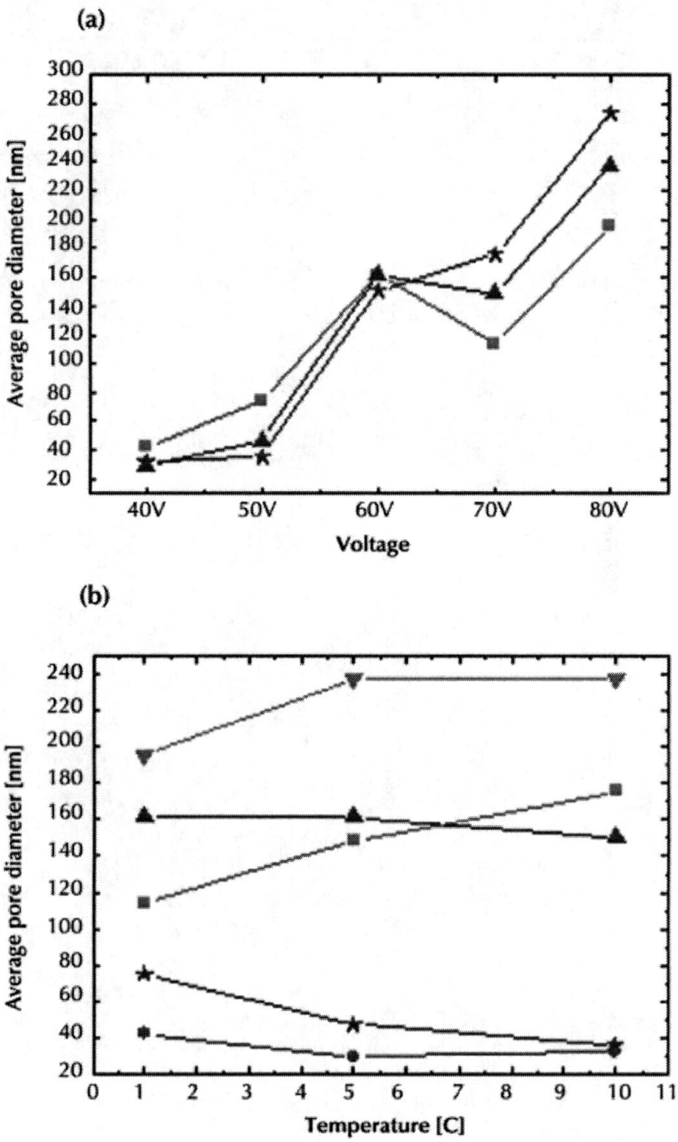

FIGURE 24 Dependence of porous diameter on temperature and voltage (a, b), respectively.

The quality of alumina templates meaning the ordering of nanopores can be improved by a number of repeated anodization processes. In Figure 25 it can be seen that after one step anodization quite well defined but randomly distributed pores are observed while after two step anodization well improved self organization is seen.

FIGURE 25 The AFM images templates prepared at 1°C, 40 V in one step and two steps anodization (a, b), respectively (P. Mazalski University of Bialsytok).

The more repetitions of anodization processes are used, the better self organization of the pores at the surface. This is already seen after two step anodization (Figure 25).

11.3.2 Single Phase Nanowires Deposition

Cu, Ni, Co, and Fe were deposited to the porous AAO of a different diameter [81]. The deposition conditions influence the quality and length of obtained nanowires. Such structures can be obtained *via* constant current deposition method when well distributed nanowires standing perpendicularly to the surface plane are relatively easily obtained. The acceleration current method in most cases (for long enough time of deposition) ends up with wires which are randomly oriented at the surface due to strong magnetic interaction between them. Such examples are depicted in Figure 26.

FIGURE 26 The AFM images of wires obtained by AC platting (a) and DC deposition (b). (P. Mazalski University of Bialsytok).

Perfect adjustment of deposition time can help to obtain the wires with tunable length, which can influence magnetic properties of the wires. The properties of nanowires are strongly dependent on the dimension of the structures. We believe that below some diameter limit the shape anisotropy will switch the easy magnetization direction from along the main axis to the perpendicular orientation [81].

11.3.3 Core-shell Nanowires

Via electrodeposition of the materials in pores wires of certain diameter and length even more than 1 μm can be obtained. Using wetting deposition process followed by thermal crystallization nanotubes can be expected [82].

Core-shell nanowires were obtained in the first step by wetting deposition method followed in the second step by thermal decomposition, and in the next step by electrodeposition. Details of the method will be presented elsewhere [83].

For elongated structures AFM images were taken, which is depicted in Figure 25. Top view shows the difference between tubes and wires (Figure 27 (a) and (b)). The tubes were obtained after dipping the matrix in the solution and afterwards the matrixes were dried in the oven when crystallization took place [84]. As a result the oxide tubes were obtained. The thickness of the walls in tubes is roughly 30–40 nm (in case of 120 nm pore). Such structures can be in the next step filled with another material in electrodeposition process.

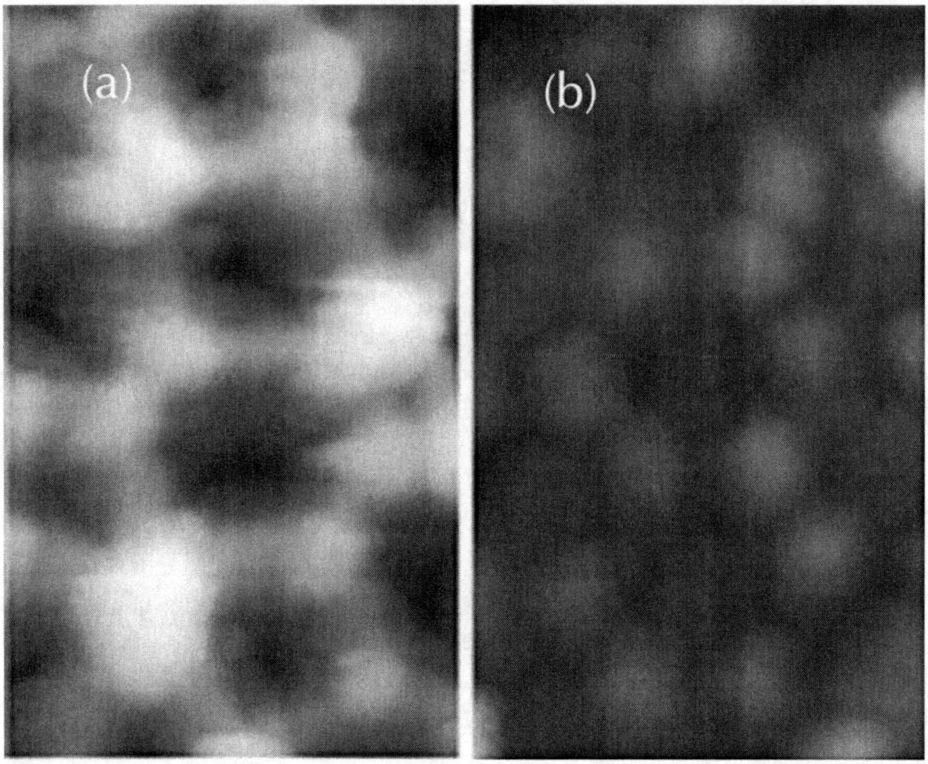

FIGURE 27 The AFM images of: top view (a) tubes, (b) nanowires, side view (c) tubes, and (d) nanowires (P. Mazalski University of Bialsytok).

The success of electrodepositing is in the first step proved by the behavior of the solution in the presence of the external magnetic field. The obtained empty tubes were in most cases non magnetic and then the tubes were not sensitive to the presence of the magnet. Electrodeposition causes that the filled tubes start to react weakly to the magnetic field (the interior of the tube was filled by magnetic material, for example Fe). The nanowires much stronger interact with the magnet and can be easily separated from the solution (see Figure 27).

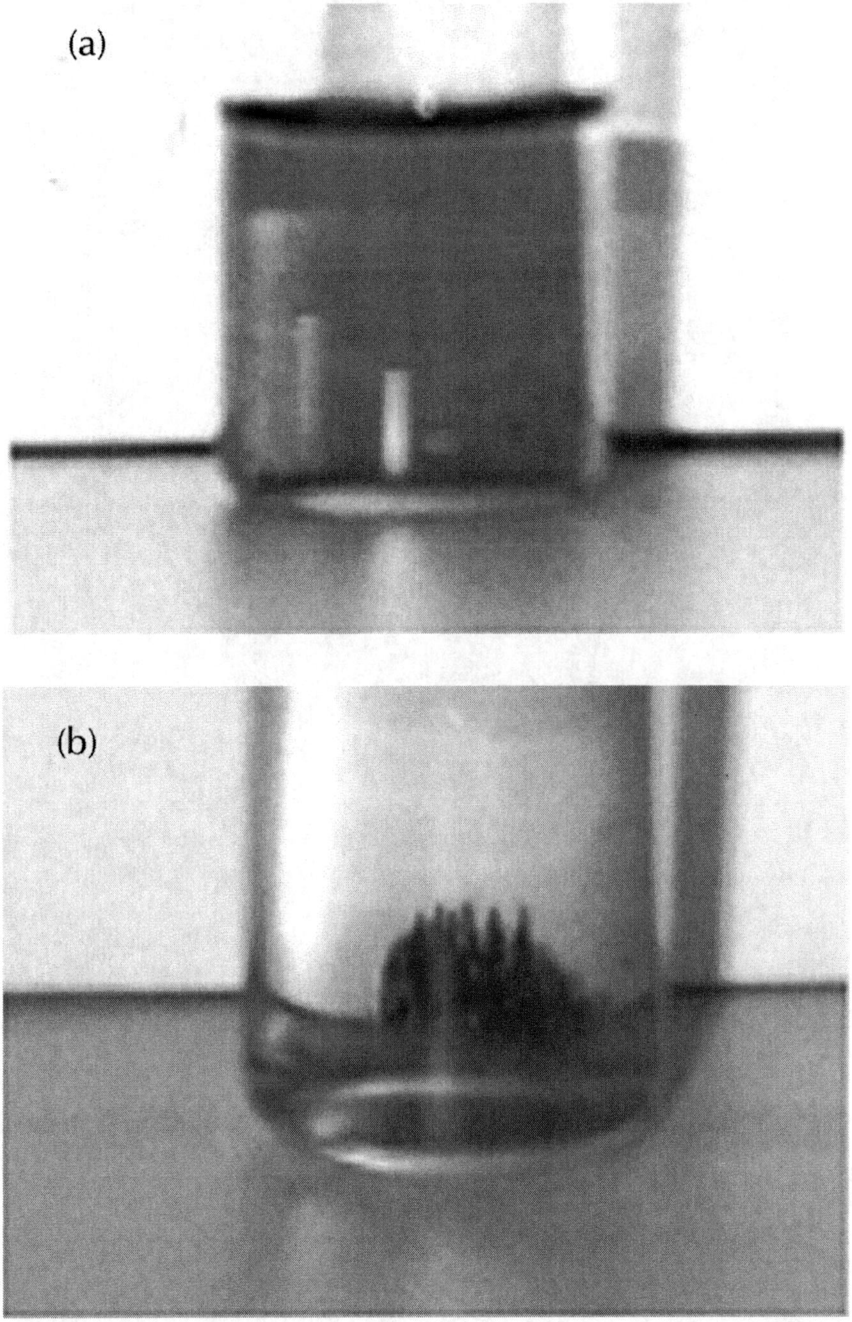

FIGURE 28 *(Continued)*

FIGURE 28 Digital camera view of the solution and powder without (a, c) and in (b, d) the external magnetic field, respectively.

As it can be seen, the tubes in the solution in the presence of the external magnetic field do not react to the field at all (Figure 28(a)). There are even not creating agglomerates since the tubes do not interact between each other (no magnetic interaction). The situation is different when the tubes are filled with magnetic material. In such a case the core-shell wires start to react to the external magnetic field and agglomerate close to the glass wall (Figure 28(b)). For the separated powder nanowires not influenced by external magnetic field a random cluster form is seen at the bottom of the flask (Figure 28(c)). After exposing them to external magnetic field, all wires strongly order along the magnetic field direction, which is depicted in Figure 28(d).

To conclude this part: core-shell nanowires can be obtained by wetting deposition method combined with electrodeposition in porous matrix. Single phase nanowires can be deposited by AC or DC deposition method. All elongated structures were obtained in porous alumina oxide matrix.

11.4 SURFACE FUNCTIONALIZATION

Magnetic nanoparticles such as iron oxides: γ-Fe_2O_3, α-Fe_2O_3 and in particular Fe_3O_4 has unique chemical and magnetic properties different from those observed in bulk materials. The features: such as low toxicity, controllable particle size, biocompatibility or hydrophobicity/hydrophilicity causes that they are the hope for new application in sciences like pharmacy and specially medicine [85, 49]. In order to allow them for chemical reaction with biomolecules, surface of magnetic nanoparticles should be functionalized with organic compounds. What is more, these types of nanoparticles have tend to agglomerate due to Vander Waals forces or magnetic interaction and they are also susceptible to air oxidation, so it is very important to stabilized NPs *via* proper media or surfactants [86, 87, 88]. To modified superparamagnetic nanoparticles different compounds are used, especially one with amine (-NH_2), hydroxide (-OH), carboxylate (-COOH) or phosphate functional groups which allow immobilization of biomolecules (Figure 29). Effectiveness of modification methods are based on physical adsorption, noncovalent and covalent immobilization [87, 89, 90, 91]. The magnetic nanoparticles could also be functionalized by coating them by inorganic metal particles such as silver or gold what has recently become particularly popular [87]. Gold and silver have also the advantage as high affinity for thiols and disulfides [92, 93, 85]. These features make that researchers have successfully bound thiol modified on Ag, Au surface and immobilized DNA, enzymes and so forth [85,49].

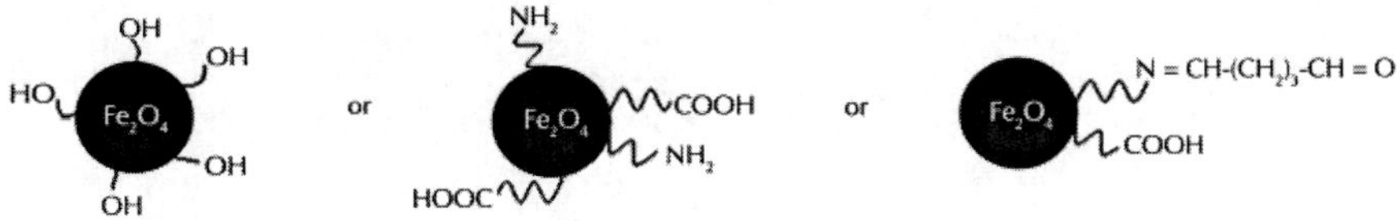

FIGURE 29 Examples of surface functionalization of nanoparticles.

Nanowires are another important nanomaterials in nanotechnology science. The integration of these structures with biological molecules fabricate a new class of materials called hybrid system, which exhibit attractive electronic, magnetic and structural properties [94, 95]. Functionalization of nanowires is also possible similarly as in magnetite NPs case with different biomolecules through various chemical bonds [96, 97, 98]. What is more, self-assembled monolayers created by directly bonded metallic nanostructured with biological particles are possible [99]. These monolayers can be modified in the next step by directly a new covalent or noncovalent bond [90, 100]. Especially the latter interactions are very important in biological system where we have to deal with the network system of hydrogen bonds [101]. There is a large number of studies on chemical modification of surfaces: thiols have huge affinity to gold surface, siloxanes, phosphonates, and carboxylic acids will readily bond to a metal oxide surface (Figure 30). Nevertheless, other ligands are also possible to use which have different character in relation to modified surfaces [102].

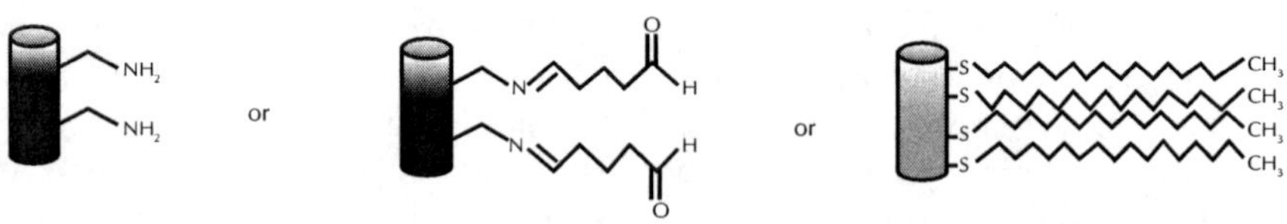

FIGURE 30 Examples of surface functionalization of nanowires.

11.5 COMPOSITES

New approaches of developing composites were attempted when nanomaterials started to be key players in material science. A tremendous number of novel composites with unique properties have been made in the last decades [103]. Recent achievements in producing nanostructured materials have stimulated researchers to build up multifunctional materials with nanomaterials as building blocks. Such structures combine the properties of their constituents but quite often develop new ones, typical only for them. Recent enthusiasm in nanotechnology influences also rapid development of a new generation of composite materials. Current interest in nanocomposites is connected with biomaterials, IT materials, electronic materials, and so on. Scientists and engineers seek to make practical materials and devices out of nanoconstituents. There is still a huge lack of knowledge in the area how nanostructures influence bulk properties.

Using the building blocks described in this paragraph, a number of composites can be obtained. The properties of such heterostructures can combine properties of elements. And this is actually the case. *Via* linkage of magnetic particles with biomolecules, the composite which possesses magnetic properties and at the same time can interact with for example human body (Figure 31). The use of external magnetic field quite often helps to get rid of organic materials that are toxic for healthy cells.

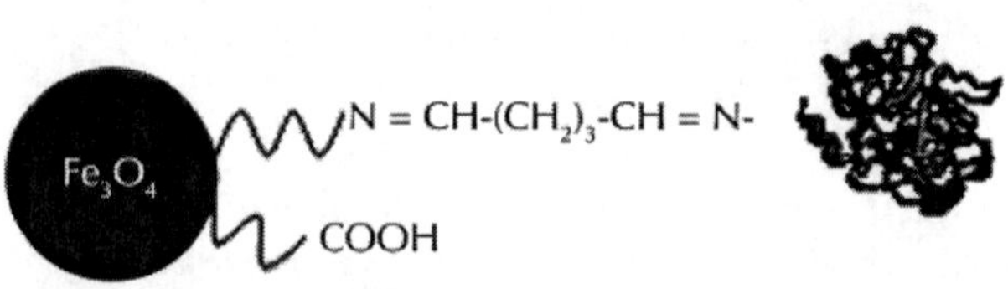

FIGURE 31 Composite magnetic nanoparticle biomolecules.

The introduction of the third part as, for example fullerene, to the system adds electronic properties to the composite. Their schematic representation is depicted in Figure 32. Similarly, the substitution of fullerene by carbon nanotubes introduces electric properties to the system and such composite can be transferred *via* external magnetic field to, for example the human brain where biomolecules and nanotubes repair the neuron connection.

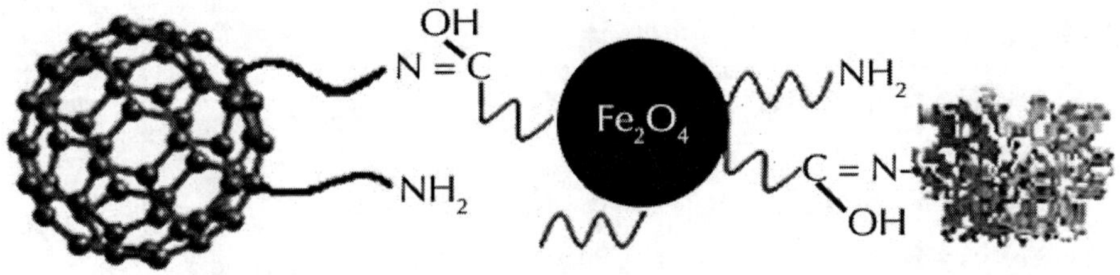

FIGURE 32 Composite fullerene magnetic nanoparticle biomolecules.

Other types of composite were these which had elongated nanostructures as nanowires or nanotubes in the structure instead of nanoparticles. The nanostructures can be made of pure metal or metal oxide and have magnetic and at the same time conducting properties or selectively only one of them. Such elements can be also connected with biomolecules directly or *via* linkage chemistry. Schematic examples are sketched in Figure 33. Changing the composition of nanotubes, the needed one can be enhanced and weakened useless for a particular application.

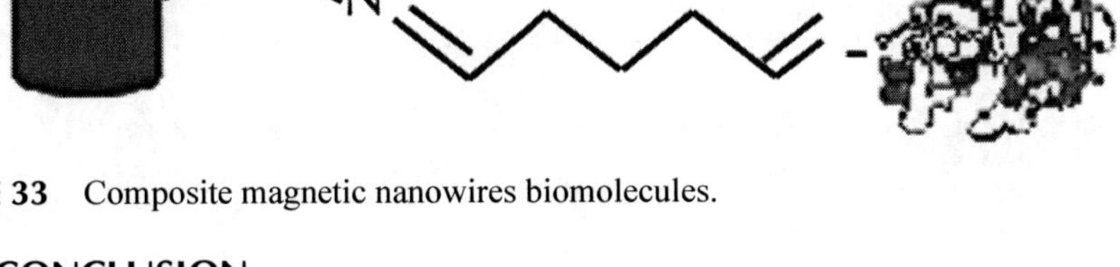

FIGURE 33 Composite magnetic nanowires biomolecules.

11.6 CONCLUSION

The possibilities in constructing new nanomaterials are huge. There are enormous number of structures which are described in the literature. However, there is also a gigantic number of unknown structures and a lot of hard work must be put to define and describe their properties. Nanotechnology brought us many new tools to design materials with application in modern IT technology, electronics, optics, and especially in biology and medicine. The application of nanostructures in bioscience allows us to follow and understand processes taking place in particular cells. Provided tools allow us to comprehend many mechanisms occurring at a cellular level. Nevertheless, the important part of the studies on nanotechnology should be devoted to environment safety and hazard. Large enthusiasm is dedicated to new materials discovered every day but only few of us think critical of all nanomaterials since nobody knows when we break the safety level and whether we will not destroy our environment *via* nanotechnological worldwide applications.

KEYWORDS

- **Accelerating current**
- **Massart's method**
- **Mössbauer spectra**
- **Nanoparticles**
- **Nanotechnology**
- **Transmission electron microscopic**

ACKNOWLEDGMENT

This work was partly supported by Polish National Grant No. N N204 246435. I would like to thank my students for laboratory assistance, my colleagues from the Institute of Chemistry and Physics Department for apparatus support.

REFERENCES

1. Abbas, K. A., Saleh, A. M., Mahamed, A., and Mohd. Azhan, N. *J. of Food Agriculture & Environment*, 7, 1417 (2009).
2. Lines, M. G. *J. All. Comp.*, **449**, 242 (2008).
3. Asiyanabola, B. and Soboyejo, W. *J. Surgical Education,* **65**, 155 (2008).
4. Pokropivny, V. V. and Skorokhod, V. V. *Physica. E*, **40**, 2521 (2008).
5. Poole, C. P. and Owens, F. J. *Introduction to nanotechnology*. Willey, New Jersey (2003).
6. Fedleim, D. L. and Foss, C. A. *Metal nanoparticles synthesis, characterization, and applications*. New York (2002).
7. Tao, Y., Endo, M., and Kaneko, K. *Recent Patents on Chemical Engineering*, **1**, 192 (2008).
8. Zhang, S. *Materials Today*, **6**, 20 (2003).
9. Mijetovic, D., Eijkel, J. T. C., and van den Berg, A. *Lab Chip*, **5**, 492 (2005).
10. Pitkethly, M. *Nano Today*, **3**, 6 (2008).
11. Salerno, M., Landoni, P., and Verganti, R. *Technological Forecasting & Social Change*, **75**, 1202 (2008).
12. Christian, P., Von der Kammer, F., Baalousha, M., and Hofmann, Th. *Ecotoxicology*, **17**, 326 (2008).
13. Abad, A., Corma, A., and Garcia, H. *Pure Apl. Chem.*, **79**, 1847 (2007).
14. Daigle, J., Ch., J. P., and Claverie, J. *Nanomater*, **1**, (2008).
15. Vakhroucher, A. V., and Liponov, A. M. *Comp. Mathematics & Mathematical Phys.*, **47**, 1702 (2007).
16. Papaefthymiou, G. C. *Nano Today*, **4**, 438 (2009).
17. Massart, R. and Cabuil, V. *J Chem. Phys.*, **84**, 967 (1987).
18. Kalska-Szostko, B., Zubowska, M., and Satula, D. *Acta Physica Polonica A*, **109** 365–369 (2006).

19. Morup, S., Topsoe, H., and Lipka, J. *J. de Phys.*, **37**, C6-287 (1976).
20. Sun, S., Zeng, H., Robinson, D. B., Raoux, S., Rice, P. M., Wang, S. X., and Li, G. *J. Am. Chem. Soc.*, **126**, 273 (2004).
21. Hargrove, S. R. and Kündig, W. *Solid. State commun.*, **8**, 303 (1970).
22. Korecki, J., Handke, B., Spiridis, N., Ślęzak, T., Flis-Kabulska, I., and Haber, J. *Thin Solid Films*, **412**, 14 (2002).
23. Kalska, B., Paggel, J. J., Fumagalli, P., Satula, D., Hilgendorff, M., and Giersig, M. *J. Appl. Phys.*, **95**, 1343 (2004).
24. Winkler, W. *Phys. Stat. Sol.*, **84**, 193 (1984).
25. Satuła, D., Kalska-Szostko, B., Szymański, K., Dobrzyński, L., and Kozubowski, J. *Acta Physica Polonica A*, pp. 1515 (2009).
26. Krupyanskii, Yu. and Suzdalev, I. P. *J. de Phys.*, **35**, C6-407 (1974).
27. Novakova, A. A., Lanchinskaya, V. Yu., Volkov, A. V., Gendler, T. S., Kiseleva, T. Yu., Moskvina, M. A., and Zezin, S. B. *J. Mag. Magn. Mater.*, **354**, 258–259 (2003).
28. Togawa, T., Sano, T., Wada, Y., Yamamato, T., Tsuji, M., and Tamanra, Y. *Solid Stat. Ion.*, **89**, 279 (1996).
29. Petitto, S. C., Tanawar, K. S., Ghose, S. K., Eng, P. J., and Trainor, T. P. *Surf. Scien.*, **604**, 1082 (2010).
30. Manova, E., Tsoncheva, T., Estournes, C. l., Paneva, D., Tenchev, K., Mitov, I., and Petrov, L. *Appl. Catal. A*, **300**, 170 (2006).
31. Mathew, D. S. and Juang, R. S. *Chem. Eng. J.*, **129**, 51 (2007).
32. Sorescu, M., Grabias, A., Tarabasanu-Mihaila, D., and Diamandescu, L. *J. Appl. Phys.*, **91**, 8135 (2002).
33. Sorescu, M., Mihaila-Tarabasanu, D., and Diamandescu, L. *Appl.Phys.Lett.*, **72**, 2047 (1998).
34. Ślawska-Waniewska, A., Diduch, P., Greneche, J. M., and Fannin, P. C. *J. Mag. Magn. Mater.*, **227**, 215–216 (2000).
35. Ngo, A. T., Bonville, P., and Pileni, M. P. *Eur. Phys. J.B.*, **9**, 583 (1999).
36. Bakuzis, A. F., Morris, P. C., and Pelegrini, F. *J. Appl. Phys.*, **85**, 7480 (1999).
37. Caizer, C. *J. Mag. Magn. Mater.*, **251**, 304 (2002).
38. Singhal, S., Singh, J., Barthwal, S. K., and Chandra, K. *J. Solid Stat. Chem.*, **178**, 3183 (2005).
39. Carp, O., Patron, L., and Reller, A. *Mater. Chem. Phys.*, **101**, 142 (2007).
40. Menzel, M., Sepelak, V., and Becker, K. D. *Sol. Stat. Ionic*, **663**, 141–142 (2001).
41. Li, F. S., Wang, L., Wang, J. B., Zhou, Q. G., Zhou, X. Z., Kunkel, H. P., and Williams, G. *J. Magn. Mag. Mater.*, **268**, 332 (2004).
42. Kalska-Szostko, B., Brancewicz, E., Orzechowska, E., Mazalski, P., and Wojciechowski, T. *Material Science Forum*, **674**, 231 (2011a).
43. Congiu, F., Concas, G., Ennas, G., Falqui, A., Fiorani, D., Marongui, G., Marras, S., Spano, G., and Testa, A. M. *J. Magn. Mag. Mater.*, **1561**, 272–276 (2004).
44. Choi, E. J., Ahn, Y., Kim, S., An, D. H., Kang, K. U., Lee, B. G., Beak, S., and Oak, H. N. *J. Magn. Mag. Mater.*, **262**, L198 (2003).
45. Li, X. and Kutal, Ch. *J. All. Comp.*, **349**, 264 (2003).
46. Sepelak, V., Baabe, D., Mienert, D., Schultze, D., Krumeich, F., Litterst, F. J., and Becker, K. D. *J. Mag. Magn. Mater.*, **257**, 337 (2003).
47. Nikiforov, M. P., Chernysheva, M. V., Eliseev, A. A., Lukashin, A. V., Tretyakov, Yu. D., Maksimov, Yu. V., Suzdalev, I. P., and Goernert, P. *Mat. Sci. Eng. B*, **109**, 226 (2004).
48. Serna, C. J., Bodker, F., Morup, S., Morale, M. P., Sandiumenge, F., and Veintemillas-Verdaguer, S. *Solid State Commun.*, **118**, 437 (2001).
49. Storhoff, J. J., Elghanian, R., and Mucic, R. C. *J. Am. Chem. Soc.*, **120** (1998).
50. Verwey, E. *Nature*, **44**, 327 (1939).
51. Zboril, R., Mashlan, M., Barcowa, K., and Vujtek, M. *Hyp. Inter.*, **139/140**, 597 (2002).
52. Morich, M., Weneda, K., and Schurer, P. J. *J. de Phys.*, **37**, C6-301 (1976).
53. Chroni, A., Frei, E. H., and Schieber, M. *J. Phys. Chem. Solids*, **23**, 545 (1962).

54. Bielański, A. "*Podstawy chemii nieorganicznej*" tom 2, Wydawnicto Naukowe PWN p. 928 (2002).
55. Kalska-Szostko, B., Cydzik, M., Satula, D., and Giersig, M. *Acta Phys. Polonica*, **119**, 15 (2011).
56. Heyeon, T., Lee, S. S., Park, J., Chung, Y., and Na, H. B. *J. Am. Chem. Soc.*, **123**, 12789 (2001).
57. Greenwood, N. N., and Gibb, T. C. *Mössbauer Spectroscopy*. Chapman and Hall Ltd, London, p. 258 (1971).
58. Barcova, K., Mashalan, M., Zboril, R., and Martenec, P. *J. Radioanalytical and Nuclear Chem.*, **255**, 529 (2003).
59. Zhang, G., Tan, J., Jin, Chen. Ch., J., Li, X., and Li, Y. *Proceedings of the 3rd Environmentl Physics Conference*, Aswan, Egypt (February 19-23, 2008).
60. Didukh, P., Greneche, J. M., Ślawska-Waniewska, A., Fannin, P. C., and Casas, L. *J. Mag. Magn. Mat.*, **613**, 242–245 (2002).
61. Bodker, F., Hansen, M. F., Koch, Ch. B., Lefmann, K., and Morup, S. *Phys. Rev. B*, **61**, 6826 (2000).
62. Wagner, J., Autenrieth, T., and Hempelmann, R. *J. Mag. Magn. Mater.*, **252**, 4 (2002).
63. Philipse, A. P., von Bruggen, M. P. B., and Pathmamanoharan, Ch. *Langmir*, **10**, 92 (1994).
64. Kalska-Szostko, B. et al. Will be published (2012a).
65. Tronc, E., Prane, P., Jolivet, J. P., Fiorani, D., Testa, A. M., Cherkaoui, R., Nogues, M., and Dormann, J. L. *Nanostructured Mater.*, **6**, 945 (1995).
66. Tronc, E., Ezzir, A., Cherkaoui, R., Chaneac, C., Nogues, M., Kachkachi, H., Fiorani, D., Testa, A. M., Greneche, J. M., and Jolivet, J. P. *J. Mag. Magn. Mater.*, **221**, 63 (2000).
67. Kalska-Szostko, B. et al. Will be published (2012b).
68. Kalska-Szostko, B. et al. Submitted for publication (2012).
69. Cho, S. J., Shahin, A. M., Long, G. J et al. *Chem. Mater.*, **18**, 960–967 (2006).
70. Shinjo, T. *Surface Science*, **438**, 329–335 (1999).
71. Paniago, R., Lupez, J. L., and Pfannes, H. D. *J. Magn. Mag. Mater.*, **1776**, 226–230 (2001).
72. Cowburn, R. and Welland, M. *Science*, **287**, 1466 (2000).
73. Khizroev, S., Kryder, M., Litvinov, D., and Thomson, D. *Appl. Phys. Lett.*, **81**, 2256 (2002).
74. Goldstein, A., Gelb, M., and Yager, P. *J. Control Release*, **70** 125 (2001).
75. Haberzettl, C. A. *Nanotechnology*, **13**, R6 (2002).
76. Romero, P. G. and Sanchez, C. *Functional Hybrid Material*. Wiley Weinheim, p. 86 (2003).
77. Wu, Y., Xiang, J., Yang, C., Lu, W., and Liber, M. C. *Nature*, **430**, 61 (2004).
78. Kim, B. H., Jung, J. H., Hong, S. H., Joo, J., Epstein, A. L., Mizoguchi, K., Kim, J. W., and Choi, H. J. *Macromolecules*, **35**, 1419 (2002).
79. Huynh, W. U., Dittmer, J. J., and Alivisatos, A. P. *Science*, **295**, 2425 (2002).
80. Kalska-Szostko, B., Brancewicz, E., Olszewski, W., Szymański, K., Mazalski, P., and Sveklo, J. *Acta Phys. Pol. A*, **115**, 542 (2009) (2009a).
81. Kalska-Szostko, B., Brancewicz, E., Olszewski, W., Szymański, K., Mazalski, P., and Sveklo, J. *Solid State. Phenom.*, **151**, 190 (2009b).
82. Kalska-Szostko B., et al. *Mater. Chem. Phys.*, (2011b).
83. Kalska-Szostko, B. et al. *Curr. Appl. Phys.*, (2011c).
84. Kalska-Szostko, B., et al. Submitted for publication (2012c).
85. Pradhan, A., Jones, R. C., Caruntu, D. et al. *Ultrason. Sonochem.*, **15**, 891 (2008).
86. Can, K., Mustafa, O., and Mustafa, E. *Colloid Surf.B*, **71**, 154–159 (2009).
87. Haratifar, E., Shahverdi, H. R., and Shakibaie, M. *J.Nanomaterials*, (2009).
88. Kinoshita, T., Seino, S. et al. *J. Magn. Mag. Mater*. **293**, 106 (2005).
89. Barie, N., Rapp, M. et al. *Biosens. Bioelectron.*, **13**, 855 (1998).
90. Birenbaum, N. S., Lai, B. T., Chen, C. S., Reich, D. H., and Meyer, G. J. *Langmuir*, **19**, 9580–9582 (2003).
91. Hwang, I., Baek, K., Jung, M., Kim, Y. et al. *J. Am. Chem. Soc.*, **129**, 4270 (2007).
92. Kinoshita, T., Seino, S., Okitsu, K. et al. *J.All.Comp.*, **359**, 46 (2003).
93. Brust, M., Walker, M., Bethell, D. et al. *J. Chem. Soc. Chem. Commun.*, p. 801 (1994).

94. Contreras, R., Sahlin, H., and Frangos, J. A. *J Biomed Mater Res A*., **80**(2), 480–5 (2007).
95. Martin, C. R. *Science*, **266**, 1961–1966 (1994).
96. Ren, Q., Zhao, Y. P., Yue, J. C., and Cui, Y. B. *Biomed Microdevices*, **8**, 201–208 (2006).
97. Skinner, K., Dwyer, C., and Washburn, S. *Nano Letters*, **6**(12), 2758–2762 (2006).
98. Bauer, L. A., Reich, D. H., and Meyer, G. J. *Langmuir*, **19**, 7043–7048 (2003).
99. Prime, K. L. and Whitesides, G. M. *Science*, **252**, 1164–1167 (1991).
100. Sullivan, T. P. and Huck, W. T. S. *J. Org. Chem*., 17–29 (2003).
101. Crooks, R. M., Kepley, L. J., and Sun, L. *Langmuir*, **8**, 2101–2103 (1992).
102. Bauer, L.A., Birenbaum, N. S., and Meyer, G. J. *J. Mater. Chem*., **14**, 517–526 (2004).
103. Kelley, A. and Zweben, C. *Comprehensive composite Nanomaterials.* Elsevier, **1-6**, (2000).

12 Electrically Conductive Nanocomposites for Structural Applications

S. Mall

CONTENTS

12.1 Introduction 239
12.2 Experiments 241
    12.2.1 Test Materials 241
    12.2.2 Specimen Details 243
    12.2.3 Test Details 245
12.3 Discussion and Results 245
    12.3.1 Lightening Strike Effect on Compression Strength 245
    12.3.2 Emi Shielding Effectiveness 247
    12.3.3 Damage Mechanisms 249
12.4 Conclusion 252
Keywords 253
References 253

12.1 INTRODUCTION

Since the advent of laminated fiber reinforced composite materials about four decades ago, their applications in the load bearing structures have been wide spread especially in aerospace systems. The next notable evolution in aerospace systems requires advanced and enabling structures that are lighter, more reliable, less expensive, survivable, and satisfy the multifunctional design goals. New materials are needed to satisfy these competing and demanding requirements. Recent advances in nanocomposites have shown promise of fulfilling these multifunctional roles. A polymer reinforced with only a few percent of electrically conductive nanofibers can make them electrically conductive and hence, suitable for several applications. Two notable examples are space bus structures and aircraft structures. The space structures require the capability to discharge electrostatic potentials, should provide sufficient conductivity for electrostatic painting, and should shield from the radio frequency interference. Current carbon based polymeric composites suffer from low electrical conductivity leading to low attenuation of high frequency electromagnetic radiation and higher susceptibility to electrostatic discharge (ESD) induced damage. Therefore, space structures made from conventional polymeric composites require additional labor intensive manufacturing steps to provide the electrical ground path, which increase mass and add extra costs in the space structures [1-3]. Thus, conductive nanocomposites are extremely useful for space applications not only to provide electromagnetic interference (EMI) and ESD protection but also due to their light weight and high strength/stiffness. Further, nonconductive composite materials offer little or no protection against lightning strikes when used in aircraft structures [4, 5]. Lightning strikes break fibers and disintegrate the resin that holds the fibers and lamina together. The current state of the art lightning protection method involves metallic meshes (or other devices) which are co-cured with conventional polymeric composites [6]. This prevents concentrated damage from the lightning strike by dispersing it around the airframe. However, metallic meshes increase weight and may corrode over time. Further, they do not fully protect the composite, and are hard to repair. Therefore, conductive nanocomposites are also extremely desirable for aircraft structures to provide protection against the lighting strike.

This study therefore, investigated nickel nanostrands (NiNS) based conductive nanocomposites for their possible applications in the space structures for EMI shielding and in the aircraft structures for protection against the lightening strike. Lightweight satellite enclosure is an example for the former application and aircraft's wing is an example for the latter application. Although the constituents and conductive nano reinforcements were slightly different for these two quite different applications, the basic composite was a polymeric matrix composite with carbon fibers containing nickel nanostrands. Further, these applications are primarily subjected to different external loading conditions so the performances of these conductive composites were characterized under different loading conditions. For the space applications and monotonic tensile loading condition was employed where its effect on the EMI shielding effectiveness of conductive nanocomposites was evaluated. On the other hand, monotonic compressive loading condition was employed because the upper side of aircraft's wing is prone to lightening strike and subjected to compressive loads.

This study investigated NiNS based conductive nanocomposites for the compressive strength degradation when subjected to a simulated lightening strike and EMI shielding effectiveness when subjected to the monotonic tensile load. The addition of NiNS in carbon fibers reinforced polymeric composite demonstrated practically no degradation of EMI shielding effectiveness, when measured in terms of db, under the tensile loading condition up to the final fracture. Further, there was no change in the ultimate tensile strength (UTS) of the parent polymeric composite from the addition of nanostrands. On the other hand, there was considerable reduction in the ultimate compressive strength of the parent polymeric composite from the addition of nanostrands when subjected to the lightening strike. It reduced the compressive strength of the laminate by about 70%, and this was due to widespread damage across the laminate thickness from the lightening strike due to higher electrical conductivity of nickel nanostrands.

12.2 EXPERIMENTS

12.2.1 Test Materials

The test material for the EMI shielding effectiveness was M55/RS-3 composite consisting of carbon fibers (M55) reinforced in toughened polycyanate (RS-3) resin. Also, a thin film consisting of NiNS in toughened polycyanate (RS-3) resin was used. One laminate was without nanostrand which will be referred to as the control laminate. The other one had NiNS film (100 g/m^2) at the both outside surfaces (top and bottom). Details of these both laminates are shown in Figure 1. The NiNS is strands of 50 ~ 1,000 nm diameter nickel particles linked in chains, microns to millimeters in length [7]. Details of NiNS are shown in Figure 2. Panels of size 15 × 15 cm were fabricated. A quasi-isotropic lay-up of $[0/90/\pm 45]_s$ having eight plies was used to fabricate both laminates using the unidirectional prepreg of M55/RS-3 and NiNS TM film. These laminates were prepared by curing in an autoclave.

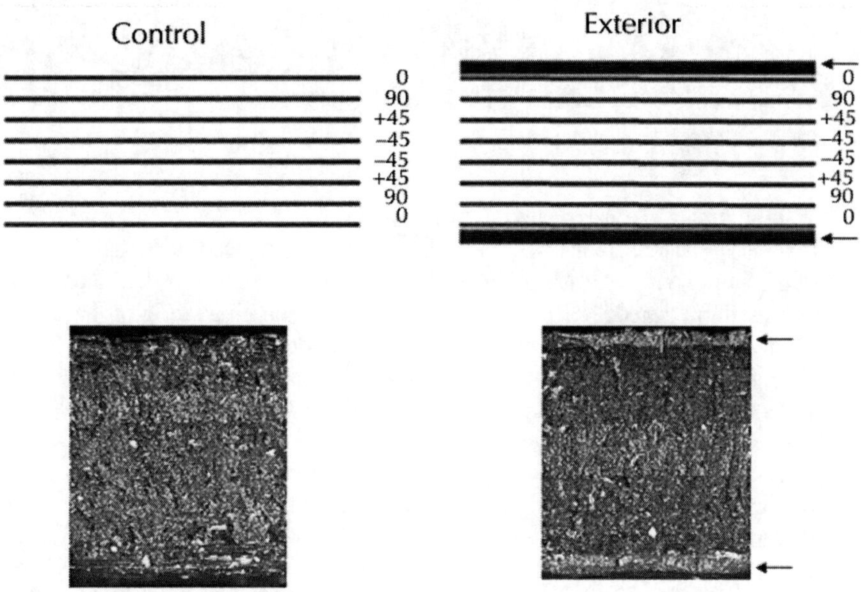

FIGURE 1 Details of laminate with and without NiNS layers (a) schematic view and (b) cross-sectional view (arrows show the nanostrands layer).

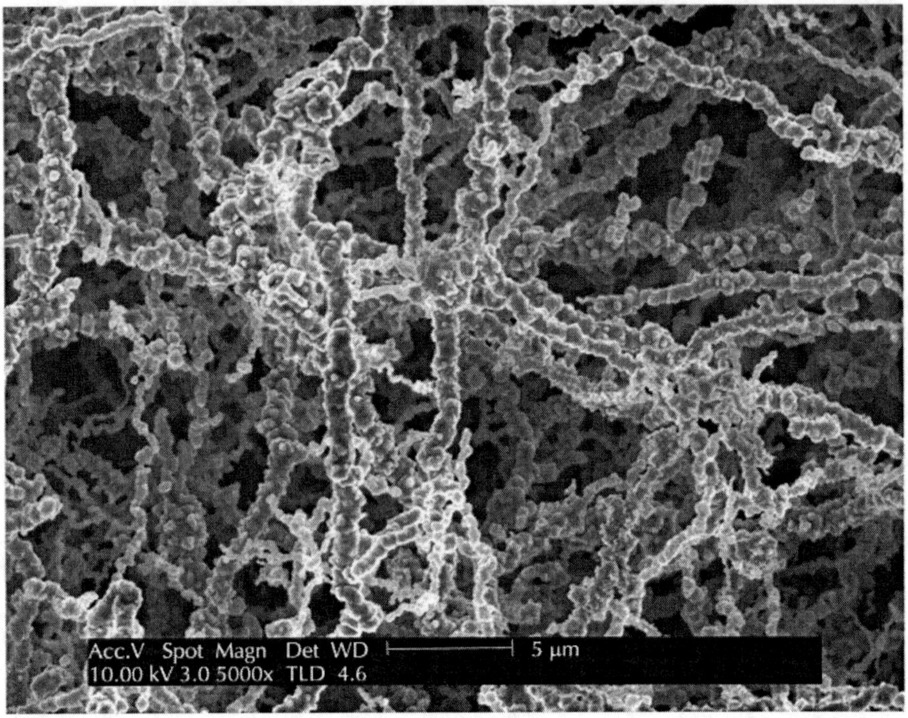

FIGURE 2 Nickel nanostrands.

For lightening strike purpose, the test material was AS4/862 composite consisting of four plies of carbon fiber (AS4 5HS 6K) fabrics reinforced in EPON 862. Also it consisted of a single ply of nickel coated carbon woven fabric (NiCCF) of AS4 5HS 12K carbon fiber embedded in EPON 862. The nickel coating, an interphase between the carbon fibers and the resin, was intended to conduct the energy of the lightening strike. The baseline panel had thus these five plies. The second panel had one more ply to serve as the lightening strike protector consisting of NiNS besides this NiCCF ply. Panels of size 30.5 × 30.5 cm were fabricated. Details of these both panels are shown in Figure 3. For laminate processing, the NiNS veil and/or Ni-coated AS4 5HS 12K carbon fabric was placed on resin transfer mold (RTM) followed by the standard AS4 5HS 6K fabrics. The matrix material was EPON 862 with Curing Agent W which was heated and injected into the mold and the standard cure cycle for the epoxy was carried out. The resin flowed into the dry preform inside the two-sided mold which was held together by a heated press. Vacuum was pulled on the resin outlet to remove air bubbles.

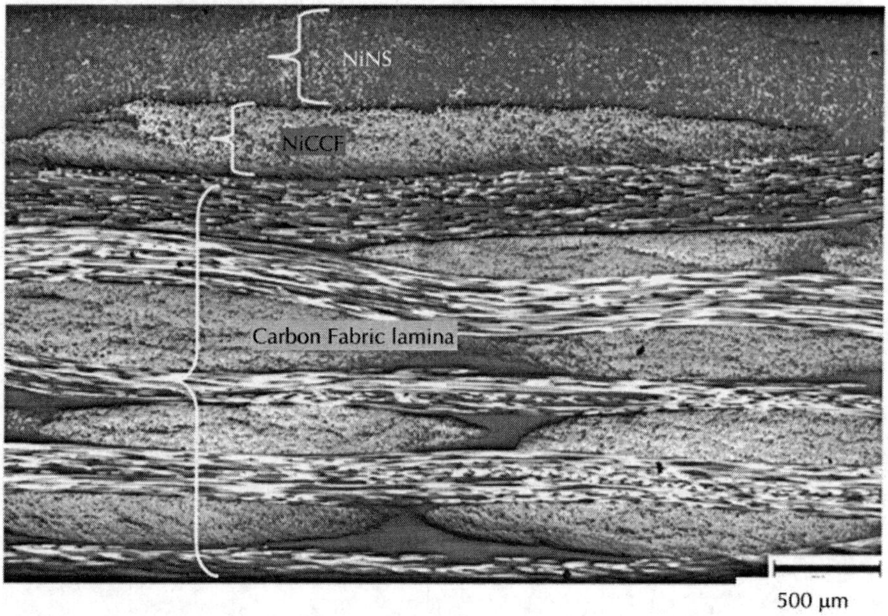

FIGURE 3 Details of panels for lightening strike protection.

12.2.2 Specimen Details

The simulated lightning strike was applied on the panels after painting with paint currently in use on military aircraft. The goal in the simulated lightening strike test was to mimic a real life lightning strike and therefore, the panel was hit with approximately 100 kA of current [8]. The panel was inspected before and after the strike using ultrasonic C-scan before and after lightning strike simulation testing. Time-of-flight (TOF) C-scan was also performed to determine the depth of composite damage after the lightning strike simulation test. The scan was done from the unstruck side of the panel, and damage was measured from the unstruck side and then subtracted from the total panel thickness to give values for the deepest damage depth for each composite panel. Figure 4 shows C-scans of both panels before and after lightning strike. The panel with NiCCF only, and with both NiNS and NiCCF had damage area of 15.8 and 19 cm^2, respectively while damage depth of 0.1 and 0.19, respectively. The volume resistivity of NiCCF paper and NiNS veil was 4.15 and 2.8 Ωcm, respectively while electrical conductivity was 4.8E-03 and 1.2E-03, respectively. Thus, NiNS was more conductive material, as well as area and depth of damage from the lightning strike was also more with nickel nanostrands.

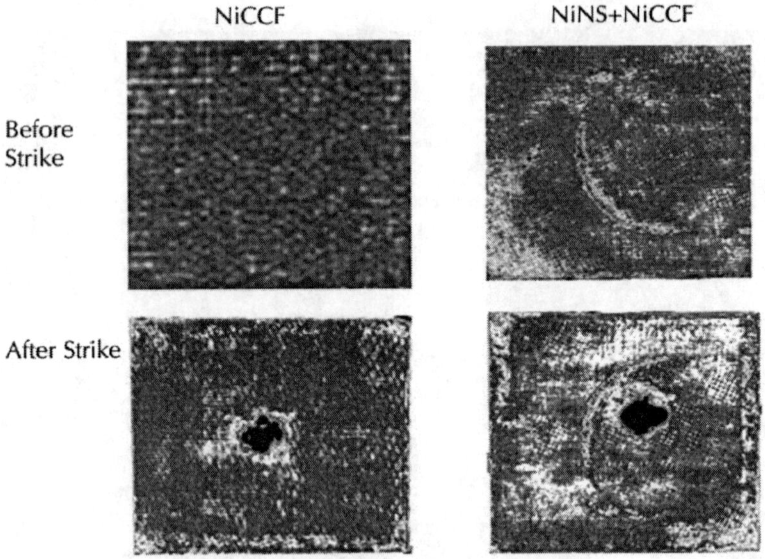

FIGURE 4 The C-scan panels before and after lightening strike.

Specimens were then machined from the panel after subjecting them to the lightening strike, and they were machined from the center and the edge of the struck panels. These locations are shown in Figure 5. These two locations had maximum and minimum damages of a panel hit by a lightning strike. The nominal size of specimens was 24 cm (length), 2.5 cm (width), and 0.26 mm (thickness). Also, specimens were machined from an unstruck panel of each system.

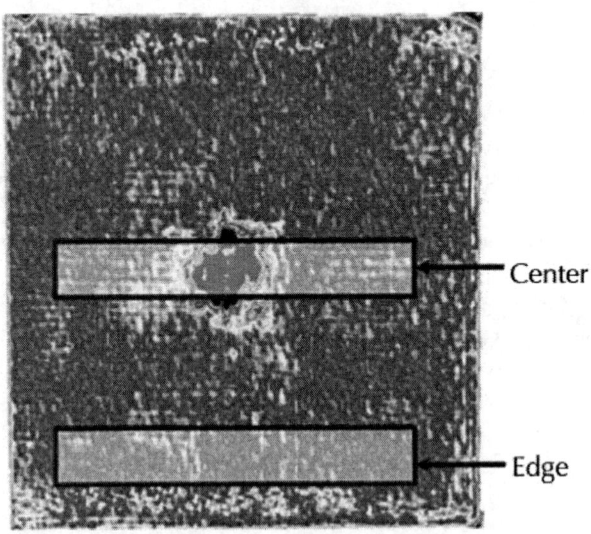

FIGURE 5 Locations of compressive strength specimens on struck panel.

For EMI shielding studies, rectangular shaped specimens of 15 (length) × 2.7 cm (width) sizes were machined from the composite panels. Glass/epoxy tabs were glued on both ends of the specimen to protect the specimen against possible damage by the grips of the test machine.

12.2.3 Test Details

The EMI shielding measurements in the terms of db were conducted before and after subjecting the tensile load to the specimen up to a certain level. This was repeated till the specimen failed. These load levels were approximately 33, 66, 80, 86, 90, 93, and 100% of the average UTS, respectively. The EMI measurements were performed before and after each tension test by the Agilent Technologies E8362B PNA Series Network Analyzer. This was done in order to determine how tensile load affects the EMI shielding effectiveness of the composite. The EMI measurements are not absolute values due to the scattering on the carbon fibers. Thus, if EMI measurements are performed at a different location on the specimen, the electromagnetic waves could be scattered differently because the fibers network act like a diffraction grating. The EMI attenuation was, therefore, measured exactly at the same spot on the specimen.

The compression test was performed on specimens before and after the lightening strike. In order to perform a compression test, it was necessary to use a fixture to prevent the specimen from buckling while under compression. The testing was conducted on a servo hydraulic test machine with hydraulic grips. All specimens were tested until their final failure which provided the ultimate compressive strength before and after the simulated lightening strike. After testing, all specimens were examined with optical and scanning electron microscopes for evaluation of damage mechanisms.

12.3 DISCUSSION AND RESULTS

12.3.1 Lightening Strike Effect on Compression Strength

The ultimate compressive strengths for the edge and center of the struck panel as well as for the unstruck specimen within each system and between both systems are compared in Figure 6. Specimen from the edge of struck specimen was not affected by the lighting's energy which was obvious from the C-scan and visual inspection. Thus the difference in the ultimate compressive strengths between the specimens from the unstuck panel and edge of struck panel could be attributed to the variation from panel to panel which is not uncommon in the making of two identical panels using the RTM process. Thus, an average of these two values may be considered as the baseline ultimate compressive strength of the panel. These average values of the ultimate compressive strength for both systems are compared in Figure 7 along with those from the center of the strike zone.

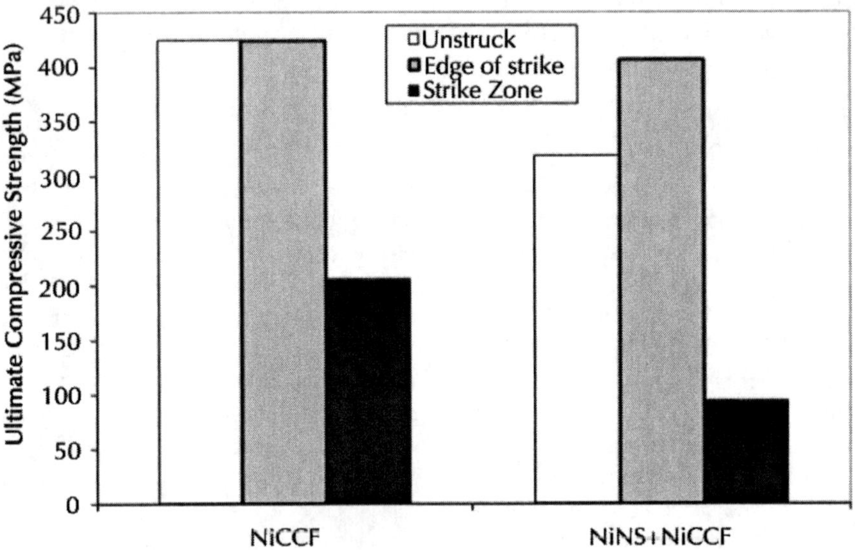

FIGURE 6 Ultimate compressive strength of unstruck, edge, and center of struck.

A comparison of ultimate compressive strength of the specimen from the center of the struck panel with the corresponding value either from the edge of the same struck panel in Figure 6 or average of the edge specimen of struck panel and specimen from unstruck panel in Figure 7 shows that there was considerable reduction after the lightening strike. It ranged from about 70% to 50%. This reduction was more (~ 70%) in the presence of NiNS veil while it was slightly less (~ 50%) without NiNS. As mentioned earlier, the panel with NiCCF only, and with both NiNS and NiCCF had damage area of 15.8 and 19 cm^2, respectively while damage depth of 0.1 and 0.19, respectively. Further, NiNS veil was more conductive material than NiCCF. This increased conductivity of NiNS increased the damage depth by about 90% while only 30% in the damage area. The compressive strength is expected to be more affected by the damage depth than the damage area. Thus, a desirable lightening protection system and/or material should disperse the damage over the wider area rather in the depth direction in order to cause minimum effect on the mechanical properties of the aerospace structures. The NiNS veil in the present form appears to be not a desirable option in this respect. Therefore, further studies are needed to devise the method (s) to utilize the present or other conductive nanocomposites as the protection against lightening strike without causing much reduction or with minimum reduction in the mechanical strength.

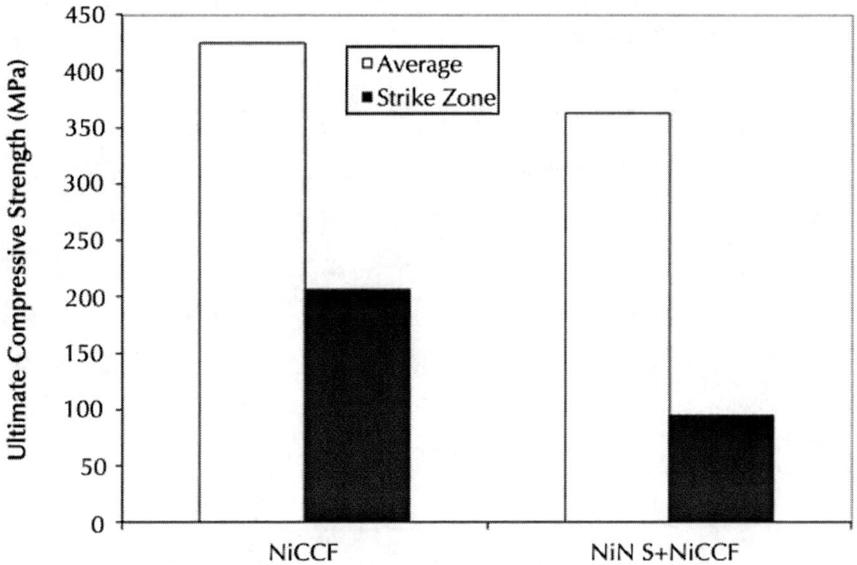

FIGURE 7 Ultimate compressive strengths of average and center of the strike zone.

12.3.2 EMI Shielding Effectiveness

Figure 8 shows the EMI attenuation values, in terms of db, as the function of the applied tensile stress level including before the application of any tensile load. These values are the average over the frequency ranging from about 8E09 to 13E10 Hz. The addition of NiNS increased the initial EMI attenuation, which is before application of any tensile load by more than 25% than the corresponding value of the control specimen without any nickel nanostrands. This increase in EMI attenuation could have been made higher by using more quantity of NiNS in composite than the one used in the present study, which were 100 g/m^2. It was not the focus of the present study, rather the focus of the present study was to investigate the effects of tensile load on EMI shielding effectiveness. Further, Figure 8 shows that there is negligible or little change in the EMI shielding attenuation with increasing tensile load up to the final fracture. This is an interesting observation because it demonstrates that the EMI shielding properties of the composite with and without nanostrands do not degrade or degrade very little under tensile loading condition up to the final fracture. As mentioned earlier, the EMI measurements in the specimens were not strictly the absolute values due to the scattering on the carbon fibers. Therefore, it is more appropriate to compare its change with increasing tensile load. This change can be looked as by normalizing the EMI data at any tensile stress level by its counterpart before the application of any tensile load. This is shown in Figure 9. EMI shielding capability in nanostrands specimen remained practically constant when compared to its initial condition. The specimen without nanostrand shows a constant decrease up to 10% before final fracture when compared to its initial value. Further, it is worthwhile to note that the addition of the nanostrands in M55/RS-3 composite did not have any detrimental effect on the UTS since both failed within ± 10% of the average strength of both panels.

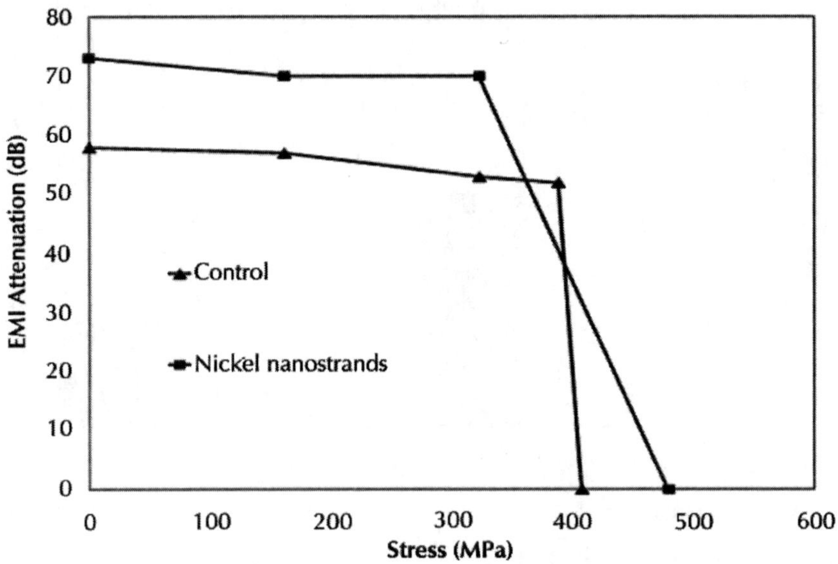

FIGURE 8 The EMI Attenuation *versus* applied tensile stress relationships.

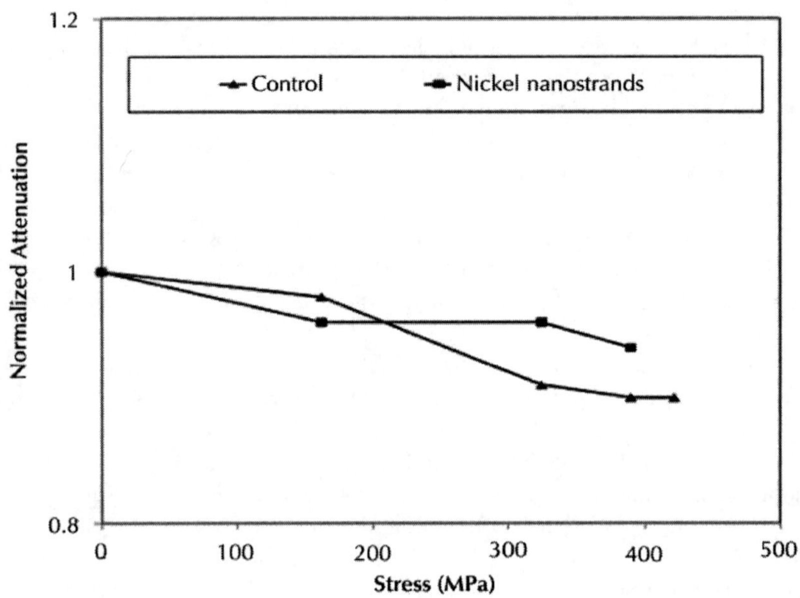

FIGURE 9 Normalized EMI attenuation *versus* applied tensile stress relationships (normalized with initial value).

12.3.3 Damage Mechanisms

EMI Specimen

The first step in the failure mechanisms in EMI specimens was the transverse matrix cracks in the 90° layers. These cracks increased in density up to a limiting value or the characteristic damage state. Thereafter, as the load increased, matrix cracks extended in ± 45° plies as well as caused the delamination along 0/90° plies, and finally precipitated the failure from the fiber breakage in 0° plies [9]. Figure 10 shows these details in a specimen without nanostrands. Figure 11 shows a fractured specimen with nanostrands in the width as well as the thickness direction. It had exactly the same damage mechanisms, that is it failed in the same manner as the specimen without nanostrands. However, it is interesting to note that nanostrands layers were intact up to the final failure of specimen as shown in Figure 12. Therefore, the panel maintained its EMI shielding effectiveness until its final failure as observed from the EMI measurements with increasing applied tensile load (Figures 8 and 9) due to these intact nanostrands.

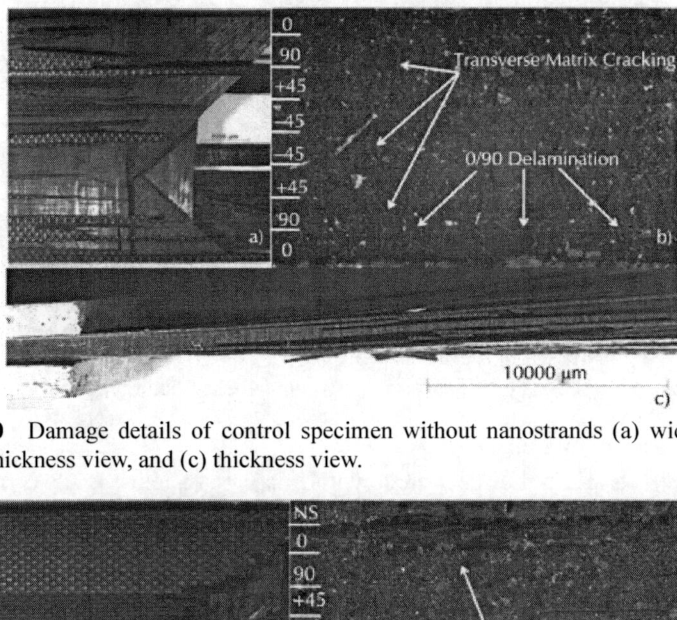

FIGURE 10 Damage details of control specimen without nanostrands (a) width view, (b) magnified thickness view, and (c) thickness view.

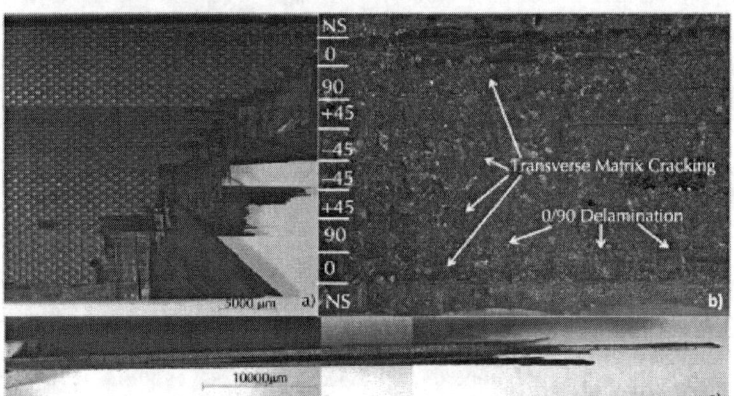

FIGURE 11 Damage details of nanostrands specimen (a) width view, (b) magnified thickness view, and (c) thickness view.

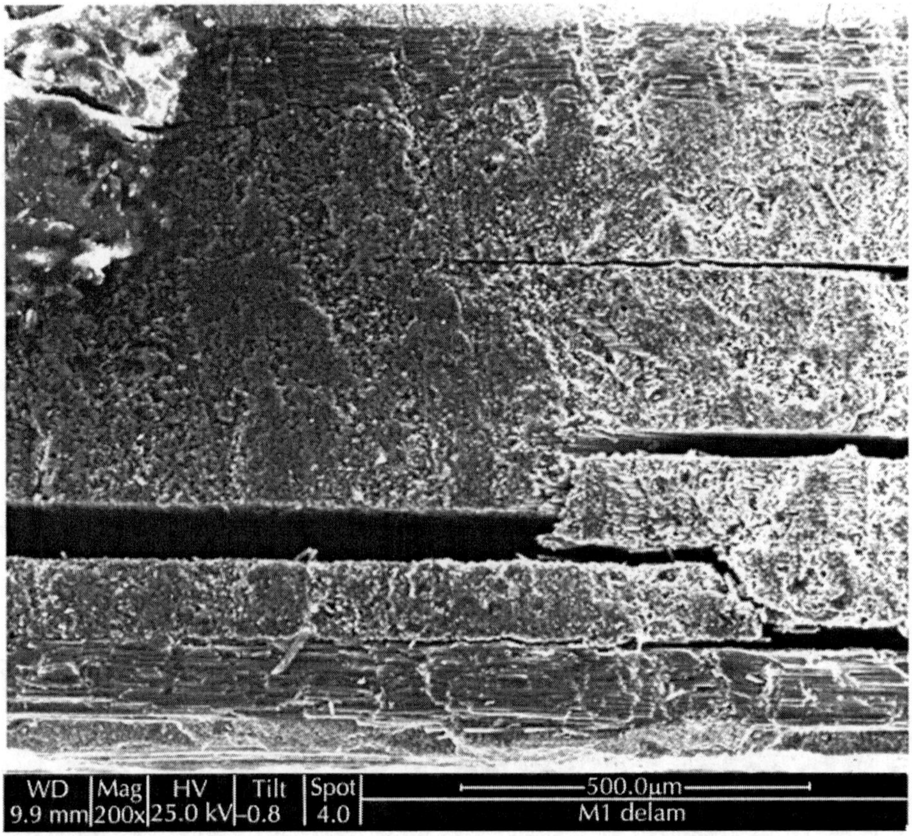

FIGURE 12 Damage details of nanostrands specimen near fracture region.

Lightening Strike Specimen

The final failure was in the shear mode with fracture at about ±45° as commonly seen in polymeric matrix composites when tested under compression [9]. Figure 13 shows a tested specimen from the unstruck panel containing a layer of nickel nanostrand. This figure shows delamination in the carbon fiber laminate, and additional delamination between the nickel nanostrand and nickel coated carbon fabric layers. A close up view of a region near the fracture region of this specimen is shown in Figure 13(b) which shows that delamination between the nickel nanostrand and nickel coated carbon fabric layer also extended in longitudinal tows of the nickel coated carbon fabric layer. However, it is interesting to note that nickel nanostrand veil itself was not damaged under the compression. The fractured region of the center specimen from the struck panel containing NiNS is shown in Figure 14. The energy from the lightening strike caused the damage and weakened the system badly that it had extensive delamination in both the carbon and nickel coated carbon fabric layers. It should be mentioned here that there was about 70% reduction in the compressive strength after the lightening strike in this case. This is a result of this extensive and wide spread delamination and damage extending in the carbon layers. One possible explanation of this behavior is that the nickel nanostrand layer acted as an insulator which trapped instead of dissipation, and directed the energy of the strike down into the remainder of the laminate causing more damage than the control. The NiNS layer was highly porous that could have trapped excessive amounts of resin in the layer which made it more resistive [7]. This increased resistivity might have made the nickel nanostrand layer act as though it was an insulator instead of the conductor which it was designed to be. The insulative layer directed the energy of the lighting strike down onto the carbon lamina that it was designed to protect.

(a)

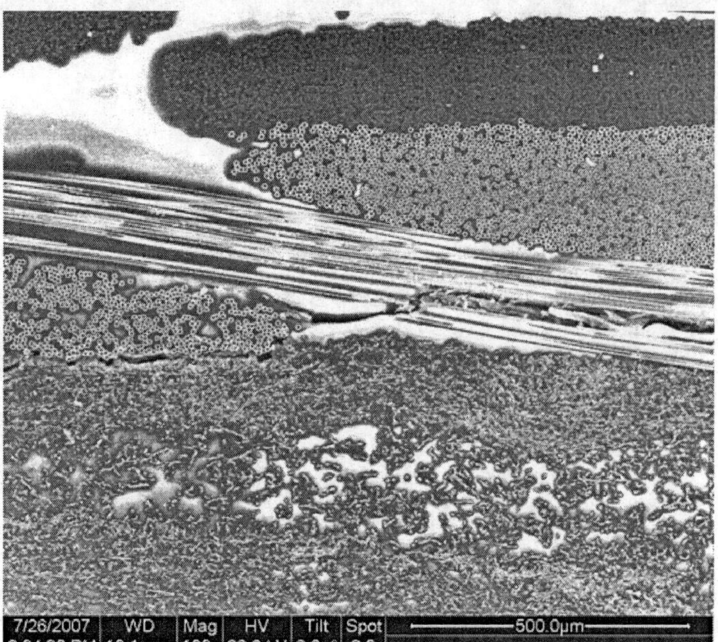

(b)

FIGURE 13 (a) Overall side view of fractured specimen from unstruck panel with nanostrands and (b) magnified view near fracture.

(a)

(b)

FIGURE 14 (a) Overall side view of fractured specimen from struck panel with nanostrands and (b) magnified view near fracture.

12.4 CONCLUSION

The conductive polymeric composites are being sought for application in aircraft structures to provide the lightning strike protection and for EMI shielding protection in space structures besides the light weight load bearing capability. This study investigated conductive nanocomposites for their compressive strength degradation after subjecting them to a simulated lightening strike and EMI shielding effectiveness when subjected to the monotonic tensile load. The following conclusions can be drawn from this study:

- The addition of the nanostrands in the M55/RS-3 composite provided better EMI shielding effectiveness without any detrimental effect on the mechanical behavior under the tensile loading condition (i.e. damage mechanisms and ultimate tensile strength). The EMI shielding effectiveness was practically constant under the tensile loading condition up to the final fracture. Failure mechanisms did not change with the inclusion of the nanostrands layers in the M55/RS-3 composite. Further, the nanostrands layers were intact up to the final composite fracture. Thus, the nanostrands are desirable option in the space composite structures.

- A considerable reduction in the compressive strength of composite material occurred from the addition of conductive NiNS when subjected to the lightening strike. This was due to increased damage depth from the increased conductivity of nickel nanostrands. A desirable lightening protection system and/or material should disperse the lightening strike damage over the wider area rather in the depth in order to cause minimum effect on the mechanical properties of the aerospace structures. The NiNS veil in the present form appears to be not a preferred option in this respect. Therefore, further studies are warranted in this direction.

KEYWORDS

- **Compressive strength**
- **Conductive nanocomposites**
- **Electromagnetic interference**
- **Lightening strike**
- **Nickel nanostrands**
- **Tensile strength**

REFERENCES

1. Bedingfield, K. L., Leach, R. D., and Alexander, M. D. *Spacecraft systems failures and anomalies attributed to the natural space environment*. NASA Reference Publication 1390 (1996).
2. Rawal, S. P. Multifunctional carbon nanocomposite coatings for space structures. Proc. AIAA/ASME/ASCE/AHS/ASC, Structures, Structural Dynamics and Material Conference, Newport, R.I (2006).
3. Rawal, S. P. and Kustas, F. M. Evaluation of carbon nanofiber-based coating and adhesives. *SAMPE Journal*, **42**, 62–70 (2006).
4. Fisher, F. A. and Plumer, J. A. *Lightning protection of aircraft*. NASA Reference, Publication, Government Printing Office, U S 1008 (1977).
5. Aircraft lightening zoning SAE Technical Standards. *SAE ARP 5414*, Detroit, USA (1999).
6. Brown, A. Evaluating lightening surface protection systems for aerospace composites. Proc. International Conference on Lightening and Static Electricity (ICOLSE). Seattle, WA. (2005).
7. Hansen, G. High aspect ratio sub-micron and nano-scale metal filaments, *SAMPE Journal*, **41**, 2–12 (2005).
8. Military Standards. MIL-STD-1757A. Lightening qualification test techniques for aerospace vehicles and hardware (1983).
9. Mall, S. *Properties and Performance of Laminated Polymer Matrix Composites*. In *Composites Engineering Handbook*. P. K. Mallick (Ed.), Mercel Dekker, New York, pp. 813–890 (1997).

13 Electrical Properties in CrO_2 Modified C-ZrO_2 of Ferromagnetic Nanocomposite

A. Sengupta, A. K. Thakur, S. Ram, and R. K. Kotnala

CONTENTS

13.1 Introduction 255
13.2 Experimental 256
13.3 Discussion and Results 256
13.4 Conclusion 260
Keywords 260
Acknowledgment 260
References 260

13.1 INTRODUCTION

A dielectric material when reinforced by ferromagnetic species often results in modified dielectric and other properties on interaction between the electric and magnetic fields [1-3]. The ferromagnetic part influences favorably the dielectric part and *vice-versa* in a single united system of a hybrid composite with tunable composite properties. Impedance studies on CrO_2 or composites in form of thin films or a powder compact have drawn great attention owing to exotic magnetization response and dielectric relaxation under alternative electromagnetic fields with controlled dielectric power loss [4, 5]. In this chapter, ZrO_2 typifies a model dielectric medium for tailoring functional ferromagnetic properties of metal oxides such as CrO_2, CoO, spinal ferrites, and so on. When used as filler in forming a composite. Owing to a high magnetic moment (1.92 μ_B per Cr^{4+} at 10 K)[6] with nearly 100% spin polarization of the conduction electrons, CrO_2 is preferred over other magnetic oxides as an additive to a ZrO_2 type of a dielectric host. We report electrical properties of CrO_2 modified ZrO_2 of a composition

$20CrO_2$-$80ZrO_2$ of a cubic (c) crystal structure in this chapter. Limited CrO_2 solubility does not permit designing a still CrO_2 rich sample without scarifying the magnetic properties [7].

The CrO_2 modified ZrO_2 with cubic (c) crystal structure and ferromagnetic properties can be useful for magnetoelectronics, spin polarizer, and magnetic sensors. In this study, we studied dielectric properties in a typical composition $20CrO_2$-$80ZrO_2$ after heating a polymer precursor at 500°C and 800°C for 2 hr in air. An average crystallite size 5 nm attained after heating at 500°C hardly grows to 8 nm when increasing heat treatment temperature to 800°C in keeping the polymorphic structure. Both the samples have ferromagnetic hysteresis loops showing 794 Oe and 211 Oe of coercivity, respectively. Temperature dependent bulk conductivity yields a typical thermoferromagnetic curve, with a Curie point T_C ~ 272°C (the sample heated at 500°C), and that is modified to a nearly paramagnetic linear curve, with T_C ~ 262°C, in prolong heating at effectively high temperature (causes a $Cr^{4+} \rightarrow Cr^{3+}$ transformation).

13.2 EXPERIMENTAL

A $20CrO_2$-$80ZrO_2$ nanocomposite was prepared by heating an aqueous polymer precursor with glycerol in air at 500–800°C for 2 hr. The precursor was obtained by dropwise addition of 17.3 ml aqueous CrO_3 (1M) to a mixed solution in $ZrOCl_2 \cdot 8H_2O$ (1M) and glycerol (2M) in equal volumes in a 138.4 ml batch with sonication at 40–50°C for 3 hr. An air dried gel at 65°C when heated in an autoclave in presence of $(NH_4)_2CO_3$ at 160°C in 1.5 atm pressure yields a stable product embedded in organic species, which could be burnt away in air at higher temperatures. Magnetic properties of nanocomposites so obtained were studied with a vibrating sample magnetometer (DMS-1600) at room temperature. Electrical properties of them were measured on cold pressed and sintered (at temperatures at which they were synthesized) powder compacts. The data were collected at selective frequencies *f* (50 Hz-100 kHz) and temperatures (30-500°C) with help of an impedance analyzer (HIOKI LCR Hi TESTER 3522-50).

13.3 DISCUSSION AND RESULTS

Two $20CrO_2$-$80ZrO_2$ powders prepared by heating a precursor at 500°C and 800°C for 2 hr in air were considered for studying the electrical and magnetic properties. Both of them have the same crystal structure of c-ZrO_2 (Fm-3m space group) but differ in the lattice parameter, that is 0.5128 nm and 0.5094 nm (i.e. comparable to the value in monolithic c-ZrO_2 of 0.5090 nm) [8], respectively, as per the X-ray diffraction (XRD). The XRD peak broadening determines an average crystallite size 5 nm after heating at 500°C that is hardly grown to 8 nm on increasing the temperature in the other sample. The values of magnetic coercivity (H_c), saturation magnetization (M_s), remanence ratio M_r/M_s, dielectric constant (ε_r), power loss (tanδ), bulk dc conductivity (σ_{dc}^{b}), and activation energy (E_a) of charge carriers studied on the two samples are given in Table 1. As an extrinsic parameter, the H_c is decreased readily from 794 Oe in the 5 nm crystallites to a value 211 Oe on growth of those to an average 8 nm value. A lowered M_s value 1.04 emu/g than the 5 nm crystallites of 1.62 emu/g indicates a considerable $Cr^{4+} \rightarrow Cr^{3+}$ (antiferromagnetic) conversion on heating at such high temperature.

TABLE 1 Magnetic, electrical, and dielectric properties in $20CrO_2$-$80ZrO_2$ nanocomposites.

Heat-treatment	H_c (Oe)	M_s (emu/g)	M_r/M_s	ε_r		Tanδ		σ_{dc}^{b} (μSm^{-1})		E_a (eV)
				A	B	A	B	A	B	B
500°C, 2 hr	794	1.62	0.253	27	47	0.08	0.04	1.22	214.26	0.65
800°C, 2 hr	211	1.04	0.038	39	50	1.06	0.32	0.66	34.23	0.95

The values A and B measured at 250°C and 350°C, while ε_r and tanδ at 100 kHz.

The Nyquist plots shown in Figure 1 for the two $20CrO_2$-$80ZrO_2$ samples at different temperatures consist of a big semicircular arc over high *f* values and a tiny semicircular arc over lower *f* values with the centers lying below the Z″ axis. The semicircle describes electrical response due to a parallel combination of a resistive and a capacitive circuit element contributed by either the grains or the grain boundaries. The first arc describes the bulk (grain) contribution while the second one the grain boundary contribution. Their intercepts on the real (Z′) axis determine the dc resistance values for the respective parts. The second semicircle grows up in the sample of larger crystallites (Figure 1(b)). Progressive shift of the intercepts of the semicircles on lower Z′ values, that is the reduction of bulk resistance with increasing the temperature, characterizes a semiconductor behavior of the samples over 320–420°C (Figure 1), that is, above the Curie point (T_C).

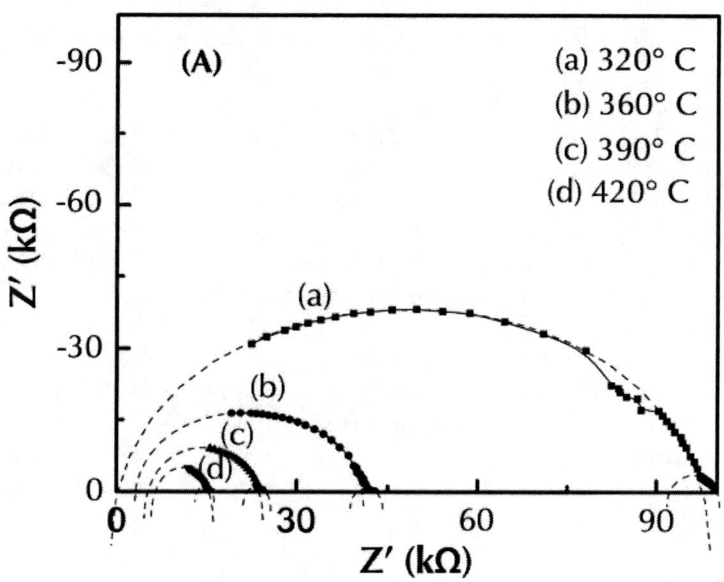

FIGURE 1 *(Continued)*

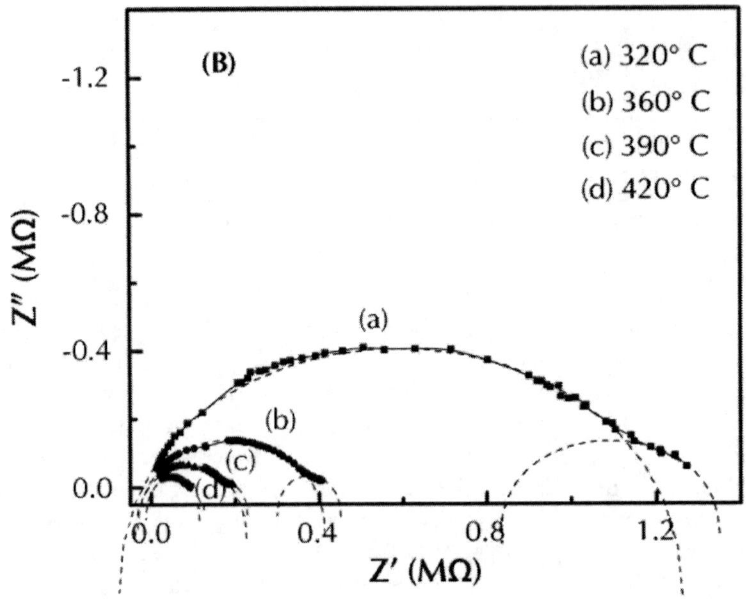

FIGURE 1 The Nyquist plots for $20CrO_2$-$80ZrO_2$ {prepared at (a) 500°C and (b) 800°C} measured at different temperatures as given in the insets.

As shown in Figure 2, temperature dependent bulk conductivity evaluated from the complex impedance spectrum exhibits a typical thermoferromagnetic curve in the sample heated at 500°C, showing a T_C ~ 272°C. The plot is modified to a nearly paramagnetic linear curve, with T_C ~ 262°C, due to a $Cr^{4+} \rightarrow Cr^{3+}$ transformation encountered in prolong heating at effectively high temperature. In the first sample, the σ_{dc}^{b} value follows the Arrhenius behavior either below 255°C or above 310°C (Figure 2(a)) with a nonlinear plot over the intermediate temperatures. The two linear regions measure the values of activation energy E_a = 0.62 and 0.65 eV, respectively. A thermally activated charge transport process describes a nearly two orders of enhanced σ_{dc}^{b} value over higher temperatures. The nonlinear region reflects in a distinct peak in the first derivative of the original plot over 260–300°C, with the maxima at ~ 272°C in the T_C point that is larger relative to a value in pure CrO_2 in the 112–127°C range [6]. The nonlinear part is narrowed down in the second sample (Figure 2(b)) with a sharp peak at ~262°C in the derivative curve in a weak residual ferromagnetism. The extended linear part is described with an enhanced E_a value 0.95 eV of a paramagnetic state.

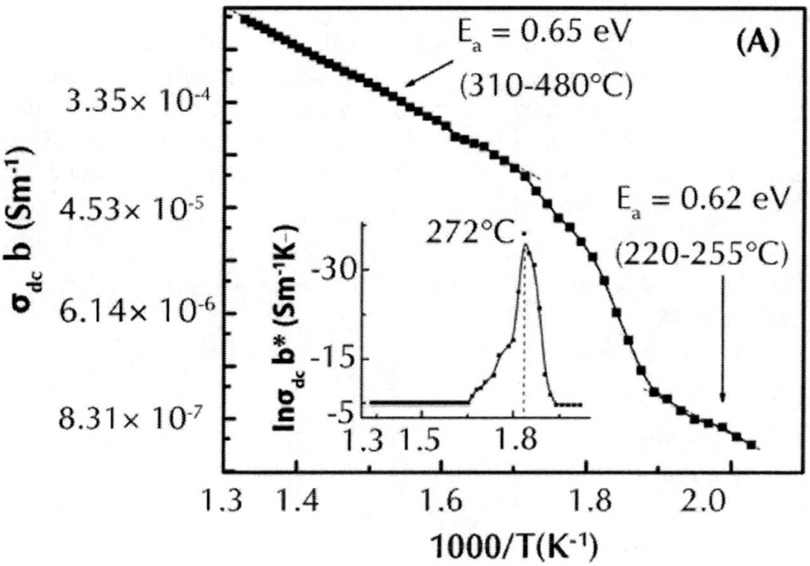

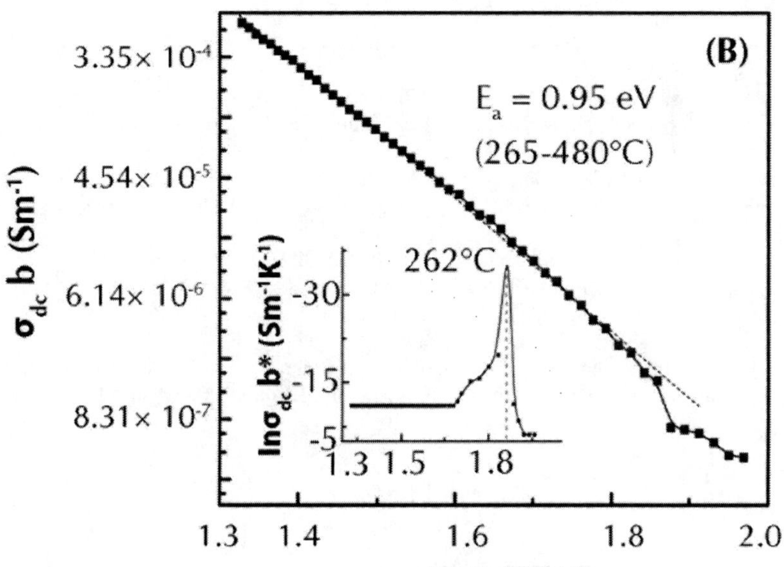

FIGURE 2 Temperature dependent bulk dc conductivity in $20CrO_2$-$80ZrO_2$ prepared by heating a precursor at (a) 500°C and (b) 800°C for 2 hr in air, with the first derivative in the insets.

13.4 CONCLUSION

A nanocomposite $20CrO_2$-$80ZrO_2$ prepared through a polymer precursor yields stable ferromagnetic material in the parent cubic crystal structure of zirconia. Useful ferromagnetic behavior does not change much unless heating the sample above 500°C in air. The results are characterized in terms of temperature dependent bulk conductivity which follows a nearly linear Arrhenius plot in the region the Curie point. The plot is influenced markedly when $Cr^{4+} \rightarrow Cr^{3+}$ transformation sets in on raising the temperature above 500°C in ambient air.

KEYWORDS

- **Dielectrics**
- **Magnetic coercivity**
- **Magnetic materials**
- **Nanocomposites**
- **Nanomaterials**

ACKNOWLEDGMENT

This work has been supported in part by the Council of Scientific and Industrial Research (CSIR), Government of India.

REFERENCES

1. Brosseau, C. and Talbot, P. *J. Appl. Phys.*, **97**, 104325 (2005).
2. Clavero, C., Sepúlveda, B., Armelles, G., Konstantinović, Z., García del Muro, M., Labarta, A., and Batlle, X. *J. Appl. Phys.*, **100**, 074320 (2006).
3. Verma, K. C., Kotnala, R. K., and Negi, N. S. *Appl. Phys. Lett.*, **92**, 152902 (2008).
4. Fu, C. M., Chao, Y. C., Hung, S. H., Lin, C. P., and Tang, J. *J. Mag. Mag. Mater.*, **282**, 283–286 (2004).
5. Singh, G. P. and Ram, S. *J. Am. Ceram. Soc.*, **91**, 322–324 2008.
6. Chioncel, L., Allmaier, H., Arrigoni, E., Yamasaki, A, Daghofer, M., Katsnelson, M. I., and Lichtenstein, A. I. *Phys. Rev. B*, **75**, 140406 (2007).
7. Ram, S. *J. Mater. Sci.*, **38**, 643–655 (2003).
8. X-ray Powder Diffraction File JCPDS–ICDD (Joint Committee on Powder Diffraction Standard International Centre for Diffraction Data, Swarthmore, PA) pp. 270–997 (1999).

14 Processable Aqueous Dispersion of Fullerene C_{60}: A Nanofluid

M. Behera, S. Ram, and H. J. Fecht

CONTENTS

14.1 Introduction 261
14.2 Experimental Details 262
14.3 Discussion and Results 262
14.4 Conclusion 266
Keywords 266
Acknowledgment 266
References 266

14.1 INTRODUCTION

Fullerenes have an important molecular system for exploiting as a modifier to engineer a variety of composite materials with functional properties. Due to unique physical and chemical properties of C_{60}, many applications have been proposed and implemented in various fields such as lubrication [1, 5], sensor [2, 3], heat transfer [4, 5], and biomedical applications [6-8] such as photodynamic therapy, drug delivery, enzyme inhibition, antioxidant, antiviral, and antibacterial activity. As C_{60} is spherical in shape, small in size, and has capability to generate reactive oxygen species (ROS), such as superoxide anion radical (O_2^-) and hydroxyl radical (OH) under photoirradiation [6, 7], it holds high potential in biological and medical applications. Extremely low solubility in most polar solvents and propensity towards aggregation restrict C_{60} of these applications. In this context, solubilization, and stabilization of C_{60} in aqueous medium has attracted great attention in scientific community. Generally, two widely used strategies have been tested to be effectively dispersing C_{60} in water: (i) surface functionalization by attaching hydrophilic functional groups [6-8] and (ii) surface alteration through stabilizing agents [9-11] (e.g. surfactants, polymers, etc.).

We describe use of a facile route to dissolve C_{60} molecules in a water medium in a significant amount in this chapter. Water soluble polymers such as

PVP, poly(vinyl alcohol) (PVA), and so on., which are well-known solubilization agents, [7, 10, 11] have been chosen in order to tailor solubility of C_{60} molecules in form of a stable liquid solution or nanofluid at room temperature. Results are characterized in terms of optical absorption, zeta potential (ξ), and dynamic light scattering (DLS).

Dispersion of fullerene C_{60} in an aqueous medium is a big challenge in processing its products and devices in selective forms such as ionic and nonionic liquids and nanofluids, or films. In this study, we used a facile route which permits C_{60} molecules dissolving in water considerably when a polymer such as poly(vinyl pyrrolidone) (PVP) present. Being soluble in water as well as many organic solvents such as butanol, ethanol, or DMF, PVP allows transferring C_{60} from an organic to an inorganic medium. As large as 90.3 μM C_{60} molecules could be solubilized in water through 120 g/l PVP molecules. Dissolved C_{60} molecules (18.4 μM) *via* 40 g/l PVP in water, reflect in a broad absorption band over 280–520 nm, reveal an average 178.3 nm of hydrodynamic length with a polydispersity index 0.349. A negatively charged surface persists with a zeta potential (-) 9.5 mV at 6.5 pH, showing C=O (PVP) nonbonding electron transfer to the C_{60} nanosurface in a weak donor-acceptor complex. Such biocompatible polymer based C_{60} nanofluids could find may useful for biological, medicinal, and other applications.

14.2 EXPERIMENTAL DETAILS

Fullerene C_{60} and PVP (K-25, 28kDa) purchased from Alfa Aeser and Aldrich respectively was used as the starting reagents. The 5.0 mg of C_{60} was dissolved in 5 ml toluene in a 1.39 mm solution while PVP was dissolved in 1-butanol of a 120 g/l PVP solution. A clear light-brown solution turned-up after 20 min sonication of a mixture of 0.2 ml C_{60} solution and 5 ml PVP at 50°C. Then, the solvent was thoroughly evaporated at 100 ± 10°C in a hot plate for a period of 10 hr. A 5.0 ml of a C_{60} nanofluid in water could be transferred on adding ultrapure water by sonication in a hot condition. That contains as large C_{60} as 90.3 μM. A similar process was followed to obtain a similar C_{60} solution (6.94 μM) in 1-butanol. Absorption spectra of the two solutions were measured on a Perkin Elmer double beam spectrophotometer (LAMBDA 1050). Particle size distribution was studied in terms of DLS using a Malvern Nano ZS instrument. The same machine was used to measure ξ-value.

14.3 DISCUSSION AND RESULTS

Figure 1 shows absorption spectrum of a C_{60} nanofluid prepared in presence of PVP molecules in water. In order to measure the spectrum, the stock sample was diluted by three times in water and then sonicated for 5 min to maintain a homogeneous dispersion. The spectrum consists of a sharp intense band at 295 nm and a broad band over 280–450 nm with average peak value 341 nm. The 295 nm band arises in the $^1A_{1g} \rightarrow {}^1T_{1u}$ (3) ($\pi \rightarrow \pi^*$) transition in C(sp^2) from PVP surface-modified C_{60} molecules whereas the broad band arises in a charge transfer (CT) band out of a C_{60}

PVP complex consisting of C_{60} molecules dispersed through PVP molecules [12, 13]. The broad band features are indicative that PVP molecules are cross-linking C_{60} molecules in a specific structure [13]. Ungurenasu and Airinei [11] observed a rather red shifted and broader band near 315 nm in 44 µM C_{60} (with 0.12 M PVP which was added through chloroform) in water in a rather different structure. A pulsed sonication we used in hot condition possibly promotes the C_{60} reordering *via* PVP molecules. In order to short out this observation, we studied absorption in C_{60} dissolved in 1-butanol (without a modifier) which does not extend such bonding. As shown in the Figure 2, a rather sharper and weaker characteristic C_{60} band stands at 330 nm along with an extremely weak band of 406 nm. The $^1A_{1g} \rightarrow {}^1T_{1u}$ (3) characteristic band at 295 nm in PVP modified C_{60} is red shifted to 330 nm. Possibly, 1-butanol is not sharing a polymer bridging with the C_{60} molecules and instead it forms merely a clathrate-like network on a C_{60} molecule [14]. An weak band centered at 406 nm is assigned to orbital forbidden singlet-singlet transition in the C(sp^2) electrons of C_{60} [12] Presence of surface modifier of PVP molecules causes amassing of C(sp^2) electrons on the C_{60} nanosurface that results in a blue-shifted band at 295 n in the quantum confinement effect [13].

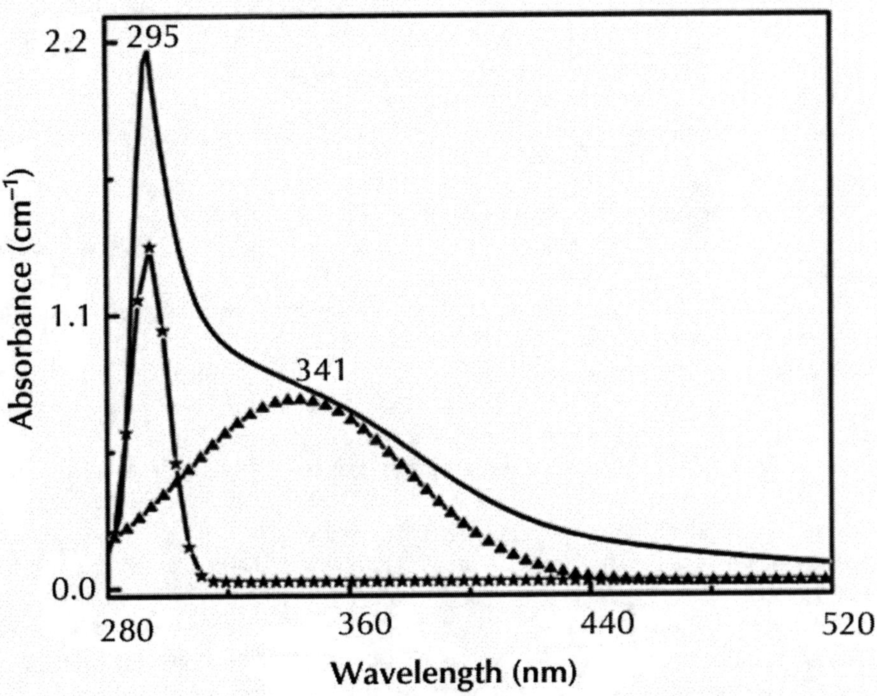

FIGURE 1 Absorption spectrum in C_{60} (18.4 µM) dissolved in water in presence of 40 g/l PVP *via* 1-butanol.

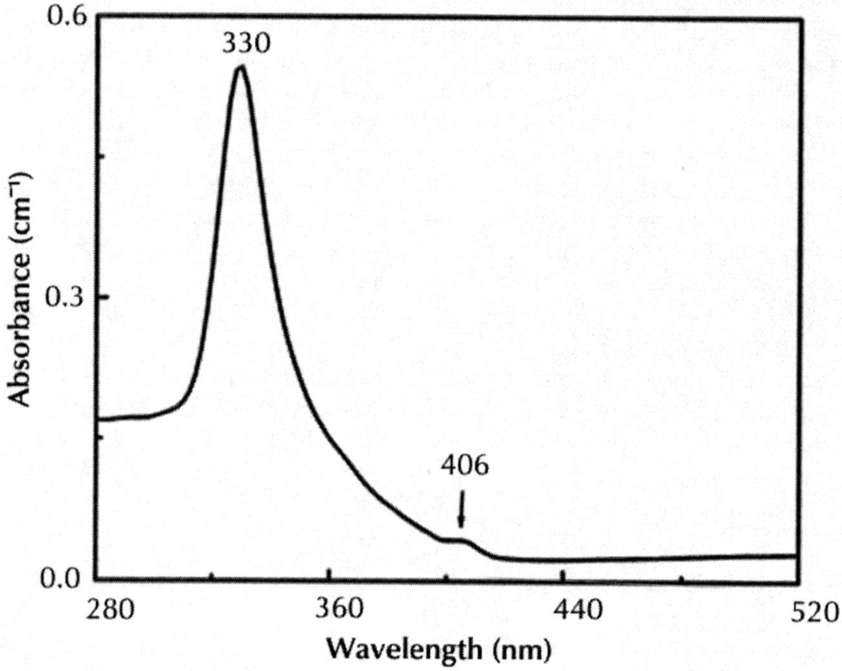

FIGURE 2 Absorption spectrum in C_{60} (6.94 μM) molecules in 1-butanol without PVP.

The ξ profile has been measured for the C_{60} nanofluid after diluting by nearly 15 times in water, such that, final sample contains 3.7 μM C_{60} and 8 g/l PVP with an average 6.5 pH. As shown in Figure 3, it consists of two distinct peaks with ξ values of (-) 14.2, and (+) 0.711 mV, with the first peak of the most prominent signal. According to it, an average ξ value is found to be (-) 9.5 mV, such that, there exists an effectively negative charge on the nanosurface of PVP encapsulated C_{60} molecules. It is a result of C=O (PVP) nonbonding electron transfer from an electron donor PVP molecule to an electron acceptor C_{60} molecule in form of a weak donor-acceptor complex. The DLS spectrum in Figure 4 consists of three distinct peaks in three hydrodynamic lengths L_{hd} of the nanofluid of 354.6, 18.6, and 7.2 nm. The three DLS peaks measure an average L_{hd} value 178.3 nm, with a polydispersity index 0.349. The first band, which is prominent, attributes to PVP encapsulated C_{60} molecules that reveal an intense ξ band. [13] The second and third bands arise in PVP molecules dispersed in solution and those which bridges a surface-interface on a C_{60} molecule, respectively. These results describe that a steric stabilization plays a major role in colloidal stability in a polymer based C_{60} nanofluids. Use of a biocompatible polymer such as PVP molecules tested in this work expands possible applications of biofriendly nanofluids for biological, medicinal, and other disciplines.

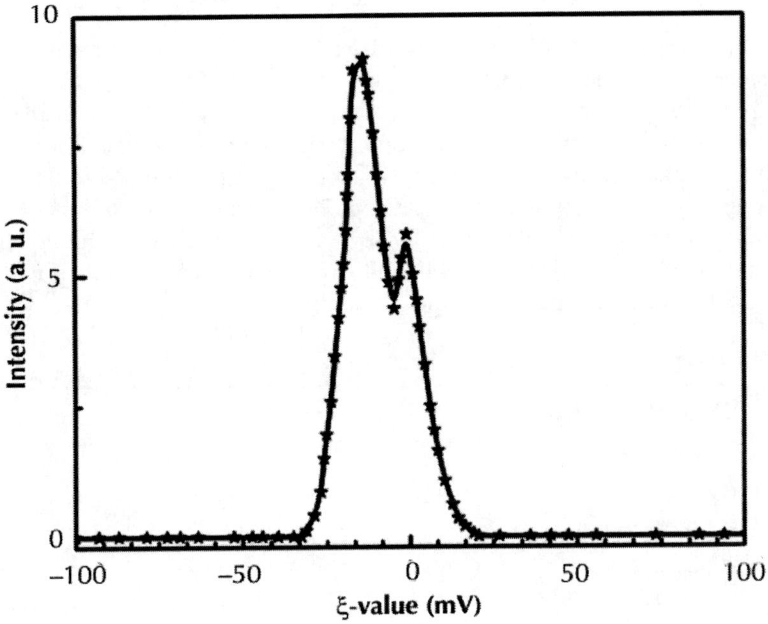

FIGURE 3 Distributions of ξ values in a C_{60} nanofluid consisting of 3.7 μM C_{60} molecules and 8.0 g/l PVP in water.

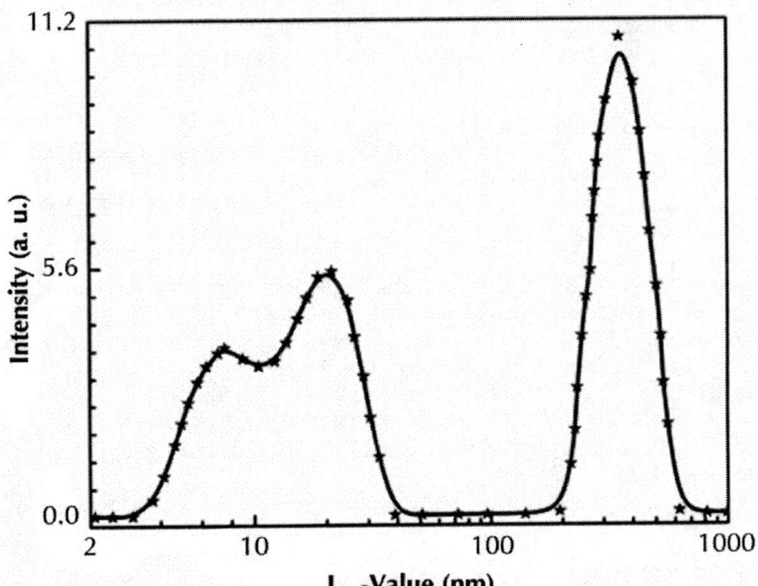

FIGURE 4 Distributions of L_{hd} values in a C_{60} nanofluid consisting of 3.7 μM C_{60} molecules and 8.0 g/l PVP in water.

14.4 CONCLUSION

A simple chemical method is used to disperse C_{60} molecules through PVP polymer molecules in a rheological nanofluid in water. Studies of the UV-visible light absorption, zeta potential, and dynamic light scattering of the nanofluid infer that PVP molecules cross-links C_{60} molecules. In a polymer complex, C_{60} molecules become easily soluble in water in a significant value, which is large as 90.3 μM at room temperature. A strong $\pi \rightarrow \pi^*$ absorption band occurs at 295 nm along with a broad CT band extending over longer wavelengths up to 520 nm. The steric stabilization plays a major role in colloidal stability in a biocompatible polymer based C_{60} nanofluids useful for biological, medicinal, and other applications.

KEYWORDS

- **Carbon chemistry**
- **Composites**
- **Hydrocolloids**
- **Nanofluids**
- **Nanosurfaces**

ACKNOWLEDGMENT

This work is supported in parts from All India Council of Technical Education, New Delhi, Silicon Institute of Technology, Bhubaneswar, and the Board of Research in Nuclear Sciences, Department of Atomic Energy (BRNS-DAE), Government of India.

REFERENCES

1. Zhang, P., Lu, J., Xue, Q., and Liu, W. *Langmuir*, **17**, 2143–2145 (2001).
2. Sudeep, P. K., Lpe, B. L., Thomas, K. G., George, M. V., Barrazouk, S., Hotchandani, S., and Kamat, P. V. *Nano Letters*, **2**, 29–35 (2002).
3. Sherigara, B. S., Kutner, W., and D'Souza, F. *Electroanalysis*, **15**, 753–772 (2003).
4. Putnam, A. S., Cahill, D. G., Braun, P. V., Ge, Z., and Shimmin, R. G. *J. Appl. Phys.*, **99**, 084308 (2006).
5. Hwang, Y., Park, H. S., Lee, J. K., and Jung, W. H. *Curr. Appl. Phys.* **6S1**, e67–e71 (2006).
6. Nakamura, E. and Isobe, H. *Acc. Chem. Res.*, **36**, 807–815 (2003).
7. Yamakoshi, Y., Umezawa, N., Ryu, A., Arakane, K., Miyata, N., Goda, Y., Masumizu, T., and Nagano, T. *J. Am. Chem. Soc.*, **125**, 12803–12809 (2003).
8. *Guldi,* D. M. and Prato, M. *Acc. Chem. Res.*, **33**, 695–703 (2000).
9. Clements, A. F, Haley, J. E., Urbas, A. M., Kost, A., Rauh, R. D., Bertone, J. F., Wang, F., Wiers, B. M., Gao, D., Stefanik, T. S., Mott, A. G., and Mackie, D. M. *J. Phys. Chem. A*, **113**, 6437–6445 (2009).
10. Yamakoshi, Y. N., Yagami, T., Fukuhara, *K.*, Sueyoshi, S., and Miyata, N. *J. Chem. Soc. Chem. Commun.*, **517**, 518 (1994).
11. Ungurenasu, C. and Airinei, A. *J. Med. Chem.*, **43**, 3186–3188 (2000).
12. Leach, S., Vervloet, M., Desprès, A., Bréheret, E., Hare, J. P., Dennis, T. J., Kroto, H. W., Taylor, R., and Walton, D. R. *M. Chem. Phys.*, **160**, 451–466 (1992).

13. Behera, M. and Ram, S. *J. Incl. Phenom. Macrocycl. Chem.*, doi: 10.1007/s10847-011-9957-y (2011).
14. Scharff, P., Risch, K., Carta-Abelmann, L., Dmytruk, I. M., Bilyi, M. M., Golub, O. A., Khavryuchenko, A. V., Buzaneva, E. V., Aksenov, V. L., Avdeev, M. V., Prylutskyy, Yu. I., and Durov, S. S. Carbon, **42**, 1203–1206 (2004).

Index

A

Adameks, G., 79–117
Arrhenius behavior, 258
Athanassiou, A., 41–50

B

Barone, A., 41–50
Behera, M., 261–267
Bello, V., 167–196
Binitha, N. N., 133–137
Biomorphic scaffolds, regeneration of loadbearing bones
alumina, 52
biohybrid, 52
biomineralization process, 52
biomorphic HA scaffolds
composite scaffolds, 74
rabbit femur, 76
bone disease, 52
bone scaffolds, mechanical biomimesis
biomechanical loads, 54
bone canaliculi, 54
central blood vessel, 54
chemical biomimesis, 54
cytoskeleton, 54
deformations act, 54
hierarchical structural, 55
load-bearing bone, 54, 55
mechanical biomimesis, 54
mechano-chemical, 54
nanometer sized crystals, 54
osteoconductive, 54
osteoinductive, 54
tensegrity, 54
trabecular bone, 54
tri-dimensional architectures, 54
bone tissue, biomorphic transformations
biomimetic hierarchical structure, 56–57
compact bone tissue and rattan wood, 56
ligneous structures, 56
mechanical load, 56
morphology, 56
osteoclasts, 56
tensegrity, 55
vascularization, 56
vascular system, 55
wood structures, 57
carbonation
biomorphic $CaCO_3$, 68
calcium oxide, 67
distribution of biomorphic, 69
kinetic of carbonation, 68
synthesize biomorphic HA, 67
carburization
biomorphic bone scaffold, 60
biomorphic CaC_2, 65
biomorphic rattan-derived carbon, 61
calcium hydride, 60–61
calcium phosphate, 60
carbon morphology, 61
carbon template, 62
carburization reaction, 65
metallic calcium, 60–61
nucleation of CaC_2, 63
SEM image of wood, 64
SEM images of biomorphic calcium carbide, 64
thermal cycle, 61, 62
vapor phase reaction, 62
vapor state, 61
wood carburization, 62
cell adhesion, 52
chemico-physical, 52
hydroxyapatite (HA), 52
ligneous templates, synthesis
biological affinity, 57
biomorphic, 57
HA, 58
carbide phases, 57
cellular structure, 57

ceramic materials, 57
consolidated oxide ceramics, 57
gas-solid reaction, 58
Haversian system, 57
inner morphology, 57
pyrolysis, 57
SEM images of, 59
sol-gel methods, 57
spongy bone, 58
mechanical load, 51
morphological, 52
osteocartilaginous regions, 52
osteoconductivity hampers, 52
osteointegration, 52
oxidation
biomorphic CaO, 66
carbon matrix, 65
schematic drawing of, 66
temperature, 66
pathologies, 52
phosphatization
calcium carbonate, 69
calcium phosphate, 69
cubic $CaCO_3$, 71
FTIR spectra, 72
XRD spectra, 72
pressure medicine, 52
scaffolds, 52
TEM micrographs of biohybrid, 53
titanium, 52
Borlaf, M., 139–146
Brookite titania nanorods, poly(Methyl Methacrylate)
AFM topographical, 45
anatase
NRs, 47
TiO2 NRs, 43
atomic force microscopy (AFM), 41
biocompatibility, 42
carbon coated Cu, 43
colloidal NRs, 41
crystalline brookite, 41
crystal structure, 42
differential scanning calorimetry (DSC), 41
measurements, 47
thermal traces, 47
fused silica, 43
heat capacity, 47
homogeneity, 48
inorganic filler, 47
Jeol JSM 1011 microscope, 43
MFP-3D AFM, 42
nanocomposites films, 42
NRs, 47
1-octadecene, 42
oleic acid, 42
oleyl amine, 42
optical microscopy inspection, 43
outer vertical bars, 48
perkin-elmer, 43
polymer nanocomposite, 41
statistical box diagrams, 48
TEM micrographs, 45–46
tetragonal anatase, 42
titania polymorphs, 42
titanium (IV) chloride, 42
titanium (IV) isopropoxide, 41, 42
transmission electron microscopy (TEM), 41
trimethylamine N-oxide dihydrate, 42

C

Caputo, G., 41–50
Chemical coprecipitation method
carbonaceous, 7
dry precipitate, 7
fatty acid, 6
ferric oxide, 7
ferrite metals, 6
hydroxides, 6
lithium
ferrite, 7
oxide, 7
Citrate precursor method
ammonium hydroxide, 8
ash-synthesized ferrite, 8
carboxylic acid, 8
carboxylic ions, 8
citric acid, 7
metal hydroxide, 8
nitrogen dioxide, 8
Colomer, M. T., 139–146

D

Dickerson, J. H., 139–146
N-Doped TiO2 photocatalyst
catalyst characterization, 134
environmental pollution, 134
methyleneblue (MB), 134
nitrogen doping, 134
N/Ti molar ratio, 134, 137
photocatalytic activity, 135
photodegradation, 133–134
pollutant degradability, 134
preparation of, 134
sol-gel method, 134
thermal treatment, 133
traditional surfactant templating method, 134
visible light activity, 134, 136–137

E

Electrically conductive nanocomposites
aerospace systems, 239
aircraft structures, 239
carbon fiber, 242
composite materials, 239
compression test, 245
compressive strength, 246
specimens, 244
conductive material, 246
conventional polymeric composites, 240
c-scan panels, 244
damage mechanisms
carbon lamina, 251
extensive delamination, 250
insulative layer, 251
intact nanostrands, 249
nickel coated carbon fabric layers, 250
nickel nanostrand, 250
NiNS layer, 251
polymeric matrix composites, 250
thickness direction, 249
unstruck panel, 251
electrical
conductivity, 239–240
ground path, 240
electromagnetic
radiation, 240
waves, 245
electromagnetic interference (EMI), 240
electrostatic discharge (ESD), 240
electrostatic potentials, 239
EMI attenuation, 245
EMI shielding effectiveness
attenuation value, 247
carbon fibers, 247
tensile load, 247
tensile stress, 248
hydraulic grips, 245
lightening protection system, 246
lightening strike
protection, 243
test, 243
mechanical strength, 246
metallic meshes, 240
monotonic tensile, 240
nickel coated carbon woven fabric (NiCCF), 242
nickel nanostrands, 240, 242
NiNS layers, 241
polycyanate (RS-3) resin, 241
quasi-isotropic lay-up, 241
resin transfer mold (RTM), 242
scanning electron microscopes, 245
servo hydraulic test machine, 245
space bus structures, 239
ultimate compressive strength, 240, 246–247
ultimate tensile strength (UTS), 240
ultrasonic C-scan, 243
Electrical properties in CrO_2. See Ferromagnetic nanocomposite

F

Fecht, H. J., 261–267
Ferromagnetic nanocomposite
activation energy, 256, 258
aqueous polymer precursor, 256
bulk DC conductivity, 256, 259
charge transport process, 258
complex impedance spectrum, 258
crystal structure, 256
cubic crystal structure, 255–256

Curie point, 256, 257
dielectric constant, 256
dielectric material, 255
electrical response, 257
electric fields, 255
exotic magnetization response, 255
grain boundary contribution, 257
hybrid composite, 255
impedance analyzer, 256
lattice parameter, 256
magnetic
coercivity, 256
fields, 255
moment, 255
metal oxides, 255
model dielectric medium, 255
Nyquist plots, 257–258
paramagnetic linear curve, 256, 258
paramagnetic state, 258
polymer precursor, 256
polymorphic structure, 256
power loss, 256
remanence ratio, 256
saturation magnetization, 256
spinal ferrites, 255
thermoferromagnetic curve, 256, 258
X-ray diffraction (XRD), 256

H

Hamada, K., 119–131

I

Imae, T., 119–131
Ion implantation
bimetallic clusters, 174
binary nanoparticles, 176
core-shell structure, 175, 176
nanocluster structures, 175
noble-metal-based systems, 175
core-satellite systems
extended plasmonic, 193
Generalized Multiparticle Mie (GMM), 192
group velocity, 192
ion beam, 189
Mie-Maxwell-Garnett (MMG), 192
nanosphere lithography, 193
optical building, 189
Raman spectroscopy, 192
satellite cluster corona, 192
scattering losses, 192
indirect synthesis, 168
linear absorption measurements
cluster composition, 178
energy-dispersive X-ray spectroscopy (EDS), 177
sub-nanometer, 178
metal nanoclusters (NCs), 167
metal nanoparticles
diffusion-controlled processes, 169
isochronal annealing, 169
nucleation, 169
optical absorption spectra, 171
precipitation processes, 169
metal NCS
cluster formation, 172
cluster surface, 174
dipolar approximation, 172
isolated clusters, 172
metallic nanoparticle, 172
optical density (OD), 174
nanocomposite Films
alcohol vapors, 183
aromatic polyimides, 180
azodicarbonamide (ADC), 184
cluster nucleation, 178
core shell particle, 181
light beam, 182
molecular dimension, 185
nucleophile, 184
plasmon band, 188
polyimide
matrix, 178
structure, 184
polymeric chain, 184
potassium polyamate, 184
silica mediated, 186
virgin polyimide film, 182
virgin vapors, 181
visible range, 186
wavelength region, 187
nonlinear optics, 167
nuclear energy, 168
thermal diffusion, 168

thermodynamic equilibrium, 168

J

Jakubowicz, J., 79–117

K

Kalska-Szostko, B., 197–237
Kotnala, R. K., 255–260
Kumbhakar, P., 19–40

M

Magnetic dye-adsorbent catalyst, magnetic photocatalyst
catalytic dye-agent, 149
comparable surface, 159
core-shell, 148
dye-adsorbent material, 149
dye-decomposition, 148
external magnetic field, 157
hydrogen titanate, 153
hydrothermal treatment, 150, 153
hysteresis loop, 151
induced magnetization (M), 151
industrial effluents, 148
magnetic ceramic particle, 149, 153
citric acid, 149
core magnetic material, 150
measurements
absorption spectra, 152
absorption spectrophotometer, 151
dye-adsorption measurements, 151
photocatalytic activity, 151
UV-radiation exposure, 152
metal-sulfide semiconductors, 148
morphological transformation, 153
nanocrystalline coating, 153
non-biodegradable pollutantss, 148
non-toxic, 148
organic synthetic dyes, 147
oxygen solubility, 147
photocatalytic degradation mechanism, 158
photo-corrosion, 148
photo-electrochemical cells, 147
photosynthetic activity, 147
powerful oxidation/reduction, 147
prehydrolized titanium (IV), 150
redox chemical reactions, 148
surface-adsorption, 158
mechanism, 159
surface-cleaning treatment, 159
XRD broad-scan spectra, 152
Mahajan, S. V., 139–146
Maity, A., 19–40
Mall, S., 239–253
Mattei, G., 167–196
Mazzoldi, P., 167–196
Mitra, A. K., 19–40
Moreno, R., 139–146
Morimoto, Y., 119–131

N

Nanotube-fullerodendron hybrids, water-dispersible carbon
amide bands, 124
Anthryl Dendron G2.0-Oac
N-dimethylformamide (DMF), 121
nitrogen atmosphere, 121
Anthryl Dendron G2.0-OH
nitrogen atmosphere, 121
carbon-based compounds, 120
carbon nanotube (CNT), 120
carboxylic acid residue, 128
chloroform, 129
electromagnetic, 120
fullerodendrimer, 120
fullerodendron
bound, 128
C60(G2.0-Oac), 122
4-dimethylaminopyridine, 123
hydroxyl-terminated fullerodendron C60(G2.0-Oh), 122
shells, 126
spectroscopies, 123
terminal hydroxyl groups, 123
hybrid material, 120
hydrophilic dendron affects, 129
hydroxyl terminals, 125
hyperbranched polymer, 120
mercury cadmium telluride (MCT), 122
methanol dispersion, 122
molecules or polymers, 120
multiwall carbon nanotubes (MWCNT), 120

MWCNT
amine-terminated, 123
hydroxyl terminated polymer, 123
polycarbonate membrane, 122
room temperature, 122
N-dimethylformamide (DMF), 120
non-covalent functionalization, 120
organized systems, 120
oxide nanoparticles, 120
polar and nonpolar mediums, 120
polymer-metal ion complex, 120
thermal gravimetric analytical (TGA), 122
transmission electron microscopic (TEM), 120

O

Oxide nanomaterials synthesis
alkoxides, 2
citrate precursor method, 2
conventional and non-conventional method, 1
conventional ceramic method
agate mortar, 5
calcined, 5
homogenization, 5
iron oxide, 5
lithium ferrite, 6
oxysalt, 5
volatile organic liquid, 5
curie temperature, 9–10
ideal crystal, 10
Doppler effect, 10
Doppler velocities, 11
electric quadrupole, 12
electron cloud, 12
electrostatic monopole, 12
fatty acid, 2
ferromagnetic spinel
oxygen ion, 8
transition metal, 8
gamma rays, 10
isomer, 12
lithium ferrites, 1, 2
lithium zinc nickel ferrites, 12
ash-synthesized, 12, 13
citric acid, 12
coulomb forces, 14
doping ion, 14
force of interaction, 14
internal magnetic field, 16
ion concentration, 14
iron nitrate, 12
lithium ferrites, 16
lithium nitrates, 12
magnetic moment, 15
metal nitrates, 12
mossbauer parameters, 16
Neel's molecular field model, 14
net internal magnetic field, 17
nickel nitrate, 12
nuclear separations, 16
octahedral site, 16
polyvinyl alcohol, 12
quadrupole splitting, 16
saturation magnetization, 14
s-electron, 16
soohoo's method, 15
spinel ferrite, 16
tetrahedral site, 16
thermal energy, 15
XRD pattern, 13
zinc nitrate, 12
Li-Zn-Ni ferrite, 2
magnetic dipole moment, 12
mössbauer isotope, 11
mössbauer spectroscopy, 10
non-conventional method
chemical coprecipitation method (See Chemical coprecipitation method)
citrate precursor method (See Citrate precursor method)
saturation magnetization, ferrites, 9
s-electron
density, 12
wave, 12
spinel structure ferrites
cubic symmetry, 4
fcc structure, 4
inverse ferrite, 5
metallic ion, 4
mixed ferrites, 5
neighboring metal ions, 4

normal ferrites, 5
oxygen ion, 3
oxygen parameter, 4
spinel structure, 4
tetrahedral ion, 4
trivalent ion, 3
synthesis methods, 2

P

Patra, N., 41–50
Pellegrini, G., 167–196
Phanjoubam, S., 1–18
Porous nanostructure ti alloys
adhesion, 81
biocompatibility, 80
corrosion resistance, 80
cytotoxicity, 80
electrochemical anodic etching
BET surface, 95
chemical composition, 91
electrolyte penetration, 90
hydrogen bubbles, 88
macropore walls, 97
micropore surface, 95
nanocrystalline sinters, 88
osseointegration time, 90
phosphorus, 91
powder metallurgy processes, 88
SEM images of Ti-6Al-4V, 90
electrochemical cathodic cap deposition
calcium phosphate, 97
layer, 99
deposition process, 110
electric field lines, 113
low cathodic potential, 99
nanocrystalline porous, 99
hydroxyapatite, 81
mechanical alloying
amorphisation, 81
amorphous structure, 85
crystallographic structure, 86
lamellar structure, 81
MBraun, 81
mechanically alloyed amorphous, 85
morphology, 85
nanocrystalline, 85
phases, 81
otropic deformations, 85
plastic deformations, 81
SEM images of lamellar, 84
sub-grains formation, 85
TEM image of nanocrystals, 86
microcrystalline Ti, 80
morphology, 80
nanocrystalline, 79, 80
osseointegration, 81
proliferation, 81
senile dementia, 80
tialv and tizrnb alloys, corrosion resistance
potentiodynamic method, 115
Tafel extrapolations, 115
Prakash, C., 1–18
Processable aqueous dispersion
absorption spectra, 262, 263–264
biocompatible polymer, 264
1-butanol, 262
charge transfer (CT), 262–263
chemical properties, 261
chloroform, 263
clathrate-like network, 263
colloidal stability, 264
composite materials, 261
distributions of, 265
donor-acceptor complex, 264
dynamic light scattering (DLS), 262
electron
acceptor, 264
transfer, 262
hydrodynamic length, 262
hydrophilic functional groups, 261
hydroxyl radical, 261
Malvern Nano ZS instrument, 262
molecular system, 261
nanofluid, 262
organic solvents, 262
particle size distribution, 262
perkin Elmer double beam spectrophotometer, 262
photodynamic therapy, 261
physical properties, 261
polydispersity index, 262
poly(vinyl alcohol) (PVA), 262
poly(vinyl pyrrolidone) (PVP), 262

quantum confinement effect, 263
reactive oxygen species (ROS), 261
singlet-singlet transition, 263
superoxide anion radical, 261
surface
alteration, 261
functionalization, 261
zeta potential, 262

R

Ram, S., 255–260, 261–267
Resmi, M. R., 133–137
Ruffini, A., 51–78

S

Salerno, M., 41–50
Sarkar, R., 19–40
Sengupta, A., 255–260
Shukla, S., 147–166
Silija, P. P., 133–137
Silver nanostructures
absorption spectra, 20
benzoic acid, 20, 21
fourier transform infrared (FTIR)
carbonyl bond frequency, 26–27
dominant role, 24
hexagonal plates, 24
high resolution transmission electron microscopy (HRTEM), 24
particle size distributions, 27–30
pure PVP, 30
SEM images, 25
surface plasmon resonance (SPR), 24
TEM images, 26
hydrothermal, 21
laser energy, 21
light induced reaction, 21
longitudinal band, 20
materials
benzoic acid, 21
methyl alcohol, 21
silver nitrate, 21
metal nanostructures, 20
nanostructures synthesized, 36
crystallographic plane, 37
cubo-octahedral, 37
ellipsoidal particles, 37
face centered cubic, 37
nanorod, 37
nucleation, 36
particle morphology, 36
polyhedron, 37
strain energy, 37
surface energy anisotropy, 37
oscillations
longitudinal modes, 21
transverse mode, 21
photoluminescence, 20
photo thermal therapy, 20
PL emission
bulk counterpart, 34
coherent oscillation, 35
efficiency, 34
hole–phonon scattering, 35
luminescence efficiency, 34
luminescent radiation, 34
optical coupling, 35
photoelectron, 34
plasmon resonance frequency, 34
quantum yield, 35
radiative transition, 34
semiconductor materials., 34
visible luminescence, 35
polyvinyl pyrolidone (PVP), 20
rod-like metal nanoparticles, optical absorption
longitudinal mode, 23
optical absorption spectrum, 23
resonance frequency, 23
silver nanorod, 23
scanning electron microscope (SEM), 22
silver nitrate, 20
spectra, 20
spherical metal nanoparticles, optical absorption
extinction coefficient, 22
metal dielectric, 22
metallic NPs, 22
surface enhanced fluorescence (SEF), 20
surface enhanced raman spectroscopy (SERS), 20

surface of NPs, 20
synthesis
benzoic acid solution, 22
metal, 20
silver NPs, 21
temperature, 21
transmission electron microscope (TEM), 22
tumor cell, 20
UV-visible absorption
absorption peak variation, 32
longitudinal plasmon band, 31
spectra, 32
synthesized NPs, 33
Single atom to nanocomposites, 227–231
assembling atomic force, 198
bottom-up approach, 198
bulk materials, 197
composites
biomaterials, 232
electronic materials, 232
heterostructures, 232
IT materials, 232
magnetic nanoparticle biomolecules, 232
practical materials, 232
core modification
alloy like composition, 203
ferrite nanoparticles, 204
hyperfine field, 207
inverse spinel structure, 203
magnetic elements, 205
magnetite substituted, 205
non collinear spin structures, 203
nonmagnetic elements, 207
nonstoichimetric ferrites, 203
quadrupole splitting, 206
stochiometric ferrites, 203
superparamagnetic, 205
crystal structure, 198
electron microscopes, 198
ferromagnetic, 203
hyperfine
field component, 202
magnetic field, 202–203
inert crystallographical structures, 200
macroscopic dimension, 197
magnetite
crystal structure, 202
nanoparticles, 201
massart's method, 200 , 204
mössbauer
spectra, 200–201
spectroscopy, 200
nanoparticle system, 200
nanopores, 198
nanowires preparation
AAO matrix, 224–227
clusters, 223
core-shell nanowires, 227231
cylinders, 223
electrochemical deposition, 223–224
electrochemical methods, 223
elongated structures, 223
hybrid nanomaterials, 223
magnetic dots, 223
memories storage media, 223
single phase nanowires deposition, 227
synthetic procedure, 223
particles inert layered structure, 199
physical methods, 200
polycrystalinity, 199
polymer matrix, 203
purposely core-shell structures
cell mitochondrial, 215
constituent layers, 220
core chemical structure, 219
core-shell nanoparticles, 219
diffractogram, 216
hyperfine field, 221, 223
intraparticle interactions, 221
lowest spectrum, 221
magnetite-silver, 221
metallic nonmagnetic shell, 221
mössbauer spectroscopy, 221
nanoparticles, 215
nanoparticle system, 221
narrow sextet, 221
nonmetallic shell, 215
policondensation process, 215
pure magnetite nanoparticles, 222
silica layer, 216
stochiometric ratio, 215

superparamagnetic fluctuation, 221
surface anisotropy, 221
uncoated magnetite, 222
quasi-zero dimensional structures, 198
spectra analyses, 200
spontaneous core-shell structures, 208
anisotropy, 214
antiferromagnetic, 213
hematite, 214
chemisorbed molecules, 214
external magnetic field, 211
fitting model, 211
macroscopic system, 209
magnetic moments, 214
magnetite nanoparticles, 210
massart's methods, 209
nanoparticle synthesis, 215
nonstechiometric magnetite, 209
orientation of magnetic moments, 214
oxidized magnetite, 209
particle morphology, 214
quadrupole moment, 212
spinel structure, 212
systems, 209
stoichiometry, 203
surface functionalization
amine, 231
biocompatibility, 231
carboxylate, 231
hydrogen bonds, 232
hydroxide, 231
metallic nanostructured, 232
noncovalent bond, 232
physical adsorption, 231
transmission electron microscopic (TEM), 200
Soibam, I., 1–18
Sprio, S., 51–78
Sugunan, S., 133–137
Suraja, P. V., 133–137

T

Tafel extrapolations, 115
Takaguchi, Y., 119–131
Tampieri, A., 51–78
Thakur, A. K., 255–260
Tio2 nanoparticle thin films
cast nanoparticle films, 139
ceramic nanoparticles (NPs), 140
crystalline materials, 139
densely-packed, 139
dynamic light scattering, 140
electrophoretic deposition (EPD), 140
facilitate tight packing, 139
hydrodynamic diameter, 140
large-scale nanoparticle ordering, 139
materials synthesis
densely packed agglomerates, 143
dispersive x-ray spectroscopy (EDS), 142–143
film cracking, 144
flake-like films, 143
isopropoxide, 141
light scattering, 141
organic surfactants, 145
suspended particles, 142
tight packing, 145
xerogel nanocrystal, 141
mechanical integrity, 139
optional sacrificial polymer layer indicated, 140
polycrystalline masses, 139
titanium dioxide (TiO_2), 140

Y

Yaakob, Z., 133–137